Nonlinear Equations and Dynamical Systems

Nonlinear Equations and Dynamical Systems

Editor: Eli Hill

MURPHY & MOORE
www.murphy-moorepublishing.com

www.murphy-moorepublishing.com

MURPHY & MOORE

Cataloging-in-Publication Data

Nonlinear equations and dynamical systems / edited by Eli Hill.
p. cm.
Includes bibliographical references and index.
ISBN 978-1-63987-726-3
1. Differential equations, Nonlinear. 2. Differentiable dynamical systems. 3. Nonlinear theories.
4. Differential equations. I. Hill, Eli.
QA372 .N66 2023
515.35--dc23

Murphy & Moore Publishing
1 Rockefeller Plaza,
New York City,
NY 10020, USA

ISBN 978-1-63987-726-3

Contents

Preface

The purpose of the book is to provide a glimpse into the dynamics and to present opinions and studies of some of the scientists engaged in the development of new ideas in the field from very different standpoints. This book will prove useful to students and researchers owing to its high content quality.

An equation whose graph solution does not form a straight line is referred to as a nonlinear equation. The variables of a nonlinear equation are either of a degree greater than one or less than one but never one. In solving a nonlinear equation by iterative method, a trial solution is assumed. Thereafter, the trial solution is substituted into the nonlinear equation to determine the error or mismatch, and the mismatch is used in some systematic manner to generate an improved estimate of the solution. Dynamical system refers to an area of study within mathematics which seeks to understand the geometrical properties of trajectories and long-term behavior. These systems use discrete mappings or ordinary differential equations to describe the evolution of a state variable over time. This book aims to shed light on some of the unexplored aspects of nonlinear equations and dynamical systems. It presents researches and studies performed by experts across the globe. The book will serve as a valuable source of reference for graduate and postgraduate students.

At the end, I would like to appreciate all the efforts made by the authors in completing their chapters professionally. I express my deepest gratitude to all of them for contributing to this book by sharing their valuable works. A special thanks to my family and friends for their constant support in this journey.

Editor

1

An Efficient Family of Optimal Eighth-Order Multiple Root Finders

Fiza Zafar [1,2,*], Alicia Cordero [2] and Juan R. Torregrosa [2]

1 Centre for Advanced Studies in Pure and Applied Mathematics, Bahauddin Zakariya University, Multan 60800, Pakistan

2 Instituto de Matemáticas Multidisciplinar, Universitat Politènica de València, 46022 València, Spain; acordero@mat.upv.es (A.C.); jrtorre@mat.upv.es (J.R.T.)

* Correspondence: fizazafar@gmail.com

Abstract: Finding a repeated zero for a nonlinear equation $f(x) = 0, f : I \subseteq R \to R$ has always been of much interest and attention due to its wide applications in many fields of science and engineering. Modified Newton's method is usually applied to solve this kind of problems. Keeping in view that very few optimal higher-order convergent methods exist for multiple roots, we present a new family of optimal eighth-order convergent iterative methods for multiple roots with known multiplicity involving a multivariate weight function. The numerical performance of the proposed methods is analyzed extensively along with the basins of attractions. Real life models from life science, engineering, and physics are considered for the sake of comparison. The numerical experiments and dynamical analysis show that our proposed methods are efficient for determining multiple roots of nonlinear equations.

Keywords: nonlinear equations; multiple zeros; optimal iterative methods; higher order of convergence

---

1. Introduction

It is well-known that Newton's method converges linearly for non-simple roots of a nonlinear equation. For obtaining multiple roots of a univariate nonlinear equation with a quadratic order of convergence, Schröder [1] modified Newton's method with prior knowledge of the multiplicity $m \geq 1$ of the root as follows:

$$x_{n+1} = x_n - m\frac{f(x_n)}{f'(x_n)}. \tag{1}$$

Scheme (1) can determine the desired multiple root with quadratic convergence and is optimal in the sense of Kung-Traub's conjecture [2] that any multipoint method without memory can reach its convergence order of at most 2^{p-1} for p functional evaluations.

In the last few decades, many researchers have worked to develop iterative methods for finding multiple roots with greater efficiency and higher order of convergence. Among them, Li et al. [3] in 2009, Sharma and Sharma [4] and Li et al. [5] in 2010, Zhou et al. [6] in 2011, Sharifi et al. [7] in 2012, Soleymani et al. [8], Soleymani and Babajee [9], Liu and Zhou [10] and Zhou et al. [11] in 2013, Thukral [12] in 2014, Behl et al. [13] and Hueso et al. [14] in 2015, and Behl et al. [15] in 2016 presented optimal fourth-order methods for multiple zeros. Additionally, Li et al. [5] (among other optimal methods) and Neta [16] presented non-optimal fourth-order iterative methods. In recent years, efforts have been made to obtain an optimal scheme with a convergence order greater than four for multiple zeros with multiplicity $m \geq 1$ of univariate function. Some of them only succeeded in developing iterative schemes of a maximum of sixth-order convergence, in the case of multiple zeros; for example, see [17,18]. However, there are only few multipoint iterative schemes with optimal eighth-order convergence for multiple zeros which have been proposed very recently.

Behl et al. [19] proposed a family of optimal eighth-order iterative methods for multiple roots involving univariate and bivariate weight functions given as:

$$\begin{aligned} y_n &= x_n - m\frac{f(x_n)}{f'(x_n)}, m \geq 1, \\ z_n &= y_n - u_n Q(h_n)\frac{f(x_n)}{f'(x_n)}, \\ x_{n+1} &= z_n - u_n t_n G(h_n, t_n)\frac{f(x_n)}{f'(x_n)}, \end{aligned} \tag{2}$$

where weight functions $Q : \mathbb{C} \to \mathbb{C}$ and $G : \mathbb{C}^2 \to \mathbb{C}$ are analytical in neighborhoods of (0) and $(0, 0)$, respectively, with $u_n = \left(\frac{f(y_n)}{f(x_n)}\right)^{\frac{1}{m}}$, $h_n = \frac{u_n}{a_1 + a_2 u_n}$ and $t_n = \left(\frac{f(z_n)}{f(y_n)}\right)^{\frac{1}{m}}$, being a_1 and a_2 complex nonzero free parameters.

A second optimal eighth-order scheme involving parameters has been proposed by Zafar et al. [20], which is given as follows:

$$\begin{aligned} y_n &= x_n - m\frac{f(x_n)}{f'(x_n)}, m \geq 1, \\ z_n &= y_n - m u_n H(u_n)\frac{f(x_n)}{f'(x_n)}, \\ x_{n+1} &= z_n - u_n t_n (B_1 + B_2 u_n) P(t_n) G(w_n)\frac{f(x_n)}{f'(x_n)}, \end{aligned} \tag{3}$$

where B_1,$B_2 \in \mathbb{R}$ are free parameters and the weight functions $H : \mathbb{C} \to \mathbb{C}$, $P : \mathbb{C} \to \mathbb{C}$ and $G : \mathbb{C} \to \mathbb{C}$ are analytic in the neighborhood of 0 with $u_n = \left(\frac{f(y_n)}{f(x_n)}\right)^{\frac{1}{m}}$, $t_n = \left(\frac{f(z_n)}{f(y_n)}\right)^{\frac{1}{m}}$ and $w_n = \left(\frac{f(z_n)}{f(x_n)}\right)^{\frac{1}{m}}$.

Recently, Geum et al. [21] presented another optimal eighth-order method for multiple roots:

$$\begin{aligned} y_n &= x_n - m\frac{f(x_n)}{f'(x_n)}, n \geq 0 \\ w_n &= x_n - m L_f(s)\frac{f(x_n)}{f'(x_n)}, \\ x_{n+1} &= x_n - m\left[L_f(s) + K_f(s, u)\right]\frac{f(x_n)}{f'(x_n)}, \end{aligned} \tag{4}$$

where $L_f : \mathbb{C} \to \mathbb{C}$ is analytic in the neighborhood of 0 and $K_f : \mathbb{C}^2 \to \mathbb{C}$ is holomorphic in the neighborhood of $(0, 0)$ with $s = \left(\frac{f(y_n)}{f(x_n)}\right)^{\frac{1}{m}}$, $u = \left(\frac{f(w_n)}{f(y_n)}\right)^{\frac{1}{m}}$.

Behl et al. [22] also developed another optimal eighth-order method involving free parameters and a univariate weight function as follows:

$$\begin{aligned} y_n &= x_n - m\frac{f(x_n)}{f'(x_n)}, \\ z_n &= y_n - m u\frac{f(x_n)}{f'(x_n)}\frac{1 + \beta u}{1 + (\beta - 2)u}, \ \beta \in \mathbb{R} \\ x_{n+1} &= z_n - u v\frac{f(x_n)}{f'(x_n)}\left(\alpha_1 + (1 + \alpha_2 v)P_f(u)\right), \end{aligned} \tag{5}$$

where α_1, $\alpha_2 \in \mathbb{R}$ are two free disposable parameters and the weight function $P_f : \mathbb{C} \to \mathbb{C}$ is an analytic function in a neighborhood of 0 with $u = \left(\frac{f(y_n)}{f(x_n)}\right)^{\frac{1}{m}}$, $v = \left(\frac{f(z_n)}{f(y_n)}\right)^{\frac{1}{m}}$.

Most recently, Behl at al. [23] presented an optimal eighth-order method involving univariate weight functions given as:

$$\begin{aligned} y_n &= x_n - m\frac{f(x_n)}{f'(x_n)}, m \geq 1, \\ z_n &= y_n - m u_n G_f(u_n)\frac{f(x_n)}{f'(x_n)}, \\ x_{n+1} &= z_n + \frac{u_n w_n}{1-w_n}\frac{f(x_n)}{f'(x_n)}\Big(H_f(u_n) + K_f(v_n)\Big), \end{aligned} \tag{6}$$

where $B_1, B_2 \in \mathbb{R}$ are free parameters and the weight functions $G_f, H_f, K_f : \mathbb{C} \to \mathbb{C}$ are analytic in the neighborhood of 0 with $u_n = \left(\frac{f(y_n)}{f(x_n)}\right)^{\frac{1}{m}}$, $v_n = \left(\frac{f(z_n)}{f(x_n)}\right)^{\frac{1}{m}}$ and $w_n = \left(\frac{f(z_n)}{f(y_n)}\right)^{\frac{1}{m}}$.

Motivated by the research going on in this direction and with a need to give more stable optimal higher-order methods, we propose a new family of optimal eighth-order iterative methods for finding simple as well as multiple zeros of a univariate nonlinear function with multiplicity $m \geq 1$. The derivation of the proposed class is based on a univariate and trivariate weight function approach. In addition, our proposed methods not only give the faster convergence but also have smaller residual error. We have demonstrated the efficiency and robustness of the proposed methods by performing several applied science problems for numerical tests and observed that our methods have better numerical results than those obtained by the existing methods. Further, the dynamical performance of these methods on the above mentioned problems supports the theoretical aspects, showing a good behavior in terms of dependence on initial estimations.

The rest of the paper is organized as follows: Section 2 provides the construction of the new family of iterative methods and the analysis of convergence to prove the eighth order of convergence. In Section 3, some special cases of the new family are defined. In Section 4, the numerical performance and comparison of some special cases of the new family with the existing ones are given. The numerical comparisons is carried out using the nonlinear equations that appear in the modeling of the predator–prey model, beam designing model, electric circuit modeling, and eigenvalue problem. Additionally, some dynamical planes are provided to compare their stability with that of known methods. Finally, some conclusions are stated in Section 5.

2. Construction of the Family

This section is devoted to the main contribution of this study, the design and convergence analysis of the proposed scheme. We consider the following optimal eighth-order class for finding multiple zeros with multiplicity $m \geq 1$:

$$\begin{aligned} y_n &= x_n - m\frac{f(x_n)}{f'(x_n)}, n \geq 0, \\ z_n &= y_n - m u_n G\left(u_n\right)\frac{f(x_n)}{f'(x_n)}, \\ x_{n+1} &= z_n - m u_n H(u_n, t_n, w_n)\frac{f(x_n)}{f'(x_n)}, \end{aligned} \tag{7}$$

where $G : \mathbb{C} \to \mathbb{C}$ and $H : \mathbb{C}^3 \to \mathbb{C}$ are analytical functions in a neighborhood of (0) and $(0, 0, 0)$, respectively being $u_n = \left(\frac{f(y_n)}{f(x_n)}\right)^{\frac{1}{m}}$, $t_n = \left(\frac{f(z_n)}{f(y_n)}\right)^{\frac{1}{m}}$ and $w_n = \left(\frac{f(z_n)}{f(x_n)}\right)^{\frac{1}{m}}$.

In the next result, we demonstrate that the order of convergence of the proposed family reaches optimal order eight.

Theorem 1. *Let us consider $x = \xi$ (say) is a zero with multiplicity $m \geq 1$ of the involved function f. In addition, we assume that $f : \mathbb{C} \to \mathbb{C}$ is an analytical function in the region enclosing the multiple zero ξ. The proposed class defined by Equation (7) has an optimal eighth order of convergence, when the following conditions are satisfied:*

$$\begin{aligned} &G(0) = 1,\ G_1 = G'(0) = 2,\ G_2 = G''(0) = 4 - \frac{G_3}{6}, G_3 = G'''(0) \\ &H_{000} = 0, H_{100} = 0, H_{010} = 1, H_{101} = 3 - \frac{G_3}{12}, H_{110} = 2 - H_{001}, \\ &H_{011} = 4, H_{020} = 1, |G_3| < \infty, |H_{001}| < \infty, \end{aligned} \tag{8}$$

where $H_{ijk} = \frac{1}{i!j!k!}\frac{\partial^{i+j+k}}{\partial u_n^i \partial t_n^j \partial w_n^k} H(u_n, t_n, w_n)|_{(u_n=0, t_n=0, w_n=0)}$ *for* $0 \leq i, k \leq 1, 0 \leq j \leq 2$.

Proof. Let us assume that $e_n = x_n - \xi$ is the error at nth step. By expanding $f(x_n)$ and $f'(x_n)$ about $x = \xi$ using Taylor series expansion, we have:

$$f(x_n) = \frac{f^{(m)}(\xi)}{m!} e_n^m \left(1 + c_1 e_n + c_2 e_n^2 + c_3 e_n^3 + c_4 e_n^4 + c_5 e_n^5 + c_6 e_n^6 + c_7 e_n^7 + c_8 e_n^8 + O(e_n^9)\right) \tag{9}$$

and:

$$\begin{aligned} f'(x_n) = \frac{f^{(m)}(\xi)}{m!} e_n^{m-1} &\Big(m + c_1(m+1)e_n + c_2(m+2)e_n^2 + c_3(m+3)e_n^3 + c_4(m+4)e_n^4 + c_5(m+5)e_n^5 \\ &+ c_6(m+6)e_n^6 + c_7(m+7)e_n^7 + c_8(m+8)e_n^8 + O(e_n^9)\Big), \end{aligned} \tag{10}$$

respectively, where $c_k = \dfrac{m!}{(m+k)!}\dfrac{f^{(m+k)}(\xi)}{f^{(m)}(\xi)}$, $k = 1, 2, 3, \ldots 8$.

By inserting the above Equations (9) and (10), in the first substep of Equation (7), we obtain:

$$y_n - \xi = \frac{c_1 e_n^2}{m} + \frac{\left(2mc_2 - (m+1)c_1^2\right)e_n^3}{m^2} + \sum_{k=0}^{4} A_k e_n^{k+4} + O(e_n^9), \tag{11}$$

where $A_k = A_k(m, c_1, c_2, \ldots, c_8)$ are given in terms of $m, c_1, c_2, c_3, \ldots, c_8$ with two explicitly written coefficients $A_0 = \frac{1}{m^3}\left\{3c_3m^2 + c_1^3(m+1)^2 - c_1c_2m(3m+4)\right\}$ and $A_1 = -\frac{1}{m^4}\left\{c_1^4(m+1)^3 - 2c_2c_1^2m(2m^2+5m+3) + 2c_3c_1m^2(2m+3) + 2m^2(c_2^2(m+2) - 2c_4m)\right\}$, etc.

With the help of Taylor series expansion and Equation (11), we get:

$$f(y_n) = f^{(m)}(\xi)e_n^{2m}\left[\frac{(\frac{c_1}{m})^m}{m!} + \frac{\left(2mc_2 - (m+1)c_1^2\right)(\frac{c_1}{m})^m e_n}{m!c_1} + \sum_{k=0}^{6} \bar{A}_k e_n^{k+2} + O(e_n^9)\right]. \tag{12}$$

Using Equations (9) and (12), we have:

$$u_n = \left(\frac{f(y_n)}{f(x_n)}\right)^{\frac{1}{m}} = \frac{c_1 e_n}{m} + \frac{\left(2mc_2 - (m+2)c_1^2\right)e_n^2}{m^2} + \tau_1 e_n^3 + \tau_2 e_n^4 + \tau_3 e_n^5 + O(e_n^6), \tag{13}$$

where $\tau_1 = \frac{1}{2m^3}\left[c_1^3(2m^2+7m+7) + 6c_3m^2 - 2c_1c_2m(3m+7)\right]$, $\tau_2 = -\frac{1}{6m^4}\left[c_1^4(6m^3+29m^2+51m+34) - 6c_2c_1^2m(4m^2+16m+17) + 12c_1c_3m^2(2m+5) + 12m^2(c_2^2(m+3) - 2c_4m)\right]$ and $\tau_3 =$

$\frac{1}{24m^5}\Big[-24m^3(c_2c_3(5m+17)-5c_5m)+12c_3c_1^2m^2(10m^2+43m+49)+12c_1m^2\Big\{c_2^2(10m^2+47m+53)-2c_4m(5m+13)\Big\}-4c_2c_1^3m(30m^3+163m^2+306m+209)+c_1^5(24m^4+146m^3+355m^2+418m+209)\Big]$.

It is clear from Equation (13) that u_n is of order one. Therefore, we can expand the weight function $G(u_n)$ in the neighborhood of origin by Taylor series expansion up to third-order terms for eighth-order convergence as follows:

$$G(u_n) \approx G(0) + u_nG'(0) + \frac{u_n^2}{2!}G''(0) + \frac{u_n^3}{3!}G'''(0). \tag{14}$$

Now, by inserting Equations (11)–(14) in the second substep of the proposed class (Equation (7)), we obtain:

$$\begin{aligned} z_n - \xi = &-\frac{(G(0)-1)c_1}{m}e_n^2 - \frac{((1+G'(0)+m-G(0)(m+3))c_1^2+2mc_2(G(0)-1))}{m^2}e_n^3 \\ &+\sum_{j=1}^{5} B_je_n^{j+3} + O(e_n^9), \end{aligned} \tag{15}$$

where $B_j = B_j(G(0), G'(0), G''(0), G'''(0), m, c_1, c_2, \ldots, c_7), j = 1, 2, 3, 4, 5$.

In order to obtain fourth-order convergence, the coefficients of e_n^2 and e_n^3 must be simultaneously equal to zero. Thus, from Equation (15), we obtain the following values of $G(0)$ and $G'(0)$:

$$G(0) = 1, G'(0) = 2. \tag{16}$$

Using Equation (16), we have:

$$z_n - \xi = -\frac{((9-G''(0)+m)c_1^2-2mc_1c_2)}{2m^3}e_n^4 + \sum_{j=1}^{4} P_je_n^{j+4} + O(e_n^9), \tag{17}$$

where $P_j = P_j(G''(0), G'''(0), m, c_1, c_2, \ldots, c_7), j = 1, 2, 3, 4$.

With the help of Equation (17) and Taylor series expansion, we have:

$$\begin{aligned} f(z_n) = f^{(m)}(\xi)e_n^{4m}\Bigg[&\frac{2^{-m}\left(\frac{(9-G''(0)+m)c_1^3-2mc_1c_2}{m^3}\right)^m}{m!} - \frac{\left(2^{-m}\left(\frac{(9-G''(0)+m)c_1^3-2mc_1c_2}{m^3}\right)^{m-1}\eta_0\right)}{3(m^3m!)}e_n \\ &+\sum_{j=1}^{7}\overline{P}_je_n^{j+1} + O(e_n^9)\Bigg], \end{aligned} \tag{18}$$

where:

$$\eta_0 = \left(124 + G'''(0) - 3G''(0)(7+3m)c_1^4 - 6m(-3G''(0)+4(7+m)c_1^2c_2 + 12m^2c_2^2 + 12m^2c_1c_3)\right)$$

and $\overline{P}_j = \overline{P}_j(G''(0), G'''(0), m, c_1, c_2, \ldots, c_7), j = 1, 2, \ldots 7$.

Using Equations (9), (12) and (18), we further obtain:

$$t_n = \left(\frac{f(z_n)}{f(y_n)}\right)^{\frac{1}{m}} = \frac{c_1^2(9-G''(0)+m)-2mc_2}{2m^2}e_n^2 + \sum_{j=1}^{4} Q_je_n^{j+2} + O(e_n^7), \tag{19}$$

where $Q_j = Q_j(G''(0), G'''(0), m, c_1, c_2, \ldots, c_6)$, $j = 1, 2, 3, 4$ and:

$$w_n = \left(\frac{f(z_n)}{f(x_n)}\right)^{\frac{1}{m}} = \frac{c_1^3(9 - G''(0) + m) - 2mc_1c_2}{2m^3} e_n^3 + \sum_{j=1}^{4} \overline{Q}_j e_n^{j+3} + O(e_n^8), \tag{20}$$

where $\overline{Q}_j = \overline{Q}_j(G''(0), G'''(0), m, c_1, c_2, \ldots, c_6)$, $j = 1, 2, 3, 4$.

Hence, it is clear from Equation (13) that t_n and w_n are of order 2 and 3, respectively. Therefore, we can expand weight function $H(u_n, t_n, w_n)$ in the neighborhood of $(0, 0, 0)$ by Taylor series expansion up to second-order terms as follows:

$$H(u_n, t_n, w_n) \approx H_{000} + u_n H_{100} + t_n H_{010} + w_n H_{001} + u_n t_n H_{110} + u_n w_n H_{101} + w_n t_n H_{110} + t_n^2 H_{020}, \tag{21}$$

where $H_{ijk} = \frac{1}{i!j!k!} \frac{\partial^{i+j+k}}{\partial u_n^j \partial t_n^j \partial w_n^k} H(u_n, t_n, w_n)|_{(0,0,0)}$, for $0 \le i, k \le 1, 0 \le j \le 2$.

Using Equations (9)–(21) in the last substep of proposed scheme (Equation (7)), we have:

$$e_{n+1} = -\frac{H_{000} c_1}{m} e_n^2 + \sum_{i=1}^{5} E_i e_n^{i+2} + O(e_n^8), \tag{22}$$

where $E_i = E_i(m, G''(0), G'''(0), H_{000}, H_{100}, H_{010}, H_{001}, H_{101}, H_{110}, H_{020}, c_1, c_2, \ldots, c_6)$, $i = 1, 2, 3, 4$.

From Equation (22), it is clear that we can easily obtain at least cubic order of convergence, for:

$$H_{000} = 0. \tag{23}$$

Moreover, $E_1 = 0$ for $H_{100} = 0$, we also have:

$$E_2 = \frac{(-1 + H_{010})\, c_1 (-9 + G''\,(0)\, c_1^2 + 2mc_2)}{2m^3}.$$

Thus, we take:

$$-1 + H_{010} = 0, \tag{24}$$

Thus, by inserting Equation (24), it results that $E_2 = 0$ and:

$$E_3 = \frac{(-2 + H_{001} + H_{110})\, c_1^2 (-9 + G''\,(0)\, c_1^2 + 2mc_2)}{2m^4}. \tag{25}$$

Therefore, by taking:

$$H_{110} = 2 - H_{001}, \tag{26}$$

we have at least a sixth-order convergence. Additionally, for $H_{020} = 1$:

$$E_4 = \frac{(-2 + 2H_{101} - G''\,(0))\, c_1^3 (-9 + G''\,(0)\, c_1^2 + 2mc_2)}{4m^5}, \tag{27}$$

which further yields:

$$H_{101} = 1 + \frac{G''\,(0)}{2}. \tag{28}$$

Finally, we take:

$$H_{011} = 4, \;\; G''\,(0) = 4 - \frac{G_3}{6}.$$

where $G_3 = G'''\,(0)$.

Then, by substituting Equations (23), (24), (26) and (28) in Equation (22), we obtain the following optimal asymptotical error constant term:

$$\begin{aligned} e_{n+1} = & \frac{c_1}{288m^7}(G_3 + 6(5+m))c_1^2 - 12mc_2)((G_3(25+m) + 2(227+90m+7m^2))c_1^4 \\ & - 2m(180 + G_3 + 24m)c_1^2c_2 + 24m^2c_2^2 + 24m^2c_1c_3)e_n^8 + O(e_n^9). \end{aligned} \tag{29}$$

Equation (29) reveals that the proposed scheme (Equation (7)) reaches optimal eighth-order convergence using only four functional evaluations (i.e., $f(x_n), f'(x_n), f(y_n)$ and $f(z_n)$) per iteration. This completes the proof. □

3. Some Special Cases of Weight Function

In this section, we discuss some special cases of our proposed class (7) by assigning different kinds of weight functions. In this regard, please see the following cases, where we have mentioned some different members of the proposed family.

Case 1: Let us describe the following polynomial weight functions directly from the hypothesis of Theorem 1:

$$\begin{aligned} G(u_n) &= 1 + 2u_n + \left(2 - \frac{G_3}{12}\right)u_n^2 + \frac{1}{6}G_3u_n^3, \\ H(u_n, t_n, w_n) &= t_n + \left(H_{001} + \left(3 - \frac{G_3}{12}\right)u_n\right)w_n + ((2 - H_{001})\,u_n + 4w_n + t_n)\,t_n, \end{aligned} \tag{30}$$

where H_{001} and G_3 are free parameters.

Case 1A: When $H_{001} = 2$, $G_3 = 0$, we obtain the corresponding optimal eighth-order iterative method as follows:

$$\begin{aligned} y_n &= x_n - m\frac{f(x_n)}{f'(x_n)}, n \geq 0, \\ z_n &= y_n - mu_n(1 + 2u_n + 2u_n^2)\frac{f(x_n)}{f'(x_n)}, \\ x_{n+1} &= z_n - mu_n(t_n^2 + w_n(2 + 3u_n + 4t_n) + t_n)\frac{f(x_n)}{f'(x_n)}. \end{aligned} \tag{31}$$

Case 2: Now, we suggest a mixture of rational and polynomial weight functions satisfying condition Equation (8) as follows:

$$\begin{aligned} G(u_n) &= \frac{1 + a_0u_n}{1 + (a_0 - 2)\,u_n + a_3u_n^2}, \\ H(u_n, t_n, w_n) &= t_n + \left(H_{001} + \left(3 - \frac{G_3}{12}\right)u_n\right)w_n + ((2 - H_{001})\,u_n + 4w_n + t_n)\,t_n, \end{aligned} \tag{32}$$

where $a_3 = -2\,(a_0 - 1) + \frac{G_3}{12}$ and a_0, G_3 and H_{001} are free parameters.

Case 2A: When $a_0 = 2, H_{001} = 2, G_3 = 12$, the corresponding optimal eighth-order iterative scheme is given by:

$$\begin{aligned} y_n &= x_n - m\frac{f(x_n)}{f'(x_n)}, n \geq 0, \\ z_n &= y_n - mu_n\left(\frac{1 + 2u_n}{1 - u_n^2}\right)\frac{f(x_n)}{f'(x_n)}, \\ x_{n+1} &= z_n - mu_n(t_n + 2\,(1 + u_n)\,w_n + (t_n + 4w_n)\,t_n)\frac{f(x_n)}{f'(x_n)}. \end{aligned} \tag{33}$$

Case 3: Now, we suggest another rational and polynomial weight function satisfying Equation (8) as follows:

$$\begin{aligned} G(u_n) &= \frac{1 + a_0 u_n}{1 + (a_0 - 2)\, u_n + a_3 u_n^2 + a_4 u_n^3}, \\ H(u_n, t_n, w_n) &= t_n + \left(H_{001} + \left(3 - \frac{G_3}{12}\right) u_n\right) w_n + \left((2 - H_{001})\, u_n + 4w_n + t_n\right) t_n, \end{aligned} \tag{34}$$

where $a_3 = -2(a_0 - 1) + \frac{G_3}{12}$, $a_4 = 2a_0 + (a_0 - 6)\,\frac{G_3}{12}$ and a_0, H_{001} and G_3 are free.

Case 3A: By choosing $a_0 = 4$, $a_3 = -5$, $a_4 = 6$, $H_{001} = 2$, $G_3 = 12$, the corresponding optimal eighth-order iterative scheme is given by:

$$\begin{aligned} y_n &= x_n - m\frac{f(x_n)}{f'(x_n)}, n \geq 0, \\ z_n &= y_n - m u_n \left(\frac{1 + 4u_n}{1 + 2u_n - 5u_n^2 + 6u_n^3}\right) \frac{f(x_n)}{f'(x_n)} \\ x_{n+1} &= z_n - m u_n (t_n + 2\,(1 + u_n)\, w_n + (t_n + 4w_n)\, t_n) \frac{f(x_n)}{f'(x_n)}. \end{aligned} \tag{35}$$

In a similar way, we can develop several new and interesting optimal schemes with eighth-order convergence for multiple zeros by considering new weight functions which satisfy the conditions of Theorem 1.

4. Numerical Experiments

This section is devoted to demonstrating the efficiency, effectiveness, and convergence behavior of the presented family. In this regard, we consider some of the special cases of the proposed class, namely, Equations (31), (33) and (35), denoted by $NS1$, $NS2$, and $NS3$, respectively. In addition, we choose a total number of four test problems for comparison: The first is a predator–prey model, the second is a beam designing problem, the third is an electric circuit modeling for simple zeros, and the last is an eigenvalue problem.

Now, we want to compare our methods with other existing robust schemes of the same order on the basis of the difference between two consecutive iterations, the residual errors in the function, the computational order of convergence ρ, and asymptotic error constant η. We have chosen eighth-order iterative methods for multiple zeros given by Behl et al. [19,23]. We take the following particular case (Equation (27)) for ($a_1 = 1$, $a_2 = -2$, $G_{02} = 2m$) of the family by Behl et al. [19] and denote it by $BM1$ as follows:

$$\begin{aligned} y_n &= x_n - m\frac{f(x_n)}{f'(x_n)}, \\ z_n &= y_n - m\,(1 + 2h_n)\, \frac{f(x_n)}{f'(x_n)} u_n, \\ x_{n+1} &= z_n - m \left(1 + t_n + t_n^2 + 3h_n^2 + h_n(2 + 4t_n - 2h_n)\right) \frac{f(x_n)}{f'(x_n)} u_n t_n. \end{aligned} \tag{36}$$

From the eighth-order family of Behl et al. [23], we consider the following special case denoted by $BM2$:

$$\begin{aligned} y_n &= x_n - m\frac{f(x_n)}{f'(x_n)}, \\ z_n &= y_n - mu_n\left(1+2u_n\right)\frac{f(x_n)}{f'(x_n)}, \\ x_{n+1} &= z_n + m\frac{u_n w_n}{1-w_n}\frac{f(x_n)}{f'(x_n)}\left(-1-2u_n-u_n^2+4u_n^3-2v_n\right). \end{aligned} \tag{37}$$

Tables 1–4 display the number of iteration indices (n), the error in the consecutive iterations $|x_{n+1}-x_n|$, the computational order of convergence $\rho \approx \frac{\log|f(x_{n+1})/f(x_n)|}{\log|(f(x_n)/f(x_{n-1}))|}$, $n \geq 1$, (the formula by Jay [24]), the absolute residual error of the corresponding function ($|f(x_n)|$), and the asymptotical error constant $\eta \approx \left|\frac{e_n}{e_{n-1}^8}\right|$. We did our calculations with 1000 significant digits to minimize the round-off error. We display all the numerical values in Tables 1–4 up to 7 significant digits with exponent. Finally, we display the values of approximated zeros up to 30 significant digits in Examples 1–4, although a minimum of 1000 significant digits are available with us.

For computer programming, all computations have been performed using the programming package *Maple* 16 with multiple precision arithmetics. Further, the meaning of $a(\pm b)$ is $a \times 10^{(\pm b)}$ in Tables 1–4.

Now, we explain the real life problems chosen for the sake of comparing the schemes as follows:

Example 1 (Predator-Prey Model). *Let us consider a predator-prey model with ladybugs as predators and aphids as preys [25]. Let x be the number of aphids eaten by a ladybug per unit time per unit area, called the predation rate, denoted by $P(x)$. The predation rate usually depends on prey density and is given as:*

$$P(x) = K\frac{x^3}{a^3+x^3}, \quad a,\ K > 0.$$

Let the growth of aphids obey the Malthusian model; therefore, the growth rate of aphids G per hour is:

$$G(x) = rx, \quad r > 0.$$

The problem is to find the aphid density x for which:

$$P(x) = G(x).$$

This gives:

$$rx^3 - Kx^2 + ra^3 = 0.$$

Let $K = 30$ aphids eaten per hour, $a = 20$ aphids and $r = 2^{-\frac{1}{3}}$ per hour. Thus, we are required to find the zero of:

$$f_1(x) = 0.7937005260x^3 - 30x^2 + 6349.604208$$

The desired zero of f_1 is 25.198420997897463295344212145564 with $m = 2$. We choose $x_0 = 20$.

Table 1. Comparison of different multiple root finding methods for $f_1(x)$.

	BM1	*BM2*	*NS1*	*NS2*	*NS3*
$\lvert x_1 - x_0\rvert$	2.064550(1)	4.789445	1.219414(1)	1.214342(1)	1.213887(1)
$\lvert f(x_1)\rvert$	1.008384(4)	4.963523	1.739946(3)	1.712863(3)	1.710446(3)
$\lvert x_2 - x_1\rvert$	1.544682(1)	4.088744(−1)	6.995715	6.944984	6.940438
$\lvert f(x_2)\rvert$	1.967429(−6)	3.035927(−7)	3.672323(−9)	6.792230(−9)	4.951247(−9)
$\lvert x_3 - x_2\rvert$	2.560869(−4)	1.005971(−4)	1.106393(−5)	1.504684(−5)	1.284684(−5)
$\lvert f(x_3)\rvert$	5.685107(−81)	6.093227(−29)	1.223217(−100)	5.427728(−98)	2.522949(−99)
η	7.900841	0.1287852	1.928645(−12)	2.780193(−12)	2.386168(−12)
ρ	7.676751	3.0078946	7.834927	7.814388	7.825421

Example 2 (Beam Designing Model). *We consider a beam positioning problem (see [26]) where an r meter long beam is leaning against the edge of the cubical box with sides of length 1 m each, such that one of its ends touches the wall and the other touches the floor, as shown in Figure 1.*

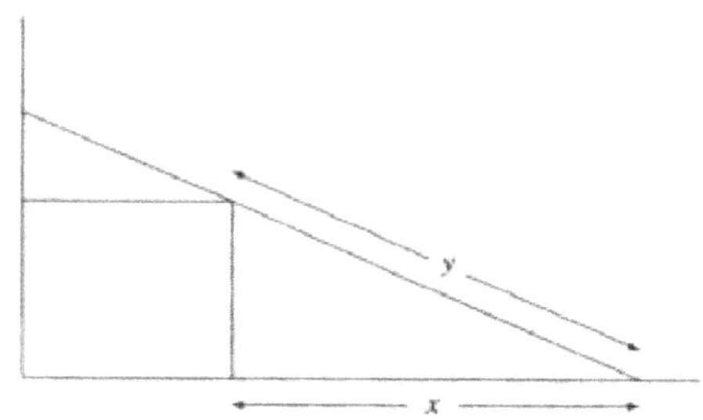

Figure 1. Beam positioning problem.

What should be the distance along the floor from the base of the wall to the bottom of the beam? Let y be the distance in meters along the beam from the floor to the edge of the box and let x be the distance in meters from the bottom of the box to the bottom of the beam. Then, for a given value of r, we have:

$$f_2(x) = x^4 + 4x^3 - 24x^2 + 16x + 16 = 0.$$

The positive solution of the equation is a double root $x = 2$. We consider the initial guess $x_0 = 1.7$.

Table 2. Comparison of different multiple root finding methods for $f_2(x)$.

	BM1	*BM2*	*NS1*	*NS2*	*NS3*
$\lvert x_1 - x_0\rvert$	1.288477	2.734437(−1)	7.427026(−1)	7.391615(−1)	7.388023(−1)
$\lvert f(x_1)\rvert$	35.99479	1.670143(−2)	5.783224	5.682280	5.672098
$\lvert x_2 - x_1\rvert$	9.884394(−1)	2.654643(−2)	4.427007(−1)	4.391589(−1)	4.388001(−1)
$\lvert f(x_2)\rvert$	3.566062(−8)	2.333107(−9)	8.652078(−11)	1.664205(−10)	1.1624462(−10)
$\lvert x_3 - x_2\rvert$	3.854647(−5)	9.859679(−6)	1.898691(−6)	2.633282(−7)	2.200800(−6)
$\lvert f(x_3)\rvert$	7.225712(−77)	5.512446(−30)	2.306147(−95)	1.620443(−92)	4.8729521(−94)
e η	4.230427(−1)	3.997726(7)	1.286982(−3)	1.903372(−3)	1.601202(−3)
ρ	7.629155	3.0090640	7.812826	7.7859217	7.800775

Example 3 (The Shockley Diode Equation and Electric Circuit). *Let us consider an electric circuit consisting of a diode and a resistor. By Kirchoff's voltage law, the source voltage drop V_S is equal to the sum of the voltage drops across the diode V_D and resistor V_R :*

$$V_s = V_R + V_D. \quad (38)$$

Let the source voltage be $V_S = 0.5$ V and from Ohm's law:

$$V_R = RI. \tag{39}$$

Additionally, the voltage drop across the diode is given by the Shockley diode equation as follows:

$$I = I_S\left(e^{\frac{V_D}{nV_T}} - 1\right), \tag{40}$$

where I is the diode current in amperes, I_S is saturation current (amperes), n is the emission or ideality constant ($1 \le n \le 2$ for silicon diode), and V_D is the voltage applied across the diode. Solving Equation (40) for V_D and using all the values in Equation (38), we obtain:

$$-0.5 + RI + nV_T \ln\left(\frac{I}{I_S} + 1\right) = 0$$

Now, for the given values of n, V_T, R and I_S, we have the following equation [27]:

$$-0.5 + 0.1I + 1.4\ln(I + 1) = 0.$$

Replacing I with x, we have

$$f_3(x) = -0.5 + 0.1x + 1.4\ln(x + 1).$$

The true root of the equation is 0.389977198390077586586453532646. *We take* $x_0 = 0.5$.

Table 3. Comparison of different multiple root finding methods for $f_3(x)$.

	BM1	*BM2*	*NS1*	*NS2*	*NS3*
$\lvert x_1 - x_0\rvert$	1.100228(−1)	1.100228(−1)	1.100228(−1)	1.100228(−1)	1.100228(−1)
$\lvert f(x_1)\rvert$	3.213611(−12)	1.902432(−10)	7.591378(−11)	4.728795(−10)	1.626799(−10)
$\lvert x_2 - x_1\rvert$	2.902439(−12)	1.718220(−10)	6.856308(−11)	4.270907(−10)	1.469276(−10)
$\lvert f(x_2)\rvert$	9.512092(−97)	6.797214(−81)	2.215753(−84)	2.393956(−77)	1.758525(−81)
$\lvert x_3 - x_2\rvert$	8.591040(−97)	6.139043(−81)	2.001202(−84)	2.162151(−77)	1.588247(−81)
$\lvert f(x_3)\rvert$	5.604505(−773)	1.805114(−644)	1.1671510(−672)	1.032863(−615)	3.278426(−649)
η	1.705849(−4)	8.081072(−3)	4.097965(−3)	1.953099(−2)	7.312887(−3)
ρ	7.999999	7.999999	7.999999	7.999999	7.999999

Example 4 (Eigenvalue Problem). *One of the challenging task of linear algebra is to calculate the eigenvalues of a large square matrix, especially when the required eigenvalues are the zeros of the characteristic polynomial obtained from the determinant of a square matrix of order greater than 4. Let us consider the following 9 × 9 matrix:*

$$A = \frac{1}{8}\begin{bmatrix} -12 & 0 & 0 & 19 & -19 & 76 & -19 & 18 & 437 \\ -64 & 24 & 0 & -24 & 24 & 64 & -8 & 32 & 376 \\ -16 & 0 & 24 & 4 & -4 & 16 & -4 & 8 & 92 \\ -40 & 0 & 0 & -10 & 50 & 40 & 2 & 20 & 242 \\ -4 & 0 & 0 & -1 & 41 & 4 & 1 & 2 & 25 \\ -40 & 0 & 0 & 18 & -18 & 104 & -18 & 20 & 462 \\ -84 & 0 & 0 & -29 & 29 & 84 & 21 & 42 & 501 \\ 16 & 0 & 0 & -4 & 4 & -16 & 4 & 16 & -92 \\ 0 & 0 & 0 & 0 & 0 & 0 & 0 & 0 & 24 \end{bmatrix}.$$

The corresponding characteristic polynomial of matrix A is given as follows:

$$f_4(x) = x^9 - 29x^8 + 349x^7 - 2261x^6 + 8455x^5 - 17663x^4 + 15927x^3 + 6993x^2 - 24732x + 12960. \tag{41}$$

The above function has one multiple zero at $\xi = 3$ of multiplicity 4 with initial approximation $x_0 = 3.1$.

Table 4. Comparison of different multiple root finding methods for $f_4(x)$.

	$BM1$	$BM2$	$NS1$	$NS2$	$NS3$
$\lvert x_1 - x_0 \rvert$	1.577283(−1)	9.9275251(−2)	1283418(−1)	1.283182(−1)	1.283180(−1)
$\lvert f(x_1) \rvert$	9.361198(−4)	2.205656(−11)	5.299339(−5)	5.281568(−5)	5.281425(−5)
$\lvert x_2 - x_1 \rvert$	5.772837(−2)	7.247474(−4)	2.834188(−2)	2.831824(−2)	2.831805(−2)
$\lvert f(x_2) \rvert$	9.059481(−49)	7.0590148(−38)	2.755794(−55)	8.779457(−55)	5.772523(−55)
$\lvert x_3 - x_2 \rvert$	3.262145(−13)	2.278878(−10)	7.661066(−15)	1.023515(−14)	9.216561(−15)
$\lvert f(x_3) \rvert$	4.543117(−408)	2.278878(−117)	4.807225(−457)	1.869778(−452)	4.077620(−454)
η	2.644775(−3)	2.264227(15)	1.840177(−2)	2.474935(−2)	2.228752(−2)
ρ	7.981915	3.000250	7.989789	7.988696	7.989189

In Tables 1–4, we show the numerical results obtained by applying the different methods for approximating the multiple roots of $f_1(x) - f_4(x)$. The obtained values confirm the theoretical results. From the tables, it can be observed that our proposed schemes $NS1$, $NS2$, and $NS3$ exhibit a better performance in approximating the multiple root of f_1, f_2 and f_4 among other similar methods. Only in the case of the example for simple zeros Behl's scheme BM1 is performing slightly better than the other methods.

Dynamical Planes

The dynamical behavior of the test functions is presented in Figures 2–9. The dynamical planes have been generated using the routines published in Reference [28]. We used a mesh of 400×400 points in the region of the complex plane $[-100, 100] \times [-100, 100]$. We painted in orange the points whose orbit converged to the multiple root and in black those points whose orbit either diverged or converged to a strange fixed point or a cycle. We worked out with a tolerance of 10^{-3} and a maximum number of 80 iterations. The multiple root is represented in the different figures by a white star.

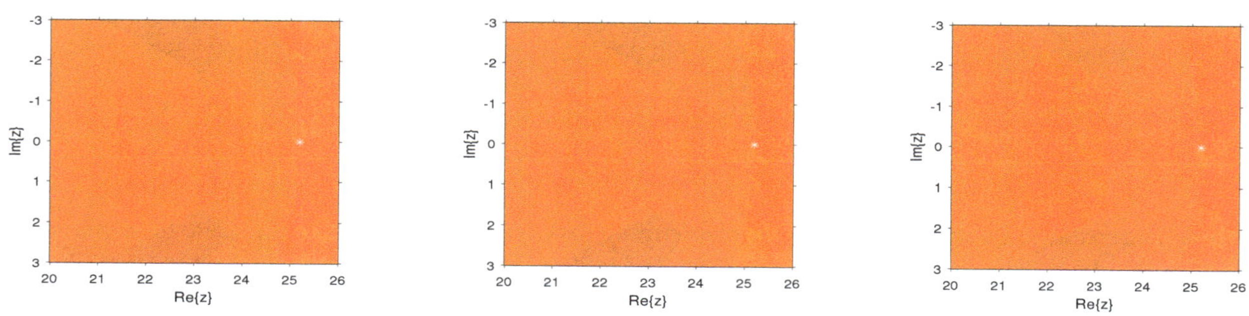

Figure 2. Dynamical planes of the methods $NS1$ (**Left**), $NS2$ (**Center**), and $NS3$ (**Right**) for $f_1(x)$.

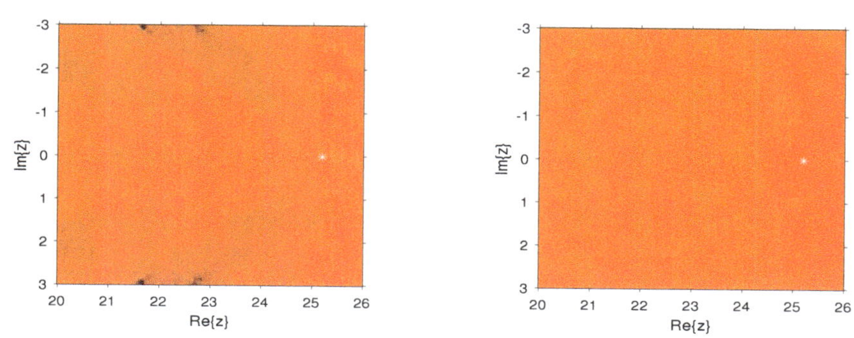

Figure 3. Dynamical planes of the methods $BM1$ (**Left**) and $BM2$ (**Right**) for $f_1(x)$.

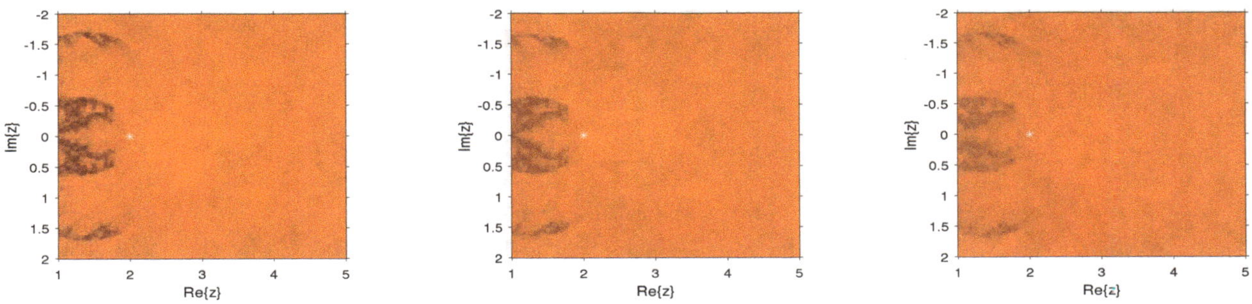

Figure 4. Dynamical planes of the methods $NS1$ (**Left**), $NS2$ (**Center**), and $NS3$ (**Right**) for $f_2(x)$.

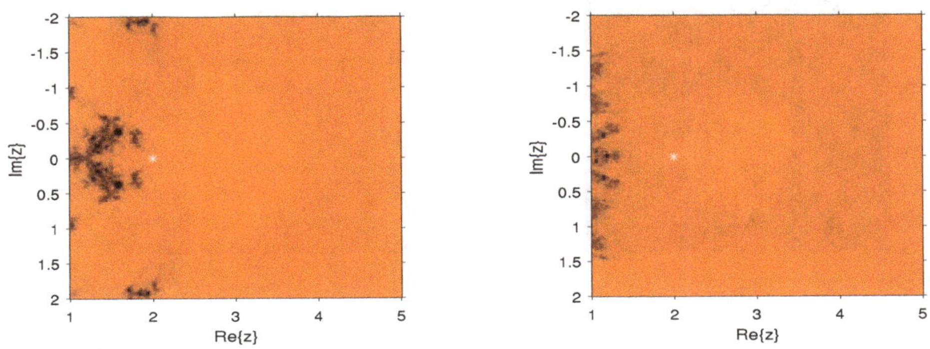

Figure 5. Dynamical planes of the methods $BM1$ (**Left**) and $BM2$ (**Right**) on $f_2(x)$.

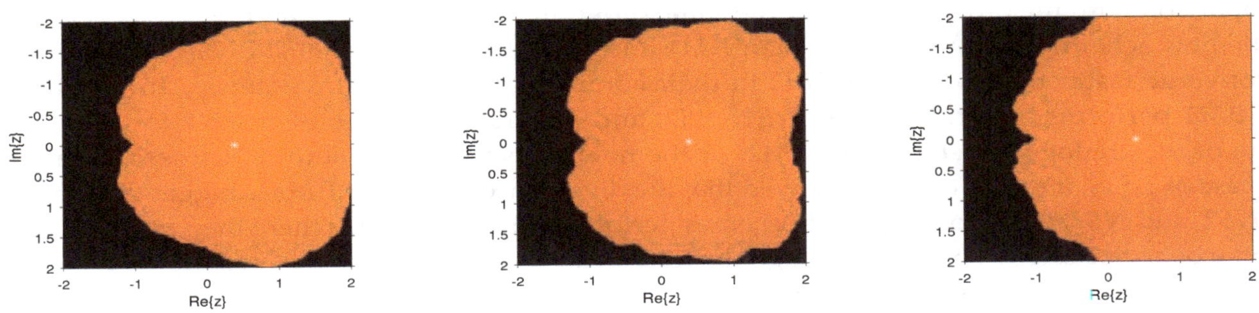

Figure 6. Dynamical planes of the methods $NS1$ (**Left**), $NS2$ (**Center**), and $NS3$ (**Right**) for $f_3(x)$.

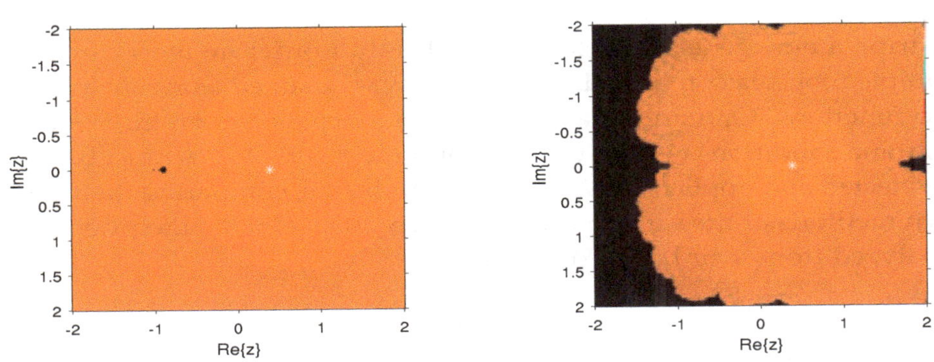

Figure 7. Dynamical planes of the methods $BM1$ (**Left**) and $BM2$ (**Right**) for $f_3(x)$.

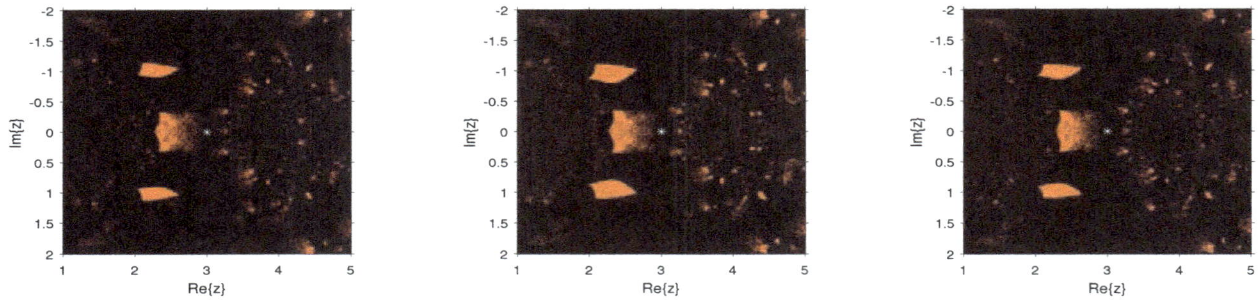

Figure 8. Dynamical planes of the methods $NS1$ (**Left**), $NS2$ (**Center**), and $NS3$ (**Right**) on $f_4(x)$.

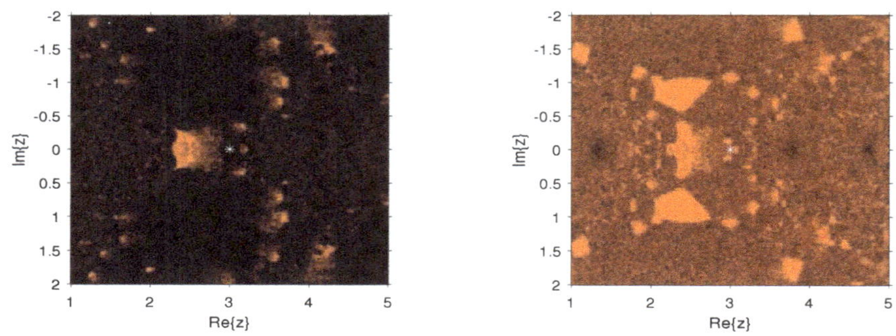

Figure 9. Dynamical planes of the methods $BM1$ (**Left**) and $BM2$ (**Right**) for $f_4(x)$.

Figures 2–9 study the convergence and divergence regions of the new schemes $NS1$, $NS2$, and $NS3$ in comparison with the other schemes of the same order. In the case of $f_1(x)$ and $f_2(x)$, we observed that the new schemes are more stable than $BM1$ and $BM2$ as they are almost divergence-free and also converge faster than $BM1$ and $BM2$ in their common regions of convergence. In the case of $f_3(x)$, $BM1$ performs better; however, $NS1$, $NS2$, and $NS3$ have an edge over $BM2$ for the region in spite of the analogous behavior to $BM2$, as the new schemes show more robustness. Similarly, in the case of $f_4(x)$, it can be clearly observed that the divergence region for $BM1$ is bigger than that for $NS1$, $NS2$, and $NS3$. Additionally, these schemes perform better than $BM2$ where they are convergent. The same behavior can be observed through the numerical comparison of these methods in Tables 1–4. As a future extension, we shall be trying to construct a new optimal eighth-order method whose stability analysis can allow to choose the optimal weight function for the best possible results.

5. Conclusions

In this manuscript, a new general class of optimal eighth-order methods for solving nonlinear equations with multiple roots was presented. This family was obtained using the procedure of weight functions with two functions: One univariate and another depending on three variables. To reach this optimal order, some conditions on the functions and their derivatives must be imposed. Several special cases were selected and applied to different real problems, comparing their performance with that of other known methods of the same order of convergence. Finally, their dependence on initial estimations was analyzed from their basins of attraction.

Author Contributions: methodology, F.Z.; writing original draft preparation, F.Z.; writing review and editing, J.R.T.; visualization, A.C.; supervision, J.R.T.

References

1. Schroder, E. Uber unendlich viele Algorithmen zur Auflosung der Gleichungen. *Math. Ann.* **1870**, *2*, 317–365. [CrossRef]
2. Kung, H.T.; Traub, J.F. Optimal order of one-point and multipoint iteration. *J. Assoc. Comput. Mach.* **1974**, *21*, 643–651. [CrossRef]
3. Li, S.; Liao, X.; Cheng, L. A new fourth-order iterative method for finding multiple roots of nonlinear equations. *Appl. Math. Comput.* **2009**, *215*, 1288–1292.
4. Sharma, J.R.; Sharma, R. Modified Jarratt method for computing multiple roots. *Appl. Math. Comput.* **2010**, *217*, 878–881. [CrossRef]
5. Li, S.G.; Cheng, L.Z.; Neta, B. Some fourth-order nonlinear solvers with closed formulae for multiple roots. *Comput. Math. Appl.* **2010**, *59*, 126–135. [CrossRef]
6. Zhou, X.; Chen, X.; Song, Y. Constructing higher-order methods for obtaining the muliplte roots of nonlinear equations. *J. Comput. Math. Appl.* **2011**, *235*, 4199–4206. [CrossRef]
7. Sharifi, M.; Babajee, D.K.R.; Soleymani, F. Finding the solution of nonlinear equations by a class of optimal methods. *Comput. Math. Appl.* **2012**, *63*, 764–774. [CrossRef]
8. Soleymani, F.; Babajee, D.K.R.; Lofti, T. On a numerical technique forfinding multiple zeros and its dynamic. *J. Egypt. Math. Soc.* **2013**, *21*, 346–353. [CrossRef]
9. Soleymani, F.; Babajee, D.K.R. Computing multiple zeros using a class of quartically convergent methods. *Alex. Eng. J.* **2013**, *52*, 531–541. [CrossRef]
10. Liu, B.; Zhou, X. A new family of fourth-order methods for multiple roots of nonlinear equations. *Non. Anal. Model. Cont.* **2013**, *18*, 143–152.
11. Zhou, X.; Chen, X.; Song, Y. Families of third and fourth order methods for multiple roots of nonlinear equations. *Appl. Math. Comput.* **2013**, *219*, 6030–6038. [CrossRef]
12. Thukral, R. A new family of fourth-order iterative methods for solving nonlinear equations with multiple roots. *J. Numer. Math. Stoch.* **2014**, *6*, 37–44.
13. Behl, R.; Cordero, A.; Motsa, S.S.; Torregrosa, J.R. On developing fourth-order optimal families of methods for multiple roots and their dynamics. *Appl. Math. Comput.* **2015**, *265*, 520–532. [CrossRef]
14. Hueso, J.L.; Martínez, E.; Teruel, C. Determination of multiple roots of nonlinear equations and applications. *J. Math. Chem.* **2015**, *53*, 880–892. [CrossRef]
15. Behl, R.; Cordero, A.; Motsa, S.S.; Torregrosa, J.R.; Kanwar, V. An optimal fourth-order family of methods for multiple roots and its dynamics. *Numer. Algor.* **2016**, *71*, 775–796. [CrossRef]
16. Neta, B. Extension of Murakami's high-order non-linear solver to multiple roots. *Int. J. Comput. Math.* **2010**, *87*, 1023–1031. [CrossRef]
17. Geum, Y.H.; Kim, Y.I.; Neta, B. A class of two-point sixth-order multiple-zero finders of modified double-Newton type and their dynamics. *Appl. Math. Comput.* **2015**, *270*, 387–400 . [CrossRef]
18. Geum, Y.H.; Kim, Y.I.; Neta, B. A sixth-order family of three-point modified Newton-like multiple-root finders and the dynamics behind their extraneous fixed points. *Appl. Math. Comput.* **2016**, *283*, 120–140. [CrossRef]
19. Behl, R.; Cordero, A.; Motsa, S.S.; Torregrosa, J.R. An eighth-order family of optimal multiple root finders and its dynamics. *Numer. Algor.* **2017**. [CrossRef]
20. Zafar, F.; Cordero, A.; Rana, Q.; Torregrosa, J.R. Optimal iterative methods for finding multiple roots of nonlinear equations using free parameters. *J. Math. Chem.* **2017**. [CrossRef]
21. Geum, Y.H.; Kim, Y.I.; Neta, B. Constructing a family of optimal eighth-order modified Newton-type multiple-zero finders along with the dynamics behind their purely imaginary extraneous fixed points. *J. Comput. Appl. Math.* **2018**, *333*, 131–156. [CrossRef]
22. Behl, R.; Zafar, F.; Alshomrani, A.S.; Junjua, M.; Yasmin, N. An optimal eighth-order scheme for multiple zeros of univariate function. *Int. J. Comput. Math.* **2018**, *15*. [CrossRef]
23. Behl, R.; Alshomrani, A.S.; Motsa, S.S. An optimal scheme for multiple roots of nonlinear equations with eighth-order convergence. *J. Math. Chem.* **2018**. [CrossRef]
24. Jay, L.O. A note on Q-order of convergence. *BIT Numer. Math.* **2001**, *41*, 422–429. [CrossRef]
25. Edelstein-Keshet, L. *Differential Calculus for the Life Sciences*; Univeristy of British Columbia: Vancouver, BC, Canada, 2017.

26. Zachary, J.L. *Introduction to Scientific Programming: Computational Problem Solving Using Maple and C*; Springer: New York, NY, USA, 2012.
27. Khoury, R. *Douglas Wilhelm Harder, Numerical Methods and Modelling for Engineering*; Springer International Publishing: Berlin, Germany, 2017.
28. Chicharro, F.I.; Cordero, A.; Torregrosa, J.R. Drawing dynamical and parameters planes of iterative families and methods. *Sci. World J.* **2013**, *2013*, 780153. [CrossRef] [PubMed]

2

On the Semilocal Convergence of the Multi–Point Variant of Jarratt Method: Unbounded Third Derivative Case

Zhang Yong [1,*], **Neha Gupta** [2], **J. P. Jaiswal** [2,*] **and Kalyanasundaram Madhu** [3]

1 School of Mathematics and Physics, Changzhou University, Changzhou 213164, China
2 Department of Mathematics, Maulana Azad National Institute of Technology, Bhopal 462003, India; neha.gupta.mh@gmail.com
3 Department of Mathematics, Saveetha Engineering College, Chennai 602105, India; kalyan742@pec.edu
* Correspondence: yzhangcczu@aliyun.com (Z.Y.); asstprofjpmanit@gmail.com (J.P.J.)

Abstract: In this paper, we study the semilocal convergence of the multi-point variant of Jarratt method under two different mild situations. The first one is the assumption that just a second-order Fréchet derivative is bounded instead of third-order. In addition, in the next one, the bound of the norm of the third order Fréchet derivative is assumed at initial iterate rather than supposing it on the domain of the nonlinear operator and it also satisfies the local ω-continuity condition in order to prove the convergence, existence-uniqueness followed by a priori error bound. During the study, it is noted that some norms and functions have to recalculate and its significance can be also seen in the numerical section.

Keywords: Banach space; semilocal convergence; ω-continuity condition; Jarratt method; error bound

MSC: 65J15; 65H10; 65G99; 47J25

---

1. Introduction

The problem of finding a solution of the nonlinear equation affects a large area of various fields. For instance, kinetic theory of gases, elasticity, applied mathematics and also engineering dynamic systems are mathematically modeled by difference or differential equations. Likewise, there are numerous problems in the field of medical, science, applied mathematics and engineering that can be reduced in the form of a nonlinear equation. Many of those problems cannot be solved directly through any of the methods. For this, we opt for numerical procedure and are able to find at least an approximate solution of the problem using various iterative methods. In this concern, Newton's method [1] is one of the best and most renowned quadratically convergent iterative methods in Banach spaces, which is frequently used by the authors as it is an efficient method and has a smooth execution. Now, consider a nonlinear equation having the form

$$L(m) = 0, \tag{1}$$

where L is a nonlinear operator defined as $L : B \subseteq \nabla_1 \to \nabla_2$, where B is a non-empty open convex domain of a Banach space ∇_1 with values in a Banach space ∇_2 which is usually known as the Newton–Kantorovich method that can be defined as

$$\begin{cases} m_0 \ \ given \ \ in \ \ B, \\ m_n = m_{n-1} - [L'(m_{n-1})]^{-1} L(m_{n-1}), \ n \in \mathbb{N}, \end{cases}$$

where $L'(m_{n-1})$ is the Fréchet derivative of L at m_{n-1}. The results on semilocal convergence have been originally studied by L.V. Kantorovich in [2]. In the early stages, he gave the method of recurrence relations and afterwards described the method of majorant principle. Subsequently, Rall in [3] and many researchers have studied the improvements of the results based on recurrence relations. A large number of researchers studied iterative methods of various order to solve the nonlinear equations extensively. The convergence of iterative methods generally relies on two types: semilocal and local convergence analysis. In the former type, the convergence of iterative methods depends upon the information available around the starting point, whereas, in the latter one, it depends on the information around the given solution.

In the literature, researchers have developed various higher order schemes in order to get better efficiency and also discussed their convergence. Various types of convergence analysis using different types of continuity conditions viz. Lipschitz continuity condition has been studied by Wang et al. in [4,5], Singh et al. in [6], and Jaiswal in [7], to name a few. Subsequently, many authors have studied the weaker continuity condition than Lipschitz namely Hölder by Hernández in [8], Parida and Gupta in [9,10], Wang and Kou in [11] are some of them. Usually, there are some nonlinear equations that neither satisfy Lipschitz nor Hölder continuity conditions; then, we need a generalized form of continuity condition such as ω-continuity, which has been studied by Ezquerro and Hernández in [12,13], Parida and Gupta in [14,15], Prashanth and Gupta in [16,17], Wang and Kou in [18–20], etc.

The algorithms having higher order of convergence plays an important role where the quick convergence is required like in the stiff system of equations. Thus, it is quite interesting to study higher order methods. In this article, we target our study on the semilocal convergence analysis using recurrence relations technique on the multi-point variant of Jarratt method when the third order Fréchet derivative becomes unbounded in the given domain.

2. The Method and Some Preliminary Results

Throughout the paper, we use the below mentioned notations:

B $\equiv$ non-empty open subset of ∇_1; $B_0 \subseteq B$ is a non-empty convex subset; ∇_1, $\nabla_2 \equiv$ Banach spaces, $U(m,b) = \{n \in \nabla_1 : \|n - m\| < b\}$, $\overline{U(m,b)} = \{n \in \nabla_1 : \|n - m\| \leq b\}$.

Here, we consider the multi-point variant of the Jarratt method suggested in [21]

$$\begin{aligned} n_n &= m_n + \tfrac{2}{3}(p_n - m_n), \\ o_n &= m_n - \Upsilon_L(m_n)\wp_n L(m_n), \\ m_{n+1} &= o_n - \left[\tfrac{3}{2}L'(n_n)^{-1}\Upsilon_L(m_n) + \wp_n\left(I - \tfrac{3}{2}\Upsilon_L(m_n)\right)\right] L(o_n), \end{aligned} \tag{2}$$

where $\Upsilon_L(m_n) = [6L'(n_n) - 2L'(m_n)]^{-1}[3L'(n_n) + L'(m_n)]$, $\wp_n = [L'(m_n)]^{-1}$, $p_n = m_n - \wp_n L(m_n)$ and I is the identity operator. In the same article for deriving semilocal convergence results, the researchers have assumed the following hypotheses:

$(A1)\|\wp_0 L(m_0)\| \leq \kappa,$
$(A2)\|\wp_0\| \leq \lambda,$
$(A3)\|L''(m)\| \leq P, m \in B,$
$(A4)\|L'''(m)\| \leq Q, m \in B,$
$(A5)\|L'''(m) - L'''(n)\| \leq \omega(\|m - n\|), \ \forall\, m, n \in B,$

where $\omega : R_+ \to R_+$, is a continuous and non-decreasing function for $m > 0$ such that $\omega(m) \geq 0$ and satisfying $\omega(\epsilon z) \leq \phi(\epsilon)\omega(z), \epsilon \in [0,1]$ and $z \in [0, +\infty)$ with $\phi : [0,1] \to R_+$, is also continuous and non-decreasing. One can realize that, if $\omega(m) = Lm$, then this condition is reduced into Lipschitz and when $\omega(m) = Lm^q$, $q \in (0,1]$ to the Hölder. Furthermore, we found some nonlinear functions which are unbounded in a given domain but seem to be bounded on a particular point of the domain.

For a motivational example, consider a function h on $(-2,2)$. We can verify the above fact by considering the following example [22]

$$h(m) = \begin{cases} m^3 ln(m^2) - 6m^2 - 3m + 8 & m \in (-2,0) \cup (0,2), \\ 0, & m = 0. \end{cases} \tag{3}$$

Clearly, we can see this fact that $h'''(m)$ is unbounded in $(-2,2)$. Hence, for avoiding the unboundedness of the function, we replace the condition $(A4)$ by the milder condition since the given example is bounded at $m = 1$. Thus, here we can assume that the norm of the third order Fréchet derivative is bounded on the initial iterate as:

$(B1) \|L'''(m_0)\| \leq \overline{A},\ m_0 \in B_0,$
where m_0 be an initial approximation. Moreover, we also assume
$(B2) \|L'''(m) - L'''(n)\| \leq \omega(\|m-n\|)\ \forall\ m,n \in B(m_0,\varepsilon),$
where $\varepsilon > 0$. For now, we choose $\varepsilon = \frac{\kappa}{\tilde{\tau}_0}$, where $\tilde{\tau}_0$ will be defined later and the rationality of this choice of such ε will be proved. Moreover, some authors have considered partial convergence conditions. The following nonlinear integral equation of mixed Hammerstein type [23]

$$m(s) = 1 + \int_0^1 G(s,t) \left(\frac{1}{2} m(t)^{\frac{5}{2}} + \frac{7}{16} m(t)^3 \right) dt, s \in [0,1], \tag{4}$$

where $m \in [0,1], t \in [0,1], G(s,t)$ is the Green function defined by

$$G(s,t) = \begin{cases} (1-s)t & t \leq s, \\ s(1-t) & s \leq t, \end{cases}$$

is an example that justified this idea which will be proved later in the numerical application section. In this study, on using recurrence relations, we first discuss the semilocal convergence of the above-mentioned algorithm by just assuming that the second-order Fréchet derivative is bounded. In addition, next, we restrict the domain of the nonlinear operator and consider the bound of the norm of the third-order Fréchet derivative on an initial iterate only rather than supposing it on the given domain of the nonlinear operator.

We start with a nonlinear operator $L : B \subseteq \nabla_1 \to \nabla_2$ and let the Hypotheses $(A1)$–$(A3)$ be fulfilled. Consider the following auxiliary scalar functions out of which Δ and Λ function are taken from the reference [21] and Γ and Θ have been recalculated:

$$\begin{aligned} \Gamma(\theta) &= 1 + \tfrac{1}{2}\tfrac{\theta}{1-\theta} + \left[1 + \tfrac{\theta}{1-\frac{2}{3}\theta}\left(1 + \tfrac{1}{2}\tfrac{\theta}{1-\theta}\right)\right] \\ &\quad \times \tfrac{\theta}{2}\left[\tfrac{1}{1-\theta} + \left(1 + \tfrac{1}{2}\tfrac{\theta}{1-\theta}\right)^2\right], \end{aligned} \tag{5}$$

$$\Delta(\theta) = \frac{1}{1 - \theta\Gamma(\theta)}, \tag{6}$$

$$\begin{aligned} \Theta(\theta) = & \left[\tfrac{\theta}{1-\frac{2}{3}\theta}\left(1 + \tfrac{1}{2}\tfrac{\theta}{1-\theta}\right) + \theta\left[1 + \tfrac{\theta}{1-\frac{2}{3}\theta}\left(1 + \tfrac{1}{2}\tfrac{\theta}{1-\theta}\right)\right]\right. \\ & + \tfrac{\theta^2}{2(1-\theta)}\left[1 + \tfrac{\theta}{1-\frac{2}{3}\theta}\left(1 + \tfrac{1}{2}\tfrac{\theta}{1-\theta}\right)\right] \\ & \left. + \tfrac{\theta}{2}\left[1 + \tfrac{\theta}{1-\frac{2}{3}\theta}\left(1 + \tfrac{1}{2}\tfrac{\theta}{1-\theta}\right)\right]^2 \Lambda(\theta)\right]\Lambda(\theta), \end{aligned} \tag{7}$$

where

$$\Lambda(\theta) \quad = \quad \frac{\theta}{2}\left[\frac{1}{1-\theta}+\left(1+\frac{1}{2}\frac{\theta}{1-\theta}\right)^2\right]. \tag{8}$$

Next, we study some of the properties of the above-stated functions. Let $k(\theta) = \Gamma(\theta)\theta - 1$. Since $k(0) = -1 < 0$ and $k(\frac{1}{2}) \approx 1.379 > 0$, then the function $k(t)$ has at least one real root in $(0, \frac{1}{2})$. Suppose γ is the smallest positive root, then clearly $\gamma < \frac{1}{2}$. Now, we begin with the following lemmas that will be used later in the main theorem(s).

Lemma 1. *Let the functions* Γ, Δ *and* Θ *be given in Equations* (5)–(7)*, respectively, and* γ *be the smallest positive real root of* $\Gamma(\theta)\theta - 1$*. Then,*

(a) $\Gamma(\theta)$ *and* $\Delta(\theta)$ *are increasing and* $\Gamma(\theta) > 1$, $\Delta(\theta) > 1$ *for* $\theta \in (0,\gamma)$,
(b) *for* $\theta \in (0,\gamma)$, $\Theta(\theta)$ *is an increasing function.*

Proof. The proof is straightforward from the expressions of Γ, Δ and Θ given in Relations (5)–(7), respectively. □

Define $\kappa_0 = \kappa, \lambda_0 = \lambda, \tau_0 = P\lambda\kappa$ and $\zeta_0 = \Delta(\tau_0)\Theta(\tau_0)$. Furthermore, we designate the following sequences as:

$$\kappa_{n+1} \quad = \quad \zeta_n\kappa_n, \tag{9}$$

$$\lambda_{n+1} \quad = \quad \Delta(\tau_n)\lambda_n, \tag{10}$$

$$\tau_{n+1} \quad = \quad P\lambda_{n+1}\kappa_{n+1} = \Delta(\tau_n)\zeta_n\tau_n, \tag{11}$$

$$\zeta_{n+1} \quad = \quad \Delta(\tau_{n+1})\Theta(\tau_{n+1}), \tag{12}$$

where $n \geq 0$. Some important properties of the immediate sequences are given by the following lemma.

Lemma 2. *If* $\tau_0 < \gamma$ *and* $\Delta(\tau_0)\zeta_0 < 1$*, where* γ *is the smallest positive root of* $\Gamma(\theta)\theta - 1 = 0$*, then we have*

(a) $\Delta(\tau_n) > 1$ *and* $\zeta_n < 1$ *for* $n \geq 0$,
(b) *the sequences* $\{\kappa_n\}, \{\tau_n\}$ *and* $\{\zeta_n\}$ *are decreasing,*
(c) $\Gamma(\tau_n)\tau_n < 1$ *and* $\Delta(\tau_n)\zeta_n < 1$ *for* $n \geq 0$.

Proof. The proof can be done readily using mathematical induction. □

Lemma 3. *Let the functions* Γ, Δ *and* Θ *be given in the Relations* (5)–(7)*, respectively. Assume that* $\alpha \in (0,1)$*, then* $\Gamma(\alpha\theta) < \Gamma(\theta)$, $\Delta(\alpha\theta) < \Delta(\theta)$ *and* $\Theta(\alpha\theta) < \alpha^2\Theta(\theta)$*, for* $\theta \in (0,\gamma)$.

Proof. For $\alpha \in (0,1)$, $\theta \in (0,\gamma)$ and by using the Equations (5)–(7), this lemma can be proved. □

3. Recurrence Relations for the Method

Here, we characterized some norms which are already derived in the reference [21] for the Method (2) and some are recalculated here.

For $n = 0$, the existence of $\wp_0$ implies the existence of p_0, n_0 and further, we have

$$\|p_0 - m_0\| \leq \kappa_0, \; \|n_0 - m_0\| \leq \frac{2}{3}\kappa_0, \tag{13}$$

i.e., p_0 and $n_0 \in U(m_0, \rho\kappa)$, where $\rho = \frac{\Gamma(\tau_0)}{1-\zeta_0}$. Let $R(m_0) = \wp_0[L'(n_0) - L'(m_0)]$ also; since $\tau_0 < 1$, we have

$$\|R(m_0)\| \leq \frac{2}{3}\tau_0, \quad \left\| \left[I + \frac{3}{2}R(m_0)\right]^{-1} \right\| \leq \frac{1}{1-\tau_0}. \tag{14}$$

Moreover,

$$\begin{aligned} \|Y_L(m_0)\| &= \left\| I - \tfrac{3}{4}\left[I + \tfrac{3}{2}R(m_0)\right]^{-1} R(m_0) \right\| \\ &\leq 1 + \left\| \tfrac{3}{4}\left[I + \tfrac{3}{2}R(m_0)\right]^{-1} \right\| \|R(m_0)\| \\ &\leq 1 + \tfrac{1}{2}\tfrac{\tau_0}{1-\tau_0}. \end{aligned} \tag{15}$$

From the second sub-step of the considered scheme, it is obvious that

$$\|o_0 - m_0\| \leq \left[1 + \frac{1}{2}\frac{\tau_0}{1-\tau_0}\right]\kappa_0. \tag{16}$$

It is similar to obtain

$$\|o_0 - p_0\| \leq \left[\frac{1}{2}\frac{\tau_0}{1-\tau_0}\right]\kappa_0. \tag{17}$$

Using the Banach Lemma, we realize that $L'(n_0)^{-1}$ exists and can be bounded as

$$\|L'(n_0)^{-1}\| \leq \frac{\lambda_0}{1 - \frac{2}{3}\tau_0}. \tag{18}$$

From Taylor's formula, we have

$$\begin{aligned} L(o_0) &= L(m_0) + L'(m_0)(o_0 - m_0) \\ &+ \int_0^1 [L'(m_0 + \theta(o_0 - m_0)) - L'(m_0)]d\theta(o_0 - m_0). \end{aligned} \tag{19}$$

From the above relation, it follows that

$$\|L(o_0)\| \leq \Lambda(\tau_0)\frac{\kappa}{\lambda}. \tag{20}$$

Though in the considered reference [21] the norm $\|m_1 - o_0\|$ has already been calculated, here we are recalculating it in a more precise way such that the recalculated norm becomes finer than the given in the reference [21] and its significance can be seen in the numerical section. The motivation for recalculating this norm has been also discussed later. From the last sub-step of the Equation (2),

$$\begin{aligned} m_1 - o_0 &= -\left[\frac{3}{2}L'(n_0)^{-1}Y_L(m_0) + \wp_n\left(I - \frac{3}{2}Y_L(m_0)\right)\right] L(o_0) \\ &= -\left[\wp_0 + \frac{3}{2}[L'(n_0)^{-1} + L'(n_0)^{-1}]\right] Y_L(m_0)L(o_0). \end{aligned}$$

On taking the norm, we have

$$\begin{aligned} \|m_1 - o_0\| \leq \tfrac{\kappa_0}{2}&\left[1 + \tfrac{\tau_0}{1-\frac{2}{3}\tau_0}\left(1 + \tfrac{1}{2}\tfrac{\tau_0}{1-\tau_0}\right)\right] \\ &\times \left[\tfrac{\tau_0}{1-\tau_0} + \tau_0\left(1 + \tfrac{1}{2}\tfrac{\tau_0}{1-\tau_0}\right)^2\right], \end{aligned} \tag{21}$$

and thus we obtain

$$\|m_1 - m_0\| \le \|m_1 - o_0\| + \|o_0 - m_0\| \le \Gamma(\tau_0)\kappa_0. \tag{22}$$

Hence, $m_1 \in U(m_0, \rho\kappa)$. Now, since the assumption $\zeta_0 < \frac{1}{\Delta(\tau_0)} < 1$, notice that $\tau_0 < \gamma$ hence $\Gamma(\tau_0) < \Gamma(\gamma)$ and it can be written as

$$\|I - \wp_0 L'(m_1)\| \le \tau_0 \Lambda(\tau_0) < 1. \tag{23}$$

Thus, $\wp_1 = [L'(m_1)]^{-1}$ exists and, by virtue of Banach lemma, it may be written as

$$\|\wp_1\| \le \frac{\lambda_0}{1 - \tau_0 \Gamma(\tau_0)} = \lambda_1.$$

Again by Taylor's expansion along o_n, we can write

$$\begin{aligned} L(m_{n+1}) &= L(o_n) + L'(p_n)(m_{n+1} - o_n) \\ &+ \textstyle\int_0^1 [L'(o_n + \theta(m_{n+1} - o_n)) - L'(p_n)] d\theta (m_{n+1} - o_n), \end{aligned} \tag{24}$$

and

$$L'(p_n) = L'(m_n) + \int_0^1 L''(m_n + \theta(p_n - m_n)) d\theta (p_n - m_n). \tag{25}$$

On using the above relation and, for $n = 0$, Equation (24) assumes the form

$$\begin{aligned} L(m_1) &= L(o_0) + L'(m_0)(m_1 - o_0) \\ &+ \int_0^1 L''(m_0 + \theta(p_0 - m_0)) d\theta (p_0 - m_0)(m_1 - o_0) \\ &+ \int_0^1 [L'(o_0 + \theta(m_1 - o_0)) - L'(p_0)] d\theta (m_1 - o_0). \end{aligned}$$

Using the last sub-step of the Scheme given in the Equation (2), the above expression can be rewritten as

$$\begin{aligned} L(m_1) &= \frac{3}{2}[L'(n_0) - L'(m_0)] L'(n_0)^{-1} \Upsilon_L(m_0) L(o_0) \\ &+ \int_0^1 L''(m_n + \theta(p_n - m_n)) d\theta (p_n - m_n)(m_{n+1} - o_n) \\ &+ \int_0^1 [L'(o_n + \theta(m_{n+1} - o_n)) - L'(p_n)] d\theta (m_{n+1} - o_n). \end{aligned}$$

In addition, thus,

$$\|L(m_1)\| \le \Theta(\tau_0)\frac{\kappa}{\lambda}. \tag{26}$$

Hence,

$$\|p_1 - m_1\| \le \Delta(\tau_0)\Theta(\tau_0)\kappa_0 = \kappa_1.$$

In addition, because $\Gamma(\tau_0) > 1$ and by triangle inequality, we find

$$\|p_1 - m_0\| \le \rho\kappa,$$

and

$$\|n_1 - m_0\| \le \|m_1 - m_0\| + \left\|\frac{2}{3}(p_1 - m_1)\right\| \le (\Gamma(\tau_0) + \zeta_0)\kappa_0 < \rho\kappa,$$

which implies $p_1, n_1 \in U(m_0, \rho\kappa)$. Furthermore, we have

$$P\|\wp_1\|\|\wp_1 L(m_1)\| \leq \Delta^2(\tau_0)\Theta(\tau_0)\tau_0 = \tau_1. \tag{27}$$

Moreover, we can state the following lemmas.

Lemma 4. *Under the hypotheses of Lemma 2, let* $\sigma = \Delta(\tau_0)\zeta_0$ *and* $\varsigma = \frac{1}{\Delta(\tau_0)}$, *we have*

$$\zeta_i \leq \varsigma\sigma^{3^n}, \tag{28}$$

$$\prod_{i=0}^{n}\zeta_i \leq \varsigma^{n+1}\sigma^{\frac{3^{n+1}-1}{2}}, \tag{29}$$

$$\kappa_n \leq \kappa\varsigma^n\sigma^{\frac{3^n-1}{2}}, \tag{30}$$

$$\sum_{i=n}^{n+m}\kappa_i \leq \kappa\varsigma^n\sigma^{\frac{3^n-1}{2}}\left(\frac{1-\varsigma^{m+1}\sigma^{\frac{3^n(3^m+1)}{2}}}{1-\varsigma\sigma^{3^n}}\right), \tag{31}$$

where $n \geq 0$ *and* $m \geq 1$.

Proof. In order to prove this lemma, first, we need to derive

$$\zeta_n \leq \varsigma\sigma^{3^n}.$$

We will prove it by executing the induction. By Lemma 3 and since $\tau_1 = \sigma\tau_0$, hence for $n = 1$,

$$\zeta_1 = \Delta(\sigma\tau_0)\Theta(\sigma\tau_0) < \sigma^2\zeta_0 < \varsigma\sigma^{3^1}.$$

Let it be true for $n = k$, then

$$\zeta_k \leq \varsigma\sigma^{3^k}, k \geq 1.$$

Now, we will prove it for $n = k+1$. Thus,

$$\zeta_{k+1} < \Delta(\sigma\tau_k)\Theta(\sigma\tau_k) < \varsigma\sigma^{3^{k+1}}.$$

Therefore, $\zeta_n \leq \varsigma\sigma^{3^n}$ is true for $n \geq 0$. Making use of this inequality, we have

$$\prod_{i=0}^{k}\zeta_i \leq \prod_{i=0}^{k}\varsigma\sigma^{3^i} = \varsigma^{k+1}\prod_{i=0}^{k}\sigma^{3^i} = \varsigma^{k+1}\sigma^{\frac{3^{k+1}-1}{2}}, k \geq 0.$$

By making use of the above-derived inequality in the Relation (9), we have

$$\kappa_n = \zeta_{n-1}\kappa_{n-1} = \zeta_{n-1}\zeta_{n-2}\kappa_{n-2} = \cdots = \kappa_0\prod_{i=0}^{n-1}\zeta_i \leq \kappa\varsigma^n\sigma^{\frac{3^n-1}{2}}, n \geq 0.$$

With the evidence that $0 < \varsigma < 1$ and $0 < \sigma < 1$, we can say that $\kappa_n \to 0$ as $n \to \infty$. Let us denote

$$\varpi = \sum_{i=k}^{k+m}\varsigma^i\sigma^{\frac{3^i}{2}}, k \geq 0, m \geq 1.$$

The above equation may also be rewritten in the following form

$$\begin{aligned} \varpi &\leq \varsigma^k \sigma^{\frac{3^k}{2}} + \varsigma\sigma^{3^k} \sum_{i=k}^{k+m-1} \varsigma^i \sigma^{\frac{3^i}{2}} \\ &= \varsigma^k \sigma^{\frac{3^k}{2}} + \varsigma\sigma^{3^k} \left(\varpi - \varsigma^{k+m} \sigma^{\frac{3^{k+m}}{2}} \right), \end{aligned}$$

and then it becomes

$$\varpi < \varsigma^k \sigma^{\frac{3^k}{2}} \left(\frac{1 - \varsigma^{m+1} \sigma^{\frac{3^k(3^m+1)}{2}}}{1 - \varsigma\sigma^{3^k}} \right).$$

Moreover,

$$\sum_{i=k}^{k+m} \kappa_i \leq \sum_{i=k}^{k+m} \kappa\varsigma^i \sigma^{\frac{3^{i}-1}{2}} \leq \kappa\varsigma^k \sigma^{\frac{3^k-1}{2}} \left(\frac{1 - \varsigma^{m+1} \sigma^{\frac{3^k(3^m+1)}{2}}}{1 - \varsigma\sigma^{3^k}} \right).$$

□

Lemma 5. *Let the hypotheses of Lemma 2 and the conditions $(A1)$–$(A3)$ hold; then, the following conditions are true for all $n \geq 0$:*

$$\begin{aligned} &(i)\wp_n = [L'(m_n)]^{-1} exists\ and\ \|\wp_n\| \leq \lambda_n, \\ &(ii)\|\wp_n L(m_n)\| \leq \kappa_n, \\ &(iii)P\|\wp_n\|\|\wp_n L(m_n)\| \leq \tau_n, \\ &(iv)\|p_n - m_n\| \leq \kappa_n, \\ &(v)\|m_{n+1} - m_n\| \leq \Gamma(\tau_n)\kappa_n, \\ &(vi)\|m_{n+1} - m_0\| \leq \rho\kappa, where\ \rho = \tfrac{\Gamma(\tau_0)}{1-\zeta_0}. \end{aligned} \tag{32}$$

Proof. By using the mathematical induction of Lemma 4, we can prove $(i) - (v)$ for $n \geq 0$. Now, for $n \geq 1$, by making use of Relation (31) and the above results, we get

$$\|m_{n+1} - m_0\| \leq \sum_{i=0}^{n} \|m_{i+1} - m_i\| < \rho\kappa.$$

□

Lastly, the following lemma can be proved in a similar way of the article by Wang and Kou [22].

Lemma 6. *Let $\rho = \frac{\Gamma(\tau_0)}{1-\zeta_0}$ and $\Delta(\tau_0)\zeta_0 < 1$ and $\tau_0 < \gamma$, where γ is the smallest positive root of $\Gamma(\theta)\theta - 1 = 0$; then, $\rho < \frac{1}{\tau_0}$.*

4. Semilocal Convergence When L''' Condition Is Omitted

In the ensuing section, our objective is to prove the convergence of the Algorithm mentioned in the Equation(2) by assuming the Hypotheses $(A1)$–$(A3)$ only. Furthermore, we will find a ball with center m_0 and of radius $\rho\kappa$ in which the solution exists and will be unique as well together with which we will define its error bound.

Theorem 1. *Suppose $L : B \subseteq \nabla_1 \to \nabla_2$ is a continuously second-order Fréchet differentiable on B. Suppose the hypotheses (A1)–(A3) are true and $m_0 \in B$. Assume that $\tau_0 = P\lambda\kappa$ and $\zeta_0 = \Delta(\tau_0)\Theta(\tau_0)$ satisfy $\tau_0 < \gamma$ and $\Delta(\tau_0)\zeta_0 < 1$, where γ is the smallest root of $\Gamma(\theta)\theta - 1 = 0$ and Γ, Δ and Θ are defined by Equations (5)–(7), respectively. In addition, suppose $\overline{U(m_0, \rho\kappa)} \subseteq B$, where $\rho = \frac{\Gamma(\tau_0)}{1-\zeta_0}$. Then, initiating* *with m_0, the iterative sequence $\{m_n\}$ creating from the Scheme given in the Equation (2) converges to a zero m^* of $L(m) = 0$ with $m_n, m^* \in \overline{U(m_0, \rho\kappa)}$ and m^* is an exclusive zero of $L(m) = 0$ in $U(m_0, \frac{2}{P\lambda} - \rho\kappa) \cap B$. Furthermore, its error bound is given by*

$$\|m_n - m^*\| \leq \Gamma(\tau_0)\kappa\varsigma^n\sigma^{\frac{3^n-1}{2}}\left(\frac{1}{1-\varsigma\sigma^{3^n}}\right), \tag{33}$$

where $\sigma = \Delta(\tau_0)\zeta_0$ and $\varsigma = \frac{1}{\Delta(\tau_0)}$.

Proof. Clearly, the sequence $\{m_n\}$ is well established in $\overline{U(m_0, \rho\kappa)}$. Now,

$$\begin{aligned} \|m_{k+l} - m_k\| &\leq \textstyle\sum_{i=k}^{k+l-1} \|m_{i+1} - m_i\| \\ &\leq \Gamma(\tau_0)\kappa\varsigma^k\sigma^{\frac{3^k-1}{2}}\left(\frac{1-\varsigma^l\sigma^{\frac{3^k(3^{l-1}+1)}{2}}}{1-\varsigma\sigma^{3^k}}\right), \end{aligned} \tag{34}$$

which shows that $\{m_k\}$ is a Cauchy sequence. Hence, there exists m^* satisfying

$$\lim_{k\to\infty} m_k = m^*.$$

Letting $k = 0, l \to \infty$ in Equation (34), we obtain

$$\|m^* - m_0\| \leq \rho\kappa,$$

which implies that $m^* \in \overline{U(m_0, \rho\kappa)}$. Next, we will show that m^* is a zero of $L(m) = 0$. Because

$$\|\wp_0\| \, \|L(m_n)\| \leq \|\wp_n\| \, \|L(m_n)\|,$$

and in the above inequality by tending $n \to \infty$ and using the continuity of L in B, we find that $L(m^*) = 0$. Finally, for unicity of m^* in $U(m_0, \frac{2}{P\lambda} - \rho\kappa) \cap B$, let m^{**} be another solution of $L(m)$ in $U(m_0, \frac{2}{P\lambda} - \rho\kappa) \cap B$. Using Taylor's theorem, we get

$$0 = L(m^{**}) - L(m^*) = \int_0^1 L'((1-t\theta)m^* + \theta m^{**})d\theta(m^{**} - m^*).$$

In addition,

$$\begin{aligned} &\|\wp_0\| \left\| \int_0^1 [L'((1-\theta)m^* + \theta m^{**}) - L'(m_0)]d\theta \right\| \\ \leq\ & P\lambda \int_0^1 [(1-\theta)\|m^* - m_0\| + \theta\|m^{**} - m_0\|]d\theta \\ \leq\ & \frac{P\lambda}{2}\left[\rho\kappa + \frac{2}{P\lambda} - \rho\kappa\right] = 1, \end{aligned}$$

which implies $\int_0^1 L'((1-\theta)m^* + \theta m^{**})d\theta$ is invertible and hence $m^{**} = m^*$. □

5. Semilocal Convergence When L''' Is Bounded on Initial Iterate

In the current section, we establish the existence and uniqueness theorem of the solution based on the weaker conditions $(A1)$–$(A3)$, $(B1)$ and $(B2)$. Define the sequences as

$$\tilde{\kappa}_{n+1} = \tilde{\zeta}_n\tilde{\kappa}_n, \tag{35}$$

$$\tilde{\lambda}_{n+1} = \Delta(\tilde{\tau}_n)\tilde{\lambda}_n, \tag{36}$$

$$\tilde{\tau}_{n+1} = P\tilde{\lambda}_{n+1}\tilde{\kappa}_{n+1} = \Delta(\tilde{\tau}_n)\tilde{\zeta}_n\tilde{\tau}_n, \tag{37}$$

$$\tilde{\mu}_{n+1} = Q\tilde{\lambda}_{n+1}\tilde{\kappa^2}_{n+1} = \Delta(\tilde{\tau}_n)\tilde{\zeta^2}_n\tilde{\mu}_n, \tag{38}$$

$$\tilde{\nu}_{n+1} = \tilde{\lambda}_{n+1}\tilde{\kappa^2}_{n+1}\omega(\tilde{\kappa}_{n+1}) \leq \Delta(\tilde{\tau}_n)\phi(\tilde{\zeta}_n)\tilde{\zeta}_n^2\tilde{\nu}_n, \tag{39}$$

$$\tilde{\zeta}_{n+1} = \Delta(\tilde{\tau}_{n+1})\Theta'(\tilde{\tau}_{n+1}, \tilde{\mu}_{n+1}, \tilde{\nu}_{n+1}), \tag{40}$$

where $n \geq 0$ and $Q = \overline{A} + \omega\left(\frac{\kappa}{\tilde{\tau}_0}\right)$. Here, we assign $\tilde{\kappa}_0 = \kappa, \tilde{\lambda}_0 = \lambda, \tilde{\tau}_0 = P\lambda\kappa, \tilde{\mu}_0 = Q\lambda\kappa^2$, $\tilde{\nu}_0 = \lambda\kappa^2\omega(\kappa)$ and $\tilde{\zeta}_0 = \Delta(\tilde{\tau}_0)\Theta'(\tilde{\tau}_0, \tilde{\mu}_0, \tilde{\nu}_0)$. From Lemma (5), it is known that

$$\|m_n - m_0\| < \rho\kappa < \frac{\kappa}{\tilde{\tau}_0}.$$

Therefore, $m_n \in U(m_0, \frac{\kappa}{\tilde{\tau}_0})$. Similarly, for $t \in [0,1]$ and $n \geq 1$ and using Lemma (6), we get

$$\begin{aligned}\|m_n + st(p_n - m_n) - m_0\| &\leq \|m_n - m_0\| + \|p_n - m_n\| \\ &\leq \sum_{i=0}^{n-1} \|m_{i+1} - m_i\| + \tilde{\kappa}_n \\ &\leq \Gamma(\tilde{\tau}_0)\sum_{i=0}^{n} \tilde{\kappa}_i \leq \rho\kappa < \frac{\kappa}{\tilde{\tau}_0}.\end{aligned}$$

Therefore, $\{m_n + st(p_n - m_n)\} \in U(m_0, \frac{\kappa}{\tilde{\tau}_0})$. This shows that the choice for $\varepsilon = \frac{\kappa}{\tilde{\tau}_0}$ is relevant. Assume that there exists a root $\tilde{\tau}_0 \in (0, \gamma)$ of the equation

$$m = \left[\overline{A} + \omega\left(\frac{\kappa}{m}\right)\right]\lambda\kappa^2.$$

It is obvious that $\tilde{\mu}_0 = Q\lambda\kappa^2$, where $Q = \overline{A} + \omega\left(\frac{\kappa}{\tilde{\tau}_0}\right)$. Notice that here we don't define $\tilde{\tau}_0$ as the root of the following equation:

$$m = \left[\overline{A} + \omega\left(\frac{\Gamma(m)\kappa}{1 - \Delta(m)\Theta'(m, \tilde{\mu}_0, \tilde{\nu}_0)}\right)\right]\lambda\kappa^2.$$

It would be remembered that, for all $m \in U(m_0, \frac{\kappa}{\tilde{\tau}_0})$, we have

$$\begin{aligned}\|L'''(m)\| = \;& \|L'''(m_0)\| + \|L'''(m) - L'''(m_0)\| \\ & \leq \overline{A} + \omega(\|m - m_0\|) \\ & \leq \overline{A} + \omega\left(\frac{\kappa}{\tilde{\tau}_0}\right) = Q.\end{aligned}$$

Here, we include two auxiliary scalar functions taken from the reference [21]

$$\begin{aligned}\Theta'(\theta, \eta, \xi) \;&= \left[\frac{5}{6}\eta + \frac{(3\theta+\eta)(6\theta+2\eta)}{27-18\theta} + \frac{(2\theta+\eta)(3\theta+\eta)}{6-4\theta}\right. \\ &\left. + \frac{(2+2\theta+\eta)(3\theta+\eta)\theta}{(12-8\theta)(1-\theta)}\right]\tilde{\Lambda}(\theta, \eta, \xi) \\ &+ \frac{1}{2}\frac{\theta^2}{1-\theta}\left[\frac{9}{6-4\theta}\left(1 + \frac{1}{2}\frac{\theta}{1-\theta}\right) + \frac{3\theta}{4(1-\theta)} + \frac{1}{2}\right]\tilde{\Lambda}(\theta, \eta, \xi) \\ &+ \frac{\theta}{2}\left[\frac{9}{6-4\theta}\left(1 + \frac{1}{2}\frac{\theta}{1-\theta}\right) + \frac{3\theta}{4(1-\theta)} + \frac{1}{2}\right]^2\tilde{\Lambda}(\theta, \eta, \xi)^2,\end{aligned} \tag{41}$$

where

$$\tilde{\Lambda}(\theta,\eta,\xi) = \frac{1}{8}\frac{\theta^3}{(1-\theta)^2} + \frac{1}{12}\frac{\theta\eta}{1-\theta} + \left(D_1 + \frac{1}{3}D_2\right)\xi. \tag{42}$$

$D_1 = \int_0^1 \int_0^1 \phi(s\theta)\theta(1-\theta)dsd\theta$ and $D_2 = \int_0^1 \int_0^1 \phi\left(\frac{2}{3}s\theta\right)\theta dsd\theta$.

Using the property of the induction and from the conditions $(A1)$–$(A3)$, $(B1)$ and $(B2)$, the following relations are true for all $n \geq 0$:

$$\begin{aligned} &(i)\wp_n = [L'(m_n)]^{-1} \text{exists and } \|\wp_n\| \leq \tilde{\lambda}_n,\\ &(ii)\|\wp_n L(m_n)\| \leq \tilde{\kappa}_n,\\ &(iii)P\|\wp_n\|\|\wp_n L(m_n)\| \leq \tilde{\tau}_n,\\ &(iv)Q\|\wp_n\|\|\wp_n L(m_n)\| \leq \tilde{\mu}_n,\\ &(v)\|\wp_n\|\|\wp_n L(m_n)\|^2\omega(\|\wp_n L(m_n)\|) \leq \tilde{\nu}_n,\\ &(vi)\|m_{n+1} - m_n\| \leq \Gamma(\tilde{\tau}_n)\tilde{\kappa}_n,\\ &(vii)\|m_{n+1} - m_0\| \leq \tilde{\rho}\kappa, \text{ where } \tilde{\rho} = \frac{\Gamma(\tilde{\tau}_0)}{1-\tilde{\zeta}_0}. \end{aligned} \tag{43}$$

The second theorem of this article is based on the weaker assumptions, which is stated as:

Theorem 2. *Suppose $L : B \subseteq \nabla_1 \to \nabla_2$ is a continuously third-order Fréchet differentiable on a non-empty open convex subset $B_0 \subseteq B$. Suppose the hypotheses $(A1)$–$(A3)$, $(B1)$ and $(B2)$ are true and $m_0 \in B_0$. Assume that $\tilde{\tau}_0 = P\lambda\kappa, \tilde{\mu}_0 = Q\lambda\kappa^2, \tilde{\nu}_0 = \lambda\kappa^2\omega(\kappa)$ and $\tilde{\zeta}_0 = \Delta(\tilde{\tau}_0)\Theta'(\tilde{\tau}_0,\tilde{\mu}_0,\tilde{\nu}_0)$ satisfy $\tilde{\tau}_0 < \gamma$ and $\Delta(\tilde{\tau}_0)\tilde{\zeta}_0 < 1$, where γ is the smallest root of $\Gamma(\theta)\theta - 1 = 0$ and Γ, Δ and Θ' are defined by Equations (5), (6) and (41). In addition, suppose $\overline{U(m_0,\tilde{\rho}\kappa)} \subseteq B_0$, where $\tilde{\rho} = \frac{\Gamma(\tilde{\tau}_0)}{1-\tilde{\zeta}_0}$. Then, initiating with m_0, the iterative sequence $\{m_n\}$ created from the Scheme given in the Equation (2) converges to a zero m^* of $L(m) = 0$ with $m_n, m^* \in \overline{U(m_0,\tilde{\rho}\kappa)}$ and m^* is an exclusive zero of $L(m) = 0$ in $U(m_0, \frac{2}{P\lambda} - \tilde{\rho}\kappa) \cap B$. Furthermore, its error bound is given by*

$$\|m_n - m^*\| \leq \Gamma(\tilde{\tau}_0)\kappa\tilde{\varsigma}^n\tilde{\sigma}^{\frac{5^n-1}{4}}\left(\frac{1}{1-\tilde{\varsigma}\tilde{\sigma}^{5^n}}\right), \tag{44}$$

where $\tilde{\sigma} = \Delta(\tilde{\tau}_0)\tilde{\zeta}_0$ and $\tilde{\varsigma} = \frac{1}{\Delta(\tilde{\tau}_0)}$.

Proof. Analogous to the proof of Theorem 1. □

6. Numerical Example

Example 1. *Consider nonlinear integral equation from the reference [23] already mentioned in the introduction is given as*

$$m(s) = 1 + \int_0^1 G(s,t)\left(\frac{1}{2}m(t)^{\frac{5}{2}} + \frac{7}{16}m(t)^3\right)dt, s \in [0,1], \tag{45}$$

where $m \in [0,1], t \in [0,1]$ and G is the Green's function defined by

$$G(s,t) = \begin{cases} (1-s)t & t \leq s, \\ s(1-t) & s \leq t. \end{cases}$$

Proof. Solving Equation (45) is equivalent to find the solution for $L(m) = 0$, where $L : B \subseteq C[0,1] \to C[0,1]$:

$$[L(m)](s) = m(s) - 1 - \int_0^1 G(s,t)\left(\frac{1}{2}m(t)^{\frac{5}{2}} + \frac{7}{16}m(t)^3\right) dt, s \in [0,1].$$

The Fréchet derivatives of L are given by

$$L'(m)n(s) = n(s) - \int_0^1 G(s,t)\left(\frac{5}{4}m(t)^{\frac{3}{2}} + \frac{21}{16}m(t)^2\right) n(t)dt, n \in B,$$

$$L''(m)no(s) = -\int_0^1 G(s,t)\left(\frac{15}{8}m(t)^{\frac{1}{2}} + \frac{21}{8}m(t)\right) n(t)o(t)dt, \quad n,o \in B.$$

Using the max-norm and taking into account that a solution m^* of Equation (45) in $C[0,1]$ must satisfy

$$\|m^*\| - \frac{1}{16}\|m^*\|^{\frac{5}{2}} - \frac{7}{128}\|m^*\|^3 - 1 \leq 0,$$

i.e., $\|m^*\| \leq s_1 = 1.18771$ and $\|m^*\| \geq s_2 = 2.54173$, where s_1 and s_2 are the positive roots of the real equation $t - \frac{t^{\frac{5}{2}}}{16} - \frac{7}{128}t^3 - 1 = 0$. Consequently, if we look for a solution m^* such that $\|m^*\| \leq s_1$, we can consider $U(0,s) \subseteq C[0,1]$, where $s \in (s_1, s_2)$, as a non-empty open convex domain. We choose, for example, $s = 2$ and therefore $B = U(0,2)$. If $m_0 = 1$, then

$$\|\wp_0\| = \frac{128}{87} = \lambda, \|\wp_0 L(m_0)\| \leq \frac{15}{87} = \kappa, \|L''(m)\| \leq \frac{15\sqrt{2}}{64} + \frac{21}{32} = P.$$

Thus, $\tau_0 \approx 0.2505$. Hence, $\tau_0\Gamma(\tau_0) = 0.4068 < 1$ and $\Delta(\tau_0)\zeta_0 = 0.790 < 1$ (It is noticeable that, if we choose the function $\Gamma(m)$ from the reference [21], then we get $\Delta(\tau_0)\zeta_0 = 1.280 > 1$ which violates one of the assumed hypotheses considered in Theorem 1 and hence this motivates us to recalculate the function $\Gamma(m)$). In addition, $U(m_0, \rho\kappa) = U(1, 0.5270) \subseteq U(0,2) = B$. Thus, the conditions of Theorem 1 of Section 4 are satisfied and the nonlinear Equation (45) has the solution m^* in the region $\{u \in C[0,1] : \|u - 1\| \leq 0.5270\}$, which is unique in $\{u \in C[0,1] : \|u - 1\| < 0.8492\} \cap B$. Hence, we can deduce that the existence ball of solution based on our result is superior to that of Wang and Kou in [23], but our uniqueness ball is inferior. □

Example 2. *Now, consider another example discussed in [22] and also mentioned in the introduction, is given by*

$$h(m) = \begin{cases} m^3 ln(m^2) - 6m^2 - 3m + 8, & m \in (-2,0) \cup (0,2), \\ 0, & m = 0. \end{cases} \tag{46}$$

Proof. Taking $U(0,2) = B$. Let $m_0 = 1$ be an initial approximation. The derivatives of h are given by

$$\begin{aligned} h'(m) &= 3m^2 ln(m^2) + 2m^2 - 12m - 3, \\ h''(m) &= 6m ln(m^2) + 10m - 12, \\ h'''(m) &= 6 ln(m^2) + 22. \end{aligned}$$

Clearly, h''' is unbounded in B and does not satisfy the condition $(A4)$ but satisfies assumption $(B1)$, and we have

$$\|\wp_0\| = \frac{1}{13} = \lambda, \|\wp_0 h(m_0)\| = \frac{1}{13} = \kappa, \|h''(m)\| \leq 12 ln(4) + 32 = P.$$

$\|h'''(m_0)\| = 22$, $\|h'''(m) - h'''(n)\| \leq \frac{12}{1-\frac{13}{32+12log(4)}}|m-n|$, for all $m, n \in U\left(1, \frac{13}{32+12ln(4)}\right)$. Here, $\omega(z) = \frac{12}{1-\frac{13}{32+12log(4)}}z$ and $\phi(\epsilon) = 1$. Here, $\tau_0 \approx 0.2878$ and since $\tau_0\Gamma(\tau_0) = 0.51440 < 1$, $\Delta(\tau_0)\zeta_0 = 0.01742 < 1$. Thus, the assumptions of Theorem 2 of Section 5 are satisfied. In addition, thus, the solution lies in the ball $m \in U(1, 0.13867)$, which is unique in $U(1, 0.39592) \cap B$. Table 1 shows the comparison of error bounds for the considered Algorithm mentioned in the Equation 2 but with two different values of function $\Gamma(m)$ (One is given in the reference [21] and the other is recalculated here). This table also confirms that the value of the recalculated function is prominent.

Table 1. Comparison of the error bounds for Method 2.

n	With Recalculated $\Gamma(m)$	With $\Gamma(m)$ Calculated in [21]
1	0.00085294	0.0019139
2	1.4091×10^{-13}	2.7117×10^{-11}
3	3.1182×10^{-61}	2.0135×10^{-48}
4	2.9759×10^{-298}	5.9108×10^{-232}

□

7. Conclusions

In this contribution, we have analyzed the semilocal convergence of a well defined multi-point variant of the Jarratt method in Banach spaces. This iterative method can be used to solve various kinds of nonlinear equations that satisfy the assumed set of hypotheses. The analysis of this method has been examined using recurrence relations by relaxing the assumptions in two different approaches. In the first approach, we have softened the classical convergence conditions to the prove convergence, existence and uniqueness results together with a priori error bounds. In another way, we have assumed the norm of the third order Fréchet derivative on an initial iterate, so that it never gets unbounded on the given domain and, in addition, it satisfies the local ω-continuity condition as well. Two numerical applications are mentioned that sustain our theoretical consideration.

Author Contributions: For research All the authors have similar contribution.

Acknowledgments: The authors are grateful to the reviewers for their significant criticism which made the paper more elegant and readable.

References

1. Ortega, J.M.; Rheinboldt, W.C. *Iterative Solution of Nonlinear Equation in Several Variables*; Academic Press: New York, NY, USA; London, UK, 1970.
2. Kantorovich, L.V.; Akilov, G.P. *Functional Analysis*; Pergamon Press: Oxford, UK, 1982.
3. Rall, L.B. *Computational Solution of Nonlinear Operator Equations*; Robert E Krieger: New York, NY, USA, 1979.
4. Wang, X.; Gu, C.; Kou, J. Semilocal convergence of a multipoint fourth-order Super–Halley method in Banach spaces. *Numer. Algor.* **2011**, *56*, 497–516. [CrossRef]
5. Wang, X.; Kou, J.; Gu, C. Semilocal convergence of a sixth-order Jarratt method in Banach spaces. *Numer. Algor.* **2011**, *57*, 441–456. [CrossRef]
6. Singh, S.; Gupta, D.K.; Martínez, E.; Hueso, J.L. Semilocal convergence analysis of an iteration of order five using recurrence relations in Banach spaces. *Mediterr. J. Math.* **2016**, *13*, 4219–4235. [CrossRef]
7. Jaiswal, J.P. Semilocal convergence of an eighth-order method in Banach spaces and its computational efficiency. *Numer. Algor.* **2016**, *71*, 933–951. [CrossRef]

8. Hernández, M.A. Chebyshev's approximation algorithms and applications. *Comput. Math. Appl.* **2001**, *41*, 433–445. [CrossRef]
9. Parida, P.K.; Gupta, D.K. Recurrence relations for semilocal convergence of a Newton-like method in Banach spaces. *J. Math. Anal. Appl.* **2008**, *345*, 350–361. [CrossRef]
10. Parida, P.K.; Gupta, D.K. Semilocal convergence of a family of third-order methods in Banach spaces under Hölder continuous second derivative. *Non. Anal. Theo. Meth. Appl.* **2008**, *69*, 4163–4173. [CrossRef]
11. Wang, X.; Kou, J. Convergence for modified Halley-like methods with less computation of inversion. *J. Diff. Eqn. Appl.* **2013**, *19*, 1483–1500. [CrossRef]
12. Ezquerro, J.A.; Hernández, M.A. On the R-order of the Halley method. *J. Math. Anal. Appl.* **2005**, *303*, 591–601. [CrossRef]
13. Ezquerro, J.A.; Hernández, M.A. A generalization of the Kantorovich type assumptions for Halley's method. *Int. J. Comput. Math.* **2007**, *84*, 1771–1779. [CrossRef]
14. Parida, P.K.; Gupta, D.K. Semilocal convergence of a family of third-order Chebyshev-type methods under a mild differentiability condition. *Int. J. Comput. Math.* **2010**, *87*, 3405–3419. [CrossRef]
15. Parida, P.K.; Gupta, D.K.; Parhi, S.K. On Semilocal convergence of a multipoint third order method with R-order $(2+p)$ under a mild differentiability condition. *J. Appl. Math. Inf.* **2013**, *31*, 399–416. [CrossRef]
16. Prashanth, M.; Gupta, D.K. Convergence of a parametric continuation method. *Kodai Math. J.* **2014**, *37*, 212–234. [CrossRef]
17. Prashanth, M.; Gupta, D.K. Semilocal convergence for Super-Halley's method under ω-differentiability condition. *Jpn. J. Ind. Appl. Math.* **2015**, *32*, 77–94. [CrossRef]
18. Wang, X.; Kou, J. Semilocal convergence of multi-point improved Super-Halley-type methods without the second derivative under generalized weak condition. *Numer. Algor.* **2016**, *71*, 567–584. [CrossRef]
19. Wang, X., Kou, J. Semilocal convergence analysis on the modifications for Chebyshev–Halley methods under generalized condition. *Appl. Math. Comput.* **2016**, *281*, 243–251. [CrossRef]
20. Wang, X.; Kou, J. Semilocal convergence on a family of root-finding multi-point methods in Banach spaces under relaxed continuity condition. *Numer. Algor.* **2017**, *74*, 643–657. [CrossRef]
21. Wang, X.; Kou, J. Semilocal convergence of a modified multi-point Jarratt method in Banach spaces under general continuity condition. *Numer. Algor.* **2012**, *60*, 369–390. [CrossRef]
22. Wang, X.; Kou, J. Convergence for a class of improved sixth-order Chebyshev–Halley type method. *Appl. Math. Comput.* **2016**, *273*, 513–524. [CrossRef]
23. Wang, X.; Kou, J. Convergence for a family of modified Chebyshev methods under weak condition. *Numer. Algor.* **2014**, *66*, 33–48. [CrossRef]

3

An Efficient Iterative Method Based on Two-Stage Splitting Methods to Solve Weakly Nonlinear Systems

Abdolreza Amiri [1], Mohammad Taghi Darvishi [1,*], Alicia Cordero [2] and Juan Ramón Torregrosa [2]

1 Department of Mathematics, Razi University, Kermanshah 67149, Iran
2 Institute for Multidisciplinary Mathematics, Universitat Politècnica de València, Camino de Vera s/n, 46022 Valencia, Spain
* Correspondence: darvishi@razi.ac.ir.

Abstract: In this paper, an iterative method for solving large, sparse systems of weakly nonlinear equations is presented. This method is based on Hermitian/skew-Hermitian splitting (HSS) scheme. Under suitable assumptions, we establish the convergence theorem for this method. In addition, it is shown that any faster and less time-consuming two-stage splitting method that satisfies the convergence theorem can be replaced instead of the HSS inner iterations. Numerical results, such as CPU time, show the robustness of our new method. This method is easy, fast and convenient with an accurate solution.

Keywords: system of nonlinear equations; Newton method; Newton-HSS method; nonlinear HSS-like method; Picard-HSS method

---

1. Introduction

For $G : D \subseteq \mathbb{C}^m \longrightarrow \mathbb{C}^m$, we consider the following system of nonlinear equations:

$$G(x) = 0. \tag{1}$$

One may encounter equations like (1) in some areas of scientific computing. In particular, when the technique of finite elements or finite differences are used to discretize nonlinear boundary problems, integral equations and certain nonlinear partial differential equations. Finding the roots of systems like (1) has widespread applications in numerical and applied mathematics. There are many iterative schemes to solve (1). The most common one is the second order classical Newton's scheme, which solves (1) iteratively as

$$x^{(n+1)} = x^{(n)} - G'(x^{(n)})^{-1}G(x^{(n)}), \qquad n = 0, 1, \ldots, \tag{2}$$

where $G'(x^{(n)})$ is the Jacobian matrix of G, evaluated in the nth iteration. To avoid computation of inverse of the Jacobian matrix $G'(x)$, Equation (2) is changed to

$$G'(x^{(n)})(x^{(n+1)} - x^{(n)}) = -G(x^{(n)}). \tag{3}$$

Equation (3) is a system of linear equations. Hence, by $s^{(n)} = x^{(n+1)} - x^{(n)}$, we have to solve the following system of equations:

$$G'(x^{(n)})s^{(n)} = -G(x^{(n)}), \tag{4}$$

whence $x^{(n+1)} = x^{(n)} + s^{(n)}$. Thus, by using this approach, we have to solve a system of linear equations such as

$$Ax = b, \tag{5}$$

which we usually use an iterative scheme to solve it.

Furthermore, an inexact Newton method [1–4] is a generalization of Newton's method for solving (1), in which, at the nth iteration, the step-size $s^{(n)}$ from current approximate solution $x^{(n)}$ must satisfy a condition such as

$$\| G(x^{(n)}) + G'(x^{(n)})s^{(n)} \| \leq \eta_n \| G(x^{(n)}) \|,$$

for a "forcing term" $\eta_n \in [0, 1)$. Let us consider system (1) in which $G(x)$ can be separated into linear and nonlinear terms, Ax and $\varphi(x)$, respectively, that is

$$G(x) = \varphi(x) - Ax \text{ or } Ax = \varphi(x). \tag{6}$$

In (6), the $m \times m$ complex matrix A is a positive definite, large and sparse matrix. In addition, vector-valued function $\varphi : D \subseteq \mathbb{C}^m \longrightarrow \mathbb{C}^m$ is continuously differentiable. Furthermore, x is an m-vector and D is an open set. When the norm of linear part Ax is strongly dominant over the norm of nonlinear part $\varphi(x)$ in a specific norm, system (6) is called a weakly nonlinear system [5,6]. Bai [5] used the separability and strong dominance between the linear and the nonlinear parts and introduced the following iterative scheme

$$Ax^{(n+1)} = \varphi(x^{(n)}). \tag{7}$$

Equation (7) is a system of linear equations. When the matrix A is positive definite, Axelsson et al. [7] solved it by a class of nested iteration methods. To solve linear positive definite systems, Bai et al. [8] applied the Hermitian/skew-Hermitian splitting (HSS) iterative scheme. For solving the large sparse, non-Hermitian positive definite linear systems, Li et al. [9] used an asymmetric Hermitian/skew-Hermitian (AHSS) iterative scheme. Moreover, to improve the robustness of the HSS method, some HSS-based iterative algorithms have been introduced. Bai and Yang [10] presented Picard-HSS and HSS-like methods to solve (7), when matrix A is a positive definite matrix. Based on the matrix multi-splitting technique, block and asynchronous two-stage methods are introduced by Bai et al. [11]. The Picard circulant and skew-circulant splitting (Picard-CSCS) algorithm and the nonlinear CSCS-like iterative algorithm are presented by Zhu and Zhang [12], when the coefficient matrix A is a Teoplitz matrix. A class of lopsided Hermitian/skew-Hermitian splitting (LHSS) algorithms and a class of nonlinear LHSS-like algorithms are used by Zhu [6] to solve the large and sparse of weakly nonlinear systems.

It must be noted that system (6) is a special form of system (1). Generally, system (6) is nonlinear. If we classify Picard-HSS and nonlinear HSS-like iterative methods as Jacobian-free schemes, in many cases, they are not as successful as Jacobian dependent schemes such as the Newton method. Most of the methods for solving nonlinear systems need to compute or approximate the Jacobian matrix in the obtained points at each step of the iterative methods, which is a very time-consuming process, especially when the Jacobian matrices $\varphi'(x^{(n)})$ are dense. Therefore, introducing any scheme that does not need to compute the Jacobian matrix and can solve a wider range of problems than the existing ones is welcome. In fact, Jacobian-free methods to solve nonlinear systems are very important and form an attractive area of research.

In this paper, we present a new iterative method to solve weakly nonlinear systems. Even though the new algorithm uses some notions of mentioned algorithms, but differs from all of them because it has three important characteristics. At the first, the new algorithm is a fully Jacobian-free one. At the second, it is easy to use, and, finally, it is very successful to solve weakly nonlinear systems. The new iterative method is a synergistic combination of high order Newton-like methods and a special splitting of the coefficient matrix A in (5).

The rest of this paper has organized as follows: in the following section, we present our new algorithm. We prove convergence of our algorithm in Section 3. We apply our algorithm to solve some problems in Section 4. In Section 5, we conclude our results and give some comments and discussions.

2. The New Algorithm

In linear system $Ax = b$, we suppose that $A = H + S$, where $H = \frac{1}{2}(A + A^*)$, $S = \frac{1}{2}(A - A^*)$, and A^* is the conjugate transpose of matrix A. Hence, H and S are, respectively, Hermitian and skew-Hermitian parts of A. By an initial guess $x_0 \in \mathbb{C}^n$, and positive constants α and tol, in HSS scheme [8], one computes x_l for $l = 1, 2, \ldots$ as

$$\begin{cases} (\alpha I + H)x_{l+\frac{1}{2}} = (\alpha I - S)x_l + b, \\ (\alpha I + S)x_{l+1} = (\alpha I - H)x_{l+\frac{1}{2}} + b, \end{cases} \tag{8}$$

where I is the identity matrix. Stopping criterion for (8) is $\|b - Ax_l\| \leq \text{tol}\|b - Ax_0\|$, for known x_0 and tol.

Bai and Guo [13] used an HSS scheme as inner iterations to generate an inexact version of Newton's method as:

(1) Consider the initial guess $x^{(0)}$, α, tol and the sequence $\{l_n\}_{n=0}^{\infty}$ of positive integers.

(2) For $n = 1, 2, \ldots$ until $\|G(x^{(n)})\| \leq \text{tol}\|G(x^{(0)})\|$ do:

(2.1) Set $s_0^{(n)} = 0$.

(2.2) For $l = 1, 2, \ldots, l_n - 1$, apply Algorithm HSS as

$$\begin{cases} (\alpha I + H(x^{(n)}))s_{l+\frac{1}{2}}^{(n)} = (\alpha I - S(x^{(n)}))s_l^{(n)} - G(x^{(n)}) \\ (\alpha I + S(x^{(n)}))s_{l+1}^{(n)} = (\alpha I - H(x^{(n)}))s_{l+\frac{1}{2}}^{(n)} - G(x^{(n)}), \end{cases}$$

and obtain $s_{l_n}^{(n)}$ such that

$$\| G(x^{(n)}) + G'(x^{(n)})s_{l_n}^{(n)} \| \leq \eta_n \| G(x^{(n)}) \|, \quad \text{for some } \eta_n \in [0, 1).$$

(2.3) Set $x^{(n+1)} = x^{(n)} + s_{l_n}^{(n)}$.

In addition, to solve weakly nonlinear problems, one can use a Picard-HSS method as a simple and Jacobian-free method, which is described as follows [10].

2.1. *Picard-HSS Iteration Method*

Suppose that $\varphi : D \subset \mathbb{C}^n \to \mathbb{C}^n$ is a continuous function and $A \in \mathbb{C}^{n\times n}$ is a positive definite matrix. For an initial guess $x^{(0)}$ and for a positive integer sequence $\{l_n\}_{n=0}^{\infty}$, Picard-HSS iterative method computes $x^{(n+1)}$ for $n = 0, 1, 2, \ldots$, by using the following iterative scheme, until the stopping criterion is satisfied [10],

(1) Set $x_l^{(n)} := x^{(n)}$;

(2) For $l = 0, 1, 2, \ldots, n - 1$, obtain $x^{(n+1)}$ from solving the following:

$$\begin{cases} (\alpha I + H)x_{l+\frac{1}{2}}^{(n)} &= (\alpha I - S)x_l^{(n)} + \varphi(x^{(n)}), \\ (\alpha I + S)x_{l+1}^{(n)} &= (\alpha I - H)x_{l+\frac{1}{2}}^{(n)} + \varphi(x^{(n)}). \end{cases}$$

(3) Set $x^{(n+1)} := x_{l_n}^{(n)}$.

The numbers $l_n, n = 0, 1, 2, \ldots$ depend on the problem, so practically they are difficult to be determined in real computations. A modified form of Picard-HSS iteration scheme, called the nonlinear HSS-like method, has been presented [10] to avoid using inner iterations as follows.

2.2. Nonlinear HSS-Like Iteration Method

Obtain $x^{(n+1)}, n = 0, 1, 2, \ldots$ from the following [10], for a given $x^{(0)} \in D \subset \mathbb{C}^n$, until the stopping condition is satisfied

$$\begin{cases} (\alpha I + H)x^{(n+\frac{1}{2})} & = & (\alpha I - S)x^{(n)} + \varphi(x^{(n)}), \\ (\alpha I + S)x^{(n+1)} & = & (\alpha I - H)x^{(n+\frac{1}{2})} + \varphi(x^{(n+\frac{1}{2})}). \end{cases}$$

However, in this method, it is necessary to evaluate the nonlinear term $\varphi(x)$ at each step, which for complicated nonlinear terms $\varphi(x)$ is too costly.

2.3. Our Proposal Iterative Scheme

For solving (6) without computing Jacobian matrices, we present a new algorithm. This algorithm is a strong tool for solving weakly nonlinear problems, as Picard and nonlinear Picard algorithms, but, in comparison with Picard and nonlinear Picard algorithms, it solves a wider range of nonlinear systems. First, we change (7) as

$$Ax^{(n+1)} = Ax^{(n)} - Ax^{(n)} + \varphi(x^{(n)}) \tag{9}$$

and

$$Ax^{(n+1)} - Ax^{(n)} = -Ax^{(n)} + \varphi(x^{(n)}). \tag{10}$$

After computing $x^{(n)}$, set $b^{(n)} = \varphi(x^{(n)})$, $G_n(x) = b^{(n)} - Ax$. Then, by intermediate iterations, obtain $x^{(n+1)}$ as:

- Let $x_0^{(n)} = x^{(n)}$ and until $\|G(x_k^{(n)})\| \leq \text{tol}_n \|G(x_0^{(n)})\|$ do:

$$As_k^{(n)} = G(x_k^{(n)}), \tag{11}$$

  where $s_k^{(n)} = x_{k+1}^{(n)} - x_k^{(n)}$ (k is the counter of the number of iterations (11)).
- For solving (11), one may use any inner solver; here, we use an HSS scheme. Next, for initial value $x_0^{(n)}$ and $k = 1, 2, \ldots, k_n - 1$ until

$$\|G_n(x_k^{(n)})\| \leq \text{tol}_n \|G_n(x_0^{(n)})\|, \tag{12}$$

  apply the HSS scheme as:

  (1) Set $s_{k,0}^{(n)} = 0$.
  (2) For $l = 0, 1, 2, \ldots, l_{k_n} - 1$, apply algorithm HSS ($l$ is the counter of the number of HSS iterations):

$$\begin{cases} (\alpha I + H)s_{k,l+\frac{1}{2}}^{(n)} & = & (\alpha I - S)s_{k,l}^{(n)} + G_n(x_k^{(n)}), \\ (\alpha I + S)s_{k,l+1}^{(n)} & = & (\alpha I - H)s_{k,l+\frac{1}{2}}^{(n)} + G_n(x_k^{(n)}) \end{cases} \tag{13}$$

and obtain $s_{k,l_{k_n}}^{(n)}$ such that

$$\| G_n(x_k^{(n)}) - As_{k,l_{k_n}}^{(n)} \| \leq \eta_k^{(n)} \| G_n(x_k^{(n)}) \|, \quad \eta_k^{(n)} \in [0,1). \tag{14}$$

(3) Set $x_{k+1}^{(n)} = x_k^{(n)} + s_{k,l_{k_n}}^{(n)}$ (l_{k_n} is the required number of HSS inner iterations for satisfying (14)).

- Finally, set $x_0^{(n+1)} = x_{k_n}^{(n)}$ (k_n is the required number of iterations (11) in the nth step, for satisfying (12)), $b^{(n+1)} = \varphi(x_0^{(n+1)})$, $G_{n+1}(x) = b^{(n+1)} - Ax$ and again apply steps 3–14 in Algorithm 1 until to achieve the following stopping criterion:

$$\|Ax^{(n)} - \varphi(x^{(n)})\| \leq \text{tol} \, \| Ax^{(0)} - \varphi(x^{(0)}) \| \, .$$

Algorithm 1: JFHSS Algorithm

Input: $x^{(0)}$, tol, α, $n \leftarrow 1$
Output: The root of $Ax - \varphi(x) = 0$
1 $root \leftarrow x^{(0)}$
2 **while** $\|Ax^{(n)} - \varphi(x^{(n)})\| > \text{tol} \, \| Ax^{(0)} - \varphi(x^{(0)}) \|$ **do**
  Input: tol_n
  Set: $x_0^{(n)} = x^{(n)}$, $b^{(n)} = \varphi(x^{(n)})$ and $G_n(x) = b^{(n)} - Ax$, $k = 1$.
3   **while** $\|G_n(x_k^{(n)})\| > \text{tol}_n \|G_n(x_0^{(n)})\|$ **do**
4   **Set:** $l = 0$, $s_{k,0}^{(n)} = 0$.
5   **while** $\| G_n(x_k^{(n)}) - As_{k,l_{k_n}}^{(n)} \| > \eta_k^{(n)} \|G_n(x_k^{(n)})\|$ **do**
6   (**the HSS algorithm**)

$$(\alpha I + H)s_{k,l+\frac{1}{2}}^{(n)} = (\alpha I - S)s_{k,l}^{(n)} + G_n(x_k^{(n)}),$$
$$(\alpha I + S)s_{k,l+1}^{(n)} = (\alpha I - H)s_{k,l+\frac{1}{2}}^{(n)} + G_n(x_k^{(n)}),$$

7   **if** $\| G_n(x_k^{(n)}) - As_{k,l_{k_n}}^{(n)} \| \leq \eta_k^{(n)} \|G_n(x_k^{(n)})\|$ **then**
8         $l \leftarrow l_{k_n}$, $x_{k+1}^{(n)} = x_k^{(n)} + s_{k,l_{k_n}}^{(n)}$
9   **else**
10        $l \leftarrow l + 1$
11   **if** $\|G_n(x_k^{(n)})\| \leq \text{tol}_n \|G_n(x_0^{(n)})\|$ **then**
12      $k \leftarrow k_n$, $x_0^{(n+1)} = x_{k_n}^{(n)}$
13   **else**
14      $k \leftarrow k + 1$
15 **if** $\|Ax^{(n)} - \varphi(x^{(n)})\| \leq \text{tol} \, \| Ax^{(0)} - \varphi(x^{(0)}) \|$ **then**
16    $root \leftarrow x^{(n)}$
17 **else**
18    $n \leftarrow n + 1$, $x^{(n)} = x_0^{(n+1)}$, $b^{(n)} = \varphi(x^{(n)})$, $G_n(x) = b^{(n)} - Ax$
19 **return** $root$

We call this new method a JFHSS (Jacobian-free HSS) algorithm, and its steps are shown in Algorithm 1.

In addition, we call the intermediate iterations Newton-like iteration because this kind of iteration uses the same procedure as an inexact Newton's method, except, since the function we use here is $b^{(n)} - Ax$ for $n = 1, 2, \cdots$, we don't need to compute any Jacobian and, in fact, the Jacobian is the matrix A. For this reason, we also call this iterative method a "Jacobian-free method".

Since the JFHSS scheme uses many HSS inner iterations, one may use another splitting scheme instead of the HSS method. In fact, if any faster and less time-consuming splitting method is available that satisfies the convergence theorem, presented in the next section, then it can be used instead of the HSS algorithm. One of these methods that is proposed in [14] is GPSS (generalized positive definite and skew-Hermitian splitting) algorithm that uses a positive-definite and skew-Hermitian splitting scheme instead of a Hermitian and skew-Hermitian one. Let H and S be the Hermitian and skew-Hermitian parts of A; then, the GPSS algorithm splits A as $A = P_1 + P_2$ where P_1 and P_2 are, respectively, positive definite and skew-Hermitian matrices. In fact, we have

$$P_1 = \mathcal{D} + 2L_{\mathcal{G}}, \quad P_2 = \mathcal{K} + L_{\mathcal{G}}^* - L_{\mathcal{G}} + S, \tag{15}$$

or

$$P_1 = \mathcal{D} + 2L_{\mathcal{G}}^*, \quad P_2 = \mathcal{K} + L_{\mathcal{G}} - L_{\mathcal{G}}^* + S, \tag{16}$$

where $\mathcal{G}$ and $\mathcal{K}$ are, respectively, Hermitian and Hermitian positive semidefinite matrices of H, that is, $H = \mathcal{G} + \mathcal{K}$; in addition, $\mathcal{D}$ and $L_{\mathcal{G}}$ are the diagonal matrix and the strictly lower triangular matrices of $\mathcal{G}$, respectively (see [14]).

Thus, to solve the system of linear Equation (5) for an initial guess $x_0 \in \mathbb{C}^n$, and positive constants α and tol, the GPSS iteration scheme (until the stopping criterion is satisfied) computes x_l for $l = 1, 2, \ldots$ by

$$\begin{cases} (\overline{\alpha} I + P_1)x_{l+\frac{1}{2}} & = & (\overline{\alpha} I - P_2)x_l + b, \\ (\overline{\alpha} I + P_2)x_{l+1} & = & (\overline{\alpha} I - P_1)x_{l+\frac{1}{2}} + b, \end{cases} \tag{17}$$

where $\overline{\alpha}$ is a given positive constant and I denotes the identity matrix. In addition, if, in Algorithm 1, we use a GPSS scheme instead of an HSS one, we denote the new method by JFGPSS (Jacobian free GPSS).

3. Convergence of the New Method

As we mentioned in the first section, for solving a nonlinear system, if one can separate (1) into linear and nonlinear terms, Ax and $\phi(x)$, when Ax is strongly dominant over the nonlinear term, Picard-HSS and nonlinear HSS-like methods can solve the problem. However, in many cases, even for weakly nonlinear ones, they may fail to solve the problems. Thus, to obtain a more useful method for solving (6), based on some splitting methods, we presented a new iterative method. Now, we prove that Algorithm 1 converges to the solution of a weakly nonlinear problem (6). In the following theorem, we prove the convergence of the JFHSS scheme.

Theorem 1. *Let $x^{(0)} \in \mathbb{C}^n$ and $\varphi : D \subset \mathbb{C}^n \to \mathbb{C}^n$ be a G-differentiable function on an open set $N_0 \subset D$ on which $\varphi'(x)$ is continuous and $\max \|A^{-1}\varphi'(x)\| = L < 1$. Let us suppose that $H = \frac{1}{2}(A + A^*)$ and $S = \frac{1}{2}(A - A^*)$ are the Hermitian and skew-Hermitian parts of the positive definite matrix A and also that $\mathcal{M}$ is an upper bound for $\|A^{-1}G(x^{(0)})\|$, and l_{k_n} is the number of HSS inner iterations in which the stopping criterion (14) is satisfied,*

$$l_*^n > \left\lfloor \frac{\ln\left(\frac{(1-\eta)(1-\eta^{k_n-1})}{L} - 1\right)}{\ln(\theta)} \right\rfloor, \tag{18}$$

with $l_*^n = \liminf_{k_n \to \infty} l_{k_n}$ *for* $n = 1, 2, 3, \ldots$, η *is the tolerance in Newton-like intermediate iterations with* $L < (1-\eta)^2$ *and* $\theta = \|T\|$, *where* T *is the HSS inner iteration matrix that can be written as*

$$T = (\alpha I + S)^{-1}(\alpha I - H)(\alpha I + H)^{-1}(\alpha I - S).$$

Then, the sequence of iteration $\{x^{(k)}\}_{k=0}^{\infty}$, *which is generated by a JFHSS scheme in Algorithm 1, is well-defined and converges to* x^*, *satisfying* $G(x^*) = 0$, *and also*

$$\|x^{(n+1)} - x^{(n)}\| \leq \delta \mathcal{M} \rho^n, \tag{19}$$

$$\|x^{(n+1)} - x^{(0)}\| \leq \frac{\delta \mathcal{M}}{1-\rho}, \tag{20}$$

where $\delta = \limsup_{n\to\infty} \dfrac{(1+\theta^{l_*^n})}{1-\eta}$ *and* $\rho = \limsup_{n\to\infty} \rho_n$ *for* $\rho_n = \dfrac{(1+\theta^{l_*^n})}{1-\eta} L + \eta^{k_{n-1}}$.

Proof. Note that $\|T\| \leq \max_{\lambda_i \in \lambda(H)} \left| \dfrac{\alpha - \lambda_i}{\alpha + \lambda_i} \right| < 1$ (see [8]), where $\lambda(H)$ is the spectral radius of H and α is a positive constant in HSS inner iterations of JFHSS scheme. Based on Algorithm 1, we can express $x^{(n+1)}$ as

$$\begin{aligned}
x^{(n+1)} &= x_{k_n}^{(n)} = x_{k_n-1}^{(n)} + (I - T^{l_{k_n}}) G_n'(x_{k_n-1}^{(n)})^{-1} G_n(x_{k_n-1}^{(n)}) \\
&= x_{k_n-1}^{(n)} + (I - T^{l_{k_n}}) A^{-1} G_n(x_{k_n-1}^{(n)}) \\
&= x_{k_n-2}^{(n)} + (I - T^{l_{k_n-1}}) A^{-1} G_n(x_{k_n-2}^{(n)}) + (I - T^{l_{k_n}}) A^{-1} G_n(x_{k_n-1}^{(n)}) \\
&= x_{k_n-3}^{(n)} + (I - T^{l_{k_n-2}}) A^{-1} G_{(n)}(x_{k_n-3}^{(n)}) + (I - T^{l_{k_n-1}}) A^{-1} G_n(x_{k_n-2}^{(n)}) \\
&\quad + (I - T^{l_{k_n}}) A^{-1} G_n(x_{k_n-1}^{(n)}) \\
&= x_0^{(n)} + (I - T^{l_1}) A^{-1} G_n(x_0^{(n)}) + (I - T^{l_2}) A^{-1} G_n(x_1^{(n)}) + \cdots \\
&\quad + (I - T^{l_{k_n-2}}) A^{-1} G_n(x_{k_n-3}^{(n)}) + (I - T^{l_{k_n-1}}) A^{-1} G_n(x_{k_n-2}^{(n)}) \\
&\quad + (I - T^{l_{k_n}}) A^{-1} G_n(x_{k_n-1}^{(n)}) = x^{(n)} + \textstyle\sum_{j=1}^{k_n} (I - T^{l_j}) A^{-1} G_n(x_{j-1}^{(n)}).
\end{aligned} \tag{21}$$

In the last equality, we used $x_0^{(n)} = x^{(n)}$. If we set $\eta' = \dfrac{\eta}{\operatorname{cond}(A)}$ in (14) instead of η, where $\operatorname{cond}(A) = \|A\|\|A^{-1}\|$, then $\eta' \leq 1$. Because of (14), we have

$$\begin{aligned}
\|G_n(x_{k_n}^{(n)})\| &\leq \|G_n(x_{k_n}^{(n)}) - G_n(x_{k_n-1}^{(n)}) + G_n'(x_{k_n-1}^{(n)})(x_{k_n}^{(n)} - x_{k_n-1}^{(n)})\| \\
&\quad + \|G_n(x_{k_n-1}^{(n)}) - G_n'(x_{k_n-1}^{(n)})(x_{k_n}^{(n)} - x_{k_n-1}^{(n)})\| \\
&= \|G_n(x_{k_n-1}^{(n)}) - A(x_{k_n}^{(n)} - x_{k_n-1}^{(n)})\| \leq \eta' \|G_n(x_{k_n-1}^{(n)})\|,
\end{aligned}$$

so

$$\begin{aligned}
\|A^{-1} G_n(x_{k_n}^{(n)})\| &\leq \|A^{-1}\| \|G_n(x_{k_n}^{(n)})\| \leq \eta' \|A^{-1}\| \|G_n(x_{k_n-1}^{(n)})\| \\
&\leq \eta' \|A^{-1}\| \|A\| \|A^{-1} G_n(x_{k_n-1}^{(n)})\| = \eta \|A^{-1} G_n(x_{k_n-1}^{(n)})\|.
\end{aligned}$$

Therefore, by mathematical induction, we can obtain

$$\|A^{-1} G_n(x_{k_n}^{(n)})\| \leq \eta^{k_n} \|A^{-1} G_n(x_0^{(n)})\|. \tag{22}$$

Then, from (21), and since $\|I - T^{l_j}\| < 1 + \theta^{l_j} \leq 1 + \theta^{l_*^n}$ for $j = 1, 2, \ldots, k_n$, we have

$$\begin{aligned}\|x^{(n+1)}-x^{(n)}\| &\leq \textstyle\sum_{j=1}^{k_n}\|I-T^{l_j}\|\|A^{-1}G_n(x_{j-1}^{(n)})\| \\ &\leq (\|I-T^{l_1}\|+\eta\|I-T^{l_2}\|+\eta^2\|I-T^{l_3}\|+\cdots \\ &+\eta^{k_n-2}\|I-T^{l_{k_n-1}}\|+\eta^{k_n-1}\|I-T^{l_{k_n}}\|)\|A^{-1}G_n(x_0^{(n)})\| \\ &= (1+\eta+\eta^2+\cdots+\eta^{k_n-2}+\eta^{k_n-1})(1+\theta^{l_*^n})\|A^{-1}G_n(x_0^{(n)})\| \\ &= \frac{1-\eta^{k_n}}{1-\eta}(1+\theta^{l_*^n})\|A^{-1}G_n(x_0^{(n)})\|.\end{aligned} \tag{23}$$

Thus, from the last inequality, since $G_n(x)=b^{(n)}-Ax$, $b^{(n)}=\varphi(x^{(n)})$, we have

$$\begin{aligned}&\|x^{(n+1)}-x^{(n)}\| \leq \frac{1-\eta^{k_n}}{1-\eta}(1+\theta^{l_*^n})\|A^{-1}(b^{(n)}-Ax^{(n)})\| \\ &= \frac{1-\eta^{k_n}}{1-\eta}(1+\theta^{l_*^n})(\|A^{-1}(\varphi(x^{(n)})-\varphi(x^{(n-1)}))\|+\|A^{-1}(\varphi(x^{(n-1)})-Ax^{(n)})\|).\end{aligned} \tag{24}$$

Then, by using the multivariable Mean Value Theorem (see [15]), we can write

$$\|A^{-1}(\varphi(x^{(n)})-\varphi(x^{(n-1)}))\| \leq \max_{x\in\mathcal{S}}\|A^{-1}\varphi'(x)\|\|x^{(n)}-x^{(n-1)}\| = L\|x^{(n)}-x^{(n-1)}\|,$$

where $\mathcal{S}=\{x\ :\ x=tx^{(n)}+(1-t)x^{(n-1)},\ 0\leq t\leq 1\}$. Thus,

$$\|A^{-1}(\varphi(x^{(n)})-\varphi(x^{(n-1)}))\| \leq L\|x^{(n)}-x^{(n-1)}\|. \tag{25}$$

From the right-hand side of (24), using (22) for $n-1$, and (25), we have

$$\begin{aligned}&\|x^{(n+1)}-x^{(n)}\| \leq \\ &= \frac{1-\eta^{k_n}}{1-\eta}(1+\theta^{l_*^n})(L\|x^{(n)}-x^{(n-1)}\|+\|A^{-1}G_{n-1}(x^n)\|) \\ &\leq \frac{1-\eta^{k_n}}{1-\eta}(1+\theta^{l_*^n})(L\|x^{(n)}-x^{(n-1)}\|+\eta^{k_{n-1}}\|A^{-1}G_{n-1}(x_0^{(n-1)})\|).\end{aligned} \tag{26}$$

If in the last inequality of (26), from (23), we use $\|x^{(n)}-x^{(n-1)}\| \leq \frac{1-\eta^{k_{n-1}}}{1-\eta}(1+\theta^{l_*^{n-1}})\|A^{-1}G_n(x_0^{(n-1)})\|$, then

$$\begin{aligned}&\|x^{(n+1)}-x^{(n)}\| \leq \\ &\frac{1-\eta^{k_n}}{1-\eta}(1+\theta^{l_*^n})(L\frac{1-\eta^{k_{n-1}}}{1-\eta}(1+\theta^{l_*^{n-1}})\|A^{-1}G_{n-1}(x_0^{(n-1)})\|+\eta^{k_{n-1}}\|A^{-1}G_{n-1}(x_0^{(n-1)})\|) \\ &\leq \frac{1-\eta^{k_n}}{1-\eta}(1+\theta^{l_*^n})(L\frac{1-\eta^{k_{n-1}}}{1-\eta}(1+\theta^{l_*^{n-1}})+\eta^{k_{n-1}})\|A^{-1}G_{n-1}(x_0^{(n-1)})\|.\end{aligned}$$

As $1-\eta^{k_n}<1$, $n=1,2,\cdots$ and by the definition of ρ and δ, we have

$$\|x^{(n+1)}-x^{(n)}\| \leq \delta\rho\|A^{-1}G_{n-1}(x_0^{(n-1)})\|. \tag{27}$$

By mathematical induction and since $\|A^{-1}G_0(x_0^{(0)})\| \leq \mathcal{M}$,

$$\|x^{(n+1)}-x^{(n)}\| \leq \delta\rho^n\|A^{-1}G_0(x_0^{(0)})\| \leq \delta\mathcal{M}\rho^n, \tag{28}$$

which yields (19). By the stopping criterion (18), we must have $\rho<1$ and then, using (19), it is easy to deduce

$$\|x^{(n+1)} - x^{(0)}\| \leq \|x^{(n+1)} - x^{(n)}\| + \|x^{(n)} - x^{(n-1)}\| + \cdots + \|x^{(1)} - x^{(0)}\| \leq \frac{\delta\mathcal{M}}{1-\rho},$$

which is the relation (20).

Thus, the sequence $\{x^{(n)}\}$ is in a ball with center $x^{(0)}$ and radius $r = \dfrac{\delta\mathcal{M}}{1-\rho}$. From (28), sequence $\{x^{(n)}\}$ also converges to its limit point x^*. From the following iteration,

$$x_1^{(n)} = x_0^{(n)} + (I - T^{l_1})A^{-1}G_n(x_0^{(n)}),$$

when $n \longrightarrow \infty$, $\|x_0^{(n)} - x^*\| \longrightarrow 0$, $\|x_1^{(n)} - x^*\| \longrightarrow 0$, $l_1 \longrightarrow \infty$. Moreover, as $\|T\| < 1$, then $T^{l_1} \to 0$ and we have

$$G(x^*) = 0,$$

which completes the proof. □

Note that, in some applications, the stopping criterion (18) may be obtained as negative; this shows that, for all $l_* \geqslant 1$, we must have $\rho < 1$.

In addition, it is easy to deduce from the above theorem that any iterative method that its iteration matrix satisfies in $\|T\| < 1$ can be used instead of the HSS method. For a JFGPSS case, the proof is similar, except, in the inner iteration, the iterative matrix is

$$T = (\overline{\alpha}I + P_2)^{-1}(\overline{\alpha}I - P_1)(\overline{\alpha}I + P_1)^{-1}(\overline{\alpha}I - P_2).$$

The following result shows the convergence of a JFGPSS algorithm:

Theorem 2. *Let $x^{(0)} \in \mathbb{C}^n$ and $\varphi : D \subset \mathbb{C}^n \to \mathbb{C}^n$ be a G-differentiable function on an open set $N_0 \subset D$, on which $\varphi'(x)$ is continuous and $\max \|A^{-1}\varphi'(x)\| = L < 1$. Let us suppose that P_1 and P_2 are generalized positive-definite and skew-Hermitian splitting parts of the positive definite matrix A as (15) and (16) and also that $\mathcal{M}$ is an upper bound for $\|A^{-1}G(x^{(0)})\|$; l_{k_n} is the number of GPSS inner iterations in which the stopping criterion (14) is satisfied,*

$$l_*^n > \left\lfloor \frac{\ln\left(\frac{(1-\eta)(1-\eta^{k_n-1})}{L} - 1\right)}{\ln(\theta)} \right\rfloor,$$

with $l_*^n = \liminf_{k_n \to \infty} l_{k_n}$ for $n = 1, 2, 3, \ldots$, η is the tolerance in Newton-like intermediate iterations with $L < (1-\eta)^2$ and $\theta = \|T\|$, where T is the GPSS inner iteration matrix that can be written as

$$T = (\overline{\alpha}I + P_2)^{-1}(\overline{\alpha}I - P_1)(\overline{\alpha}I + P_1)^{-1}(\overline{\alpha}I - P_2).$$

Then, the sequence of iteration $\{x^{(k)}\}_{k=0}^{\infty}$, generated by JFGPSS scheme in Algorithm 1, is well-defined and converges to x^*, satisfying $G(x^*) = 0$, and also

$$\|x^{(n+1)} - x^{(n)}\| \leqslant \delta\mathcal{M}\rho^n,$$
$$\|x^{(n+1)} - x^{(0)}\| \leqslant \frac{\delta\mathcal{M}}{1-\rho},$$

where $\delta = \limsup_{n\to\infty} \dfrac{(1+\theta^{l_*^n})}{1-\eta}$ and $\rho = \limsup_{n\to\infty} \rho_n$ for $\rho_n = \dfrac{(1+\theta^{l_*^n})}{1-\eta}L + \eta^{k_{n-1}}$.

Proof. Let us note that, in this theorem, we also have $\|T\| < 1$ (for more details, see [16]). The rest of the proof is similar to Theorem 1. □

In the next section, we apply our new iterative method on some weakly nonlinear systems of equations.

4. Application

Now, we use JFHSS and JFGPSS algorithms for solving some nonlinear systems. These examples show that JFHSS and JFGPSS methods perform better than nonlinear HSS-like and Picard-HSS methods.

Example 1. *Consider the following two-dimensional nonlinear convection-diffusion equation*

$$\begin{aligned} -(u_{xx}+u_{yy})+q(u_x+u_y) &= f(x,y), \quad (x,y)\in\Omega \\ u(x,y) &= h(x,y), \quad (x,y)\in\partial\Omega \end{aligned}$$

where $\Omega=(0,1)\times(0,1)$, $\partial\Omega$ *is its boundary and* q *is a positive constant for measuring the magnitude of the convection term. We solve this problem for each of the following cases:*

Case 1 $f(x,y)=e^{u(x,y)}$, $h(x,y)=0$.
Case 2 $f(x,y)=-e^{u(x,y)}-\sin(1+u_x(x,y)+u_y(x,y))$, $h(x,y)=-e^{x+y}$.

To discretize this convection-diffusion equation, for the convective term, we use a central difference method while, for the diffusion term, we use a five-point finite difference method. These yield the following nonlinear system

$$H(u)=Mu+h^2\psi(u), \tag{29}$$

where $h=\dfrac{1}{N+1}$ is the equidistance step-size with N as a known natural number and $M=A_N\otimes I_N+A_N\otimes I_N$, $B=C_N\times C_N$ with tridiagonal matrices $A_N=\text{tridiag}(-1-qh/2,2,1+qh/2)$, $C_N=\text{tridiag}(-1/h,0,1/h)$ and I_N is $N\times N$ identity matrix. For case 1, we have $\psi(u)=-\varphi(u)$ and, for case 2, $\psi(u)=\sin(1+Bu)+\varphi(u)$, where $\varphi(u)=(e^{u_1},e^{u_2},...,e^{u_n})^T$; moreover, $\otimes$ is the Kronecker product symbol, $n=N\times N$ and $\sin(u)$ means $(\sin(u_1),\sin(u_2),\cdots,\sin(u_n))^T$. To apply Picard-HSS, nonlinear HSS-like, JFHSS and JFGPSS methods for solving (29), the stopping criterion for the outer iteration in all methods is chosen as

$$\frac{\| Mu^{(n)}+h^2\psi(u^{(n)}) \|}{\| Mu^{(0)}+h^2\psi(u^{(0)}) \|}\leq 10^{-12}. \tag{30}$$

Meanwhile, the Newton-like iteration (in JFHSS and JFGPSS methods) is

$$\frac{\| G_n(u_{k_n}^{(n)}) \|}{\| G_n(u_0^{(n)}) \|}\leq 10^{-1}, \tag{31}$$

and also the stopping criterion for HSS and GPSS processes in each Newton-like inner iteration is

$$\| G_n(u_k^{(n)})-As_{k,l_{k_n}}^{(n)} \|\leq \eta\|G_n(u_k^{(n)})\|, \tag{32}$$

where $\{u^{(n)}\}$ is the sequence generated by the JFHSS method. k_n and l_{k_n} are, respectively, the number of Newton-like inner iterations and HSS and GPSS inner iterations, required for satisfying Relations (31) and (32).

Moreover, to avoid computing the Jacobian in Picard-HSS method, we propose the following stopping criterion for inner iterations

$$\| G(u^{(n)})+As_{l_n}^{(n)} \|\leq \eta\|G(u^{(n)})\|. \tag{33}$$

In order to use a JFGPSS method, we apply the following decomposition on matrix M in Equation (29),

$$P_1 = \mathcal{D} + 2L_{\mathcal{G}}, \quad P_2 = L_{\mathcal{G}}^* - L_{\mathcal{G}} + S. \tag{34}$$

In addition, $\mathcal{K} = 0$, so $\mathcal{G} = H$ is the Hermitian part of M and $S = \frac{1}{2}(M - M^*)$ is the skew Hermitian part of M.

Numerical results for $q = 1000$, $q = 2000$ and initial points $u^{(0)} = \bar{1}$, $u^{(0)} = 4 \times \bar{1}$ for both cases and $u^{(0)} = 12 \times \bar{1}$ for case 1 and $u^{(0)} = 13 \times \bar{1}$ for case 2 and different values of N for JFHSS, JFGPSS, nonlinear HSS-like and Picard-HSS schemes are reported in Tables 1 and 2. Other numerical results such as CPU-time (total CPU time), the number of outer and inner iteration steps (denoted as IT_{out} and IT_{inn}, respectively), and the norm-2 of the function at the last step (denoted by $\|F(u^{(n)})\|$) are also presented in these tables. For JFHSS and JFGPSS algorithms, the values of IT_{int} and IT_{inn} are reported. The former is the obtained number when total inner HSS or GPSS iteration is used in Newton-like iterations, divided by the sum of total Newton-like iterations, while the latter is the total number of intermediate iterations of the Newton-like method.

Except for $u^{(0)} = \bar{1}$, which is relatively close to the solution (in case 1, the real solution u is near zero and, in case 2, almost for all coordinates of the solution, u_i, $i = 1, 2, \cdots, n$, $0 \leq u_i \leq 1$), the nonlinear HSS-like method for other initial points of Tables 1 and 2 could not perform the iterations at all, but JFHSS and JFGPSS methods for all points in both cases could easily solve the problem. Picard-HSS for these three initial points could not solve the problem and, in all cases, fails to solve the problem, especially for $q > 500$.

Numerical results show that the inner iterations for both JFHSS and nonlinear HSS-like are almost the same but for JFGPSS is less than these two methods. For example, in Table 1, for $u^{(0)} = \bar{1}$, $q = 1000$ and $N = 40$, the number of inner iterations for JFHSS and JFGPSS methods are, respectively, 133 and 96 and this number for total iterations in the nonlinear HSS-like method (consider that there is only one kind of iteration in a nonlinear HSS-like method) is 127. However, the nonlinear HSS-like method needs to evaluate a greater number of the nonlinear term $\psi(u)$ than the JFHSS method (for the JFHSS method, only 12 function evaluations are required compared to 254 function evaluations for the nonlinear HSS-like method). Thus, JFHSS and JFGPSS methods can significantly reduce the computational cost of evaluation of the nonlinear term, especially when the nonlinear part is so complicated, e.g., in Example 2, the difference between the computational cost of the nonlinear HSS-like method and the JFHSS method has increased, since the problem has a more complicated nonlinear term.

It must be noted that, in the inner iteration, for solving the linear systems related to the Hermitian part (in HSS scheme) and the skew-Hermitian part (in both HSS and GPSS schemes), we have employed respectively the conjugate gradient (CG) method and the Lanczos method (for more details, see [17]).

In this example, η = tol was used for all steps; in most cases, we obtained equal Newton-like and outer iterations at each step; however, in general, choosing equal η and tol does not always lead to equal Newton-like and outer iterations. For example, in cases that nonlinearity increases (e.g., when we choose initial value $u^{(0)} = 12 \times \bar{1}$, in the first steps, the nonlinear term $h^2\psi(u)$ is so big) result in a different number of Newton-like and outer iterations. In all tables of this paper, a, b denote the number $a \cdot 10^b$.

Table 1. Results for JFHSS, JFGPSS, nonlinear HSS-like and Picard-HSS methods of Example 1, Case 1 (η = tol = 0.1).

N			30	40	60	70	80	100
$q = 1000$, $u^{(0)} = \bar{1}$	JFHSS	CPU	0.65	1.81	7.46	13.21	24.45	59.32
		IT_{out}	12	12	12	12	12	12
		IT_{int}	12	12	12	12	12	12
		IT_{inn}	9	11.08	10.75	10.75	10.41	10.91
		$\|F(u^{(n)})\|$	1.86, −11	3.35, −11	1.70, −11	3.41, −11	3.54, −11	2.43, −11
	JFGPSS	CPU	0.63	1.46	5.79	9.84	17.28	44.50
		IT_{out}	12	12	11	11	11	11
		IT_{int}	14	12	11	11	11	11
		IT_{inn}	8.78	8	7.64	7.45	7.90	8.73
		$\|F(u^{(n)})\|$	5.45, −11	1.89, −11	7.69, −11	1.02, −10	9.63, −11	5.09, −11
	Nonlinear HSS-like	CPU	0.82	2.03	8.26	14.60	24.65	61.35
		IT	129	127	123	124	128	126
		$\|F(u^{(n)})\|$	1.45, −10	1.53, −10	1.25, −10	1.10, −10	8.60, −11	8.91, −11
	Picard-HSS	-	-	-	-	-	-	-
$q = 2000$, $u^{(0)} = \bar{1}$	JFHSS	CPU	1.04	2.71	11.32	19.87	31.48	76.13
		IT_{out}	12	12	12	12	12	12
		IT_{int}	12	12	12	12	12	12
		IT_{inn}	16.08	14.67	14.25	14.17	14	14.08
		$\|F(u^{(n)})\|$	1.47, −10	9.30, −11	7.92, −11	8.80, −11	9.56, −11	6.56, −11
	JFGPSS	CPU	0.85	2.20	8.57	14.26	23.50	54.90
		IT_{out}	12	12	12	12	12	12
		IT_{int}	12	12	12	12	12	12
		IT_{inn}	14.42	12.42	10.58	10	9.84	9.91
		$\|F(u^{(n)})\|$	1.57, −10	8.49, −11	3.33, −11	2.80, −11	2.38, −11	4.23, −11
	Nonlinear HSS-like	CPU	1.32	2.94	12.11	20.51	33.88	80.97
		IT	188	172	167	166	165	165
		$\|F(u^{(n)})\|$	3.24, −10	2.50, −10	2.07, −10	2.32, −10	2.037, −10	1.81, −10
	Picard-HSS	-	-	-	-	-	-	-
$q = 1000$, $u^{(0)} = 4 \times \bar{1}$	JFHSS	CPU	0.80	2.24	9.34	14.56	23.77	60.21
		IT_{out}	12	12	12	12	12	12
		IT_{int}	12	12	12	12	12	12
		IT_{inn}	11.08	11	10.67	10.75	10.50	11.25
		$\|F(u^{(n)})\|$	1.94, −10	1.70, −10	9.33, −11	9.94, −11	1.15, −10	8.77, −11
	JFGPSS	CPU	0.56	1.51	6.55	12.47	21.03	55.50
		IT_{out}	12	12	11	12	11	11
		IT_{int}	12	12	11	12	11	11
		IT_{inn}	8.92	8.34	8.72	8.75	9.55	10.63
		$\|F(u^{(n)})\|$	9.76, −11	7.68, −11	4.60, −10	6.35, −11	3.73, −10	3.78, −10
	Nonlinear HSS-like	-	-	-	-	-	-	-
	Picard-HSS	-	-	-	-	-	-	-
$q = 2000$, $u^{(0)} = 4 \times \bar{1}$	JFHSS	CPU	0.99	2.51	11.20	19.45	32.23	77.58
		IT_{out}	12	12	12	12	12	12
		IT_{int}	12	12	12	12	12	12
		IT_{inn}	16.08	14.67	14.25	14.17	14	14.08
		$\|F(u^{(n)})\|$	5.88, −10	3.71, −10	3.20, −10	3.57, −10	3.75, −10	2.69, −10
	JFGPSS	CPU	0.85	2.20	8.58	14.02	23.22	54.94
		IT_{out}	12	12	12	12	12	12
		IT_{int}	12	12	12	12	12	12
		IT_{inn}	14.42	12.41	10.58	9.92	9.84	9.84
		$\|F(u^{(n)})\|$	6.26, −10	3.44, −10	1.31, −10	1.63, −10	1.08, −10	2.08, −10
	Nonlinear HSS-like	-	-	-	-	-	-	-
	Picard-HSS	-	-	-	-	-	-	-
$q = 1000$, $u^{(0)} = 12 \times \bar{1}$	JFHSS	CPU	0.81	2.23	10.41	18.89	31.31	81.74
		IT_{out}	12	12	12	12	12	12
		IT_{int}	14	14	14	14	14	14
		IT_{inn}	10.85	12.83	11.28	11.71	11.64	12.93
		$\|F(u^{(n)})\|$	1.47, −8	1.05, −8	7.50, −9	4.55, −9	3.08, −9	3.29, −9
	JFGPSS	CPU	0.66	1.70	7.95	14.30	25.44	63.80
		IT_{out}	12	12	12	12	12	12
		IT_{int}	14	14	14	14	14	14
		IT_{inn}	8.78	8	7.86	8.64	9.07	9.92
		$\|F(u^{(n)})\|$	8.02, −9	3.11, −8	3.40, −9	2.32, −9	1.61, −9	6.16, −10
	Nonlinear HSS-like	-	-	-	-	-	-	-
	Picard-HSS	-	-	-	-	-	-	-

Table 1. *Cont.*

N			30	40	60	70	80	100
$q = 2000$, $u^{(0)} = 12 \times \bar{1}$	JFHSS	CPU	1.06	2.90	13.55	21.72	38.94	87.18
		IT_{out}	12	12	12	12	12	12
		IT_{int}	14	14	14	13	14	13
		IT_{inn}	14.93	14.36	14.86	14.62	14.57	15
		$\|F(u^{(n)})\|$	2.06, −8	1.48, −8	9.76, −9	6.96, −9	6.58, −9	5.72, −9
	JFGPSS	CPU	0.95	2.45	10.03	17.81	29.06	69.57
		IT_{out}	12	12	12	12	12	12
		IT_{int}	14	14	14	14	14	13
		IT_{inn}	13.71	11.64	10.71	10.85	10.64	11.31
		$\|F(u^{(n)})\|$	1.91, −8	1.30, −8	6.08, −9	3.26, −9	6.35, −9	3.23, −9
	Nonlinear HSS-like	-	-	-	-	-	-	-
	Picard-HSS	-	-	-	-	-	-	-

Table 2. Results for JFHSS, JFGPSS, nonlinear HSS-like and Picard-HSS methods of Example 1, Case 2 ($\eta = \text{tol} = 0.1$).

N			30	40	60	70	80	100
$q = 1000$, $u^{(0)} = \bar{1}$	JFHSS	CPU	0.73	2.03	9.54	16.99	27.27	65.47
		IT_{out}	11	11	11	12	12	12
		IT_{int}	12	12	12	13	13	13
		IT_{inn}	11.25	11.41	11.92	11.23	10.92	12
		$\|F(u^{(n)})\|$	1.64, −10	1.18, −10	1.43, −11	1.32, −11	1.95, −11	1.56, −11
	JFGPSS	CPU	0.57	1.53	7.19	12.59	19.42	53.59
		IT_{out}	11	11	11	11	11	12
		IT_{int}	12	12	12	12	12	13
		IT_{inn}	9	8.25	8.75	8.92	8.25	9
		$\|F(u^{(n)})\|$	5.45, −11	8.19, −11	8.56, −11	7.6, −11	5.27, −11	4.81, −12
	Nonlinear HSS-like	CPU	0.82	2.30	9.86	14.38	29.31	59.91
		IT	128	128	123	124	121	126
		$\|F(u^{(n)})\|$	1.81, −10	1.43, −10	1.25, −10	1.10, −10	1.15, −10	1.06, −10
	Picard-HSS	-	-	-	-	-	-	-
$q = 2000$, $u^{(0)} = \bar{1}$	JFHSS	CPU	0.98	2.63	11.73	20.51	36.07	77.20
		IT_{out}	11	11	11	11	12	12
		IT_{int}	12	12	12	12	13	13
		IT_{inn}	16	15	14.50	14.67	14.30	14.62
		$\|F(u^{(n)})\|$	2.49, −10	2.26, −10	2.03, −10	2.30, −10	2.61, −11	1.74, −11
	JFGPSS	CPU	0.88	2.26	8.61	15.08	25.83	60.7
		IT_{out}	11	11	11	11	12	11
		IT_{int}	12	12	12	12	13	12
		IT_{inn}	14.33	12.08	10.58	10.66	9.69	11
		$\|F(u^{(n)})\|$	2.04, −10	2.91, −10	1.91, −10	1.09, −10	1.64, −11	1.72, −10
	Nonlinear HSS-like	CPU	1.15	3.85	12.52	19.61	37.70	79.26
		IT	187	171	166	166	164	164
		$\|F(u^{(n)})\|$	3.68,-10	3.07, −10	2.48, −10	2.17, −10	2.42, −10	2.13, −10
	Picard-HSS	-	-	-	-	-	-	-
$q = 1000$, $u^{(0)} = 4 \times \bar{1}$	JFHSS	CPU	0.72	2.28	9.39	16.62	28.53	67.23
		IT_{out}	11	11	11	12	11	11
		IT_{int}	12	12	12	12	12	12
		IT_{inn}	11.41	11.33	11.41	11.75	12.08	12.34
		$\|F(u^{(n)})\|$	1.62, −10	2.01, −10	1.64, −10	1.92, 10	2.89, −10	2.47, −10
	JFGPSS	CPU	0.69	1.97	8.85	16.53	26.53	70.80
		IT_{out}	11	11	11	11	12	11
		IT_{int}	12	12	12	12	13	12
		IT_{inn}	10.91	11.16	11	11.42	11.42	12.34
		$\|F(u^{(n)})\|$	2.18, −10	1.22, −10	1.21, −10	8.35, −11	1.15, −10	1.17, −10
	Nonlinear HSS-like	-	-	-	-	-	-	-
	Picard-HSS	-	-	-	-	-	-	-

Table 2. *Cont.*

N			30	40	60	70	80	100
$q = 2000$, $u^{(0)} = 4 \times \bar{1}$	JFHSS	CPU	0.97	2.59	11.06	20.24	33.75	79.80
		IT_{out}	11	11	11	11	11	11
		IT_{int}	12	12	12	12	12	12
		IT_{inn}	15.92	15.08	14.50	14.58	14.66	14.92
		$\|F(u^{(n)})\|$	9.65, −10	4.15, −10	4.31, −10	4.40, −10	3.82, −10	3.39, −10
	JFGPSS	CPU	0.88	2.18	8.75	14.60	25.05	64.79
		IT_{out}	11	11	11	11	11	11
		IT_{int}	12	12	12	12	12	12
		IT_{inn}	14.50	12.42	10.66	10.42	10.58	11.08
		$\|F(u^{(n)})\|$	5.06, −10	3.61, −10	2.53, −10	3.32, −10	2.95, −10	1.97, −10
	Nonlinear HSS-like	-	-	-	-	-	-	-
	Picard-HSS	-	-	-	-	-	-	-
$q = 1000$, $u^{(0)} = 13 \times \bar{1}$	JFHSS	CPU	0.77	2.33	10.82	19.56	32.13	85.32
		IT_{out}	12	12	12	12	12	12
		IT_{int}	14	14	14	14	14	14
		IT_{inn}	10.85	12.83	11.28	11.71	11.64	12.92
		$\|F(u^{(n)})\|$	1.44, −8	1.36, −8	7.47, −9	4.56, −9	3.8, −9	3.54, −9
	JFGPSS	CPU	0.65	1.74	8.00	14.54	25.02	64.19
		IT_{out}	12	12	12	12	12	12
		IT_{int}	14	14	14	14	14	14
		IT_{inn}	8.78	8	8.28	8.64	9.07	9.86
		$\|F(u^{(n)})\|$	8.03, −9	1.44, −8	3.35, −9	4.76, −9	1.69, −9	1.085, −9
	Nonlinear HSS-like	-	-	-	-	-	-	-
	Picard-HSS	-	-	-	-	-	-	-
$q = 2000$, $u^{(0)} = 13 \times \bar{1}$	JFHSS	CPU	1.08	2.97	11.27	22.45	39.98	89.15
		IT_{out}	12	12	12	12	12	12
		IT_{int}	14	14	14	14	14	14
		IT_{inn}	14.93	14.35	14.43	14.62	14.57	15
		$\|F(u^{(n)})\|$	2.01, −8	1.49, −8	8.73, −9	6.97, −9	6.57, −9	5.72, −9
	JFGPSS	CPU	0.99	2.41	10.15	17.98	29.33	67.45
		IT_{out}	12	12	12	12	12	12
		IT_{int}	14	14	14	14	14	13
		IT_{inn}	13.78	11.64	10.71	10.86	10.64	11.31
		$\|F(u^{(n)})\|$	1.70, −8	1.30,−8	6.02, −9	3.23, −9	6.34, −9	3.21, −9
	Nonlinear HSS-like	-	-	-	-	-	-	-
	Picard-HSS	-	-	-	-	-	-	-

The optimal value for parameter α that minimizes the boundary of spectral radius of the iteration matrices is important because it also improves the convergence speed of Picard-HSS, nonlinear HSS-like, JFHSS and JFGPSS methods. There are no general results to determine the optimal α and $\overline{\alpha}$, so we need to obtain the optimal values of parameters α and $\overline{\alpha}$ experimentally. However, Bai and Golub [8] proved that spectral radius of HSS iterative matrix that is obtained from the coefficient matrix M in (29) is bounded by $\|T\| \leq \sigma(\alpha) \equiv \max_{\lambda_i \in \lambda(H)} \left|\frac{\alpha - \lambda_i}{\alpha + \lambda_i}\right| < 1$, and the minimum of this bound is obtained when

$$\alpha = \alpha^* = \sqrt{\lambda_{min(H)} \lambda_{max(H)}},$$

where $\lambda_{min(H)}$ and $\lambda_{max(H)}$ are, respectively, the smallest and the largest eigenvalues of Hermitian matrix H. Usually, in an HSS scheme, $\alpha_{opt} \neq \alpha^* \equiv \underset{\alpha>0}{\operatorname{argmin}}\{\sigma(\alpha)\} < 1$ and $\rho(T(\alpha^*)) \geqslant \rho(T(\alpha_{opt}))$. When q or $qh/2$ is small, $\sigma(\alpha)$ is close to $\rho(T(\alpha))$ and in this case α^* is close to α_{opt} and α^* can be a good estimation for α_{opt}. However, when q or $qh/2$ is large (the skew-Hermitian part is dominant), hence $\sigma(\alpha)$ deviates too much from $\rho(T(\alpha))$, so using α^* is not useful. In this case, $\rho(T(\alpha))$ attains its minimum at α_{opt} that is far from α^*, but close to $qh/2$ (see [8]).

In the GPSS case, a spectral radius of $T(\overline{\alpha})$ is bounded by $\|V(\overline{\alpha})\|$, where $V(\overline{\alpha}) = (\overline{\alpha}I - P_1)(\overline{\alpha}I + P_1)^{-1}$. Since $\|V(\overline{\alpha})\|_2 \leqslant 1$ (see [18]), GPSS inner iterations unconditionally converge to the exact solution in each inner iteration of a JFGPSS scheme. However, when $P_1 \in \mathbb{C}^{n \times n}$ is a general positive-definite matrix, we do not have any formula to compute $\overline{\alpha}^* \equiv \underset{\overline{\alpha}>0}{\operatorname{argmin}}\{\|V(\overline{\alpha})\|\}$ that is the value that minimizes the boundary of iteration matrix $T(\overline{\alpha})$, nor do we have a formula for $\overline{\alpha}_{opt}$, the value that minimizes $\|T(\overline{\alpha})\|$.

In Table 3, the optimal values of α_{opt} and $\overline{\alpha}_{opt}$ have been written (tested and optimal α and $\overline{\alpha}_{opt}$) that are determined experimentally by using increments as 0.25. In addition, the corresponding spectral radius of the iteration matrices $T(\alpha)$ and $T(\overline{\alpha})$ for HSS and GPSS algorithms that are used as inner iterations to solve (29) are reported in this table. One can see that the spectral radius of GPSS method in all cases is smaller than HSS scheme, which results in faster convergence.

Table 3. Optimal value of α for HSS and GPSS inner iterations for different values of N and q of Example 1.

N			30	40	50	60	70	80	90	100
$q=1000$	HSS	α_{opt}	18	15	10.5	9	8	6	5.75	5.75
		$\rho(T(\alpha_{opt}))$	0.7226	0.6930	0.6743	0.6613	0.6513	0.6485	0.6459	0.6467
		α^*	0.4047	0.3062	0.2462	0.2059	0.1769	0.1551	0.1381	0.1244
		$\rho(T(\alpha^*))$	0.8971	0.9211	0.9360	0.9461	0.9535	0.9590	0.9634	0.9669
		$\frac{qh}{2}$	16.1290	12.1951	9.8039	8.1967	7.0423	6.1728	5.4945	4.9505
		$\rho(T(\frac{qh}{2}))$	0.7236	0.6974	0.6783	0.6674	0.6608	0.6574	0.6562	0.6569
	GPSS	$\overline{\alpha}_{opt}$	11.25	9.5	8.5	7.5	7	6.5	6	5.5
		$\rho(T(\overline{\alpha}_{opt}))$	0.5428	0.5140	0.5076	0.4983	0.4959	0.4902	0.4982	0.4983
$q=2000$	HSS	α_{opt}	26	22	16	13.5	12	10	8.75	8
		$\rho(T(\alpha_{opt}))$	0.7911	0.7663	0.6499	0.7399	0.0.7373	0.7302	0.7302	0.7242
		α^*	0.1638	0.0938	0.0606	0.0424	0.0313	0.0241	0.0191	0.0155
		$\rho(T(\alpha^*))$	0.9579	0.9757	0.9842	0.9889	0.9918	0.9937	0.9950	0.9959
		$\frac{qh}{2}$	32.2581	24.39	19.61	16.3934	14.0845	12.35	10.99	9.9010
		$\rho(T(\frac{qh}{2}))$	0.7953	0.77	0.7512	0.7439	0.7343	0.728	0.7282	0.7270
	GPSS	$\overline{\alpha}_{opt}$	15	13	11	10	9	8	7.5	7
		$\rho(T(\overline{\alpha}_{opt}))$	0.6424	0.6212	0.6144	0.6063	0.6036	0.6028	0.6090	0.6033

Example 2 ([10]). *We consider the two-dimensional nonlinear convection-diffusion equation*

$$-(u_{xx}+u_{yy})+qe^{x+y}(xu_x+yu_y)=ue^u+\sin(\sqrt{1+u_x^2+u_y^2}),\quad (x,y)\in\Omega,$$
$$u(x,y)=0,\quad (x,y)\in\partial\Omega,$$

where $\Omega=(0,1)\times(0,1)$, $\partial\Omega$ is its boundary and q is a positive constant for measuring magnitude of the convection term. By applying the upwind finite difference scheme on the equidistance discretization grid (stepsize $h=\frac{1}{N+1}$) with the central difference scheme to the convective term, we obtain a system of nonlinear equations in the general form (for more details, see [10])

$$H(x)=Mx-h^2\psi(x). \tag{35}$$

We have selected zero vector $u^{(0)}=\overline{0}=(0,0,\cdots,0)^T$ as the initial guess. In addition, again (31) and (32) are used respectively as the stopping criteria for the inner iterations and Newton-like iterations in the JFHSS method and (30) for outer iterations in JFHSS, Picard-HSS and nonlinear HSS-like methods. Moreover, to avoid computing Jacobian in Picard-HSS and nonlinear HSS-like methods, we used (33). Similar to Example 1, one can use other iterative methods instead of HSS in Algorithm 1, for which the spectral radius of its iteration matrix is smaller and thus results in faster convergence.

Numerical results for $N=32,48,64$, optimal α and different values of q for JFNHSS, Picard-HSS and nonlinear HSS-like schemes are reported in Table 4. In addition, we adopted the experimentally optimal parameters α to obtain the least CPU times for these iterative methods. One can see that JFHSS performs better than nonlinear HSS-like and Picard-HSS methods in all cases.

Table 4. Results of JFHSS, nonlinear HSS-like and Picard-HSS methods for Example 2 (η = tol = 0.1).

q			50	100	200	400	1200	2000
$N = 32$		α_{opt}	1.4	1.6	2.5	8	21.5	34
	JFHSS	CPU	1.23	1.42	1.29	1.53	1.71	1.86
		IT_{out}	12	12	12	12	12	12
		IT_{int}	12	12	12	12	12	12
		IT_{inn}	11.34	11.67	12	12.75	16.25	20.34
		$\|\|F(u^{(n)})\|\|$	1.54, −14	2.2, −14	1.47, −14	6.87, −15	8.22, −15	1.3, −14
	Nonlinear HSS-like	CPU	2.03	2.39	2.25	2.31	2.42	2.45
		IT	129	137	140	146	160	167
		$\|\|F(u^{(n)})\|\|$	1.1, −14	2.25, −14	2.31, −14	2.23, −14	2.4, −14	2.3, −14
	Picard-HSS	CPU	7.96	8.31	7.76	8	8.60	8.86
		IT_{out}	12	12	12	12	12	12
		IT_{inn}	121.1	131.91	126.75	145.34	146.34	147
		$\|\|F(u^{(n)})\|\|$	1.1, −14	1.24, −14	1.57, −14	1.96, −14	1.84, −14	1.6, −14
$N = 48$		α_{opt}	0.8	1.4	2.6	4.8	13	20.5
	JFHSS	CPU	5.25	5.31	5.5	5.93	6.21	6.28
		IT_{out}	12	12	12	12	12	12
		IT_{int}	12	12	12	12	12	12
		IT_{inn}	13.66	14.58	15.083	16.08	17.34	17.58
		$\|\|F(u^{(n)})\|\|$	2.42, −14	6.04, −15	6.36, −15	1.96, −14	6.15, −15	8.60, −15
	Nonlinear HSS-like	CPU	8.87	11.828	10.02	10.31	11.28	11.85
		IT	161	209	178	186	201	207
		$\|\|F(u^{(n)})\|\|$	1.5, −14	1.59, −14	1.46, −14	1.57, −14	1.615, −14	1.46, −14
	Picard-HSS	CPU	50.81	50.01	51.85	53.34	56.32	59.95
		IT_{out}	12	12	12	12	12	12
		IT_{inn}	177.16	179.1	183.50	189.34	202.75	213.25
		$\|\|F(u^{(n)})\|\|$	7.7, −15	9.67, −15	1.11, −14	1.23, −14	1.22, −14	1.26, −14
$N = 64$		α_{opt}	0.7	1	1.8	3.3	8.9	14.2
	JFHSS	CPU	21.68	18.23	18.65	19.156	20.53	21.39
		IT_{out}	12	12	12	12	12	12
		IT_{int}	12	12	12	12	12	12
		IT_{inn}	21	17.39	18.17	18.75	19.91	20.84
		$\|\|F(u^{(n)})\|\|$	1.61, −14	6.73, −15	9.15, −15	8.39, −15	7.7, −15	4.71, −15
	Nonlinear HSS-like	CPU	38.57	31.78	33.50	34.65	36.56	37.70
		IT	246	206	213	221	235	242
		$\|\|F(u^{(n)})\|\|$	1.17, −14	1.26, −14	1.26, −14	1.16, −14	1.19, −14	1.22, −14
	Picard-HSS	CPU	219.54	217.45	266.83	225.37	228.60	248.35
		IT_{out}	12	12	12	12	12	12
		IT_{inn}	219.54	248.58	230.75	252	258.75	264.50
		$\|\|F(u^{(n)})\|\|$	6.12, −15	7.7, −15	8.9, −15	1.0, −14	1.1, −14	1.1, −14

5. Conclusions

In this paper, an iterative method based on two-stage splitting methods has been proposed to solve weakly nonlinear systems and a convergence property of this method has been investigated. This method is a combination of an inexact Newton method, Hermitian and skew-Hermitian splitting (or generalized positive definite and skew-Hermitian splitting) scheme. The advantage of our new method, Picard-HSS and nonlinear HSS-like over the methods like Newton method is that they don't need explicit construction and accurate computation of the Jacobian matrix. Hence, computation works and computer memory may be saved in actual application; however, numerical results show that JFHSS and JFGPSS methods perform better than the two other ones.

Numerical results show that JFHSS and JFGPSS iteration algorithms are effective, robust, and feasible nonlinear solvers for a class of weakly nonlinear systems. Moreover, employing these algorithms to solve nonlinear systems is found to be simple, accurate, fast, flexible, convenient and have small computation cost. In addition, it must be noted that, even though our inner iteration scheme in this paper are HSS and GPSS methods, another inner iteration solver can be used subject to the condition that the iteration matrix satisfies in $||T|| < 1$.

Author Contributions: The contributions of authors are roughly equal.

Acknowledgments: The third and fourth authors have been partially supported by the Spanish Ministerio de Ciencia, Innovación y Universidades PGC2018-095896-B-C22 and Generalitat Valenciana PROMETEO/2016/089.

References

1. Shen, W.; Li, C. Kantorovich-type convergence criterion for inexact Newton methods. *Appl. Numer. Math.* **2009**, *59*, 1599–1611. [CrossRef]
2. An, H.-B.; Bai, Z.-Z. A globally convergent Newton-GMRES method for large sparse systems of nonlinear equations. *Appl. Numer. Math.* **2007**, *57*, 235-252. [CrossRef]
3. Eisenstat, S.C.; Walker, H.F. Globally convergent inexact Newton methods. *SIAM J. Optim.* **1994**, *4*, 393–422. [CrossRef]
4. Gomes-Ruggiero, M.A.; Lopes, V.L.R.; Toledo-Benavides, J.V. A globally convergent inexact Newton method with a new choice for the forcing term. *Ann. Oper. Res.* **2008**, *157*, 193–205. [CrossRef]
5. Bai, Z.-Z. A class of two-stage iterative methods for systems of weakly nonlinear equations. *Numer. Algorithms* **1997**, *14*, 295–319. [CrossRef]
6. Zhu, M.-Z. Modified iteration methods based on the Asymmetric HSS for weakly nonlinear systems. *J. Comput. Anal. Appl.* **2013**, *15*, 188–195.
7. Axelsson, O.; Bai, Z.-Z.; Qiu, S.-X. A class of nested iteration schemes for linear systems with a coefficient matrix with a dominant positive definite symmetric part. *Numer. Algorithms* **2004**, *35*, 351–372. [CrossRef]
8. Bai, Z.-Z.; Golub, G.H.; Ng, M.K. Hermitian and skew-Hermitian splitting methods for non-Hermitian positive definite linear systems. *SIAM J. Matrix Anal. Appl.* **2003**, *24*, 603–626. [CrossRef]
9. Li, L.; Huang, T.-Z.; Liu, X.-P. Asymmetric Hermitian and skew-Hermitian splitting methods for positive definite linear systems. *Comput. Math. Appl.* **2007**, *54*, 147–159. [CrossRef]
10. Bai, Z.-Z.; Yang, X. On HSS-based iteration methods for weakly nonlinear systems. *Appl. Numer. Math.* **2009**, *59*, 2923–2936. [CrossRef]
11. Bai, Z.-Z.; Migallón, V.; Penadés, J.; Szyld, D.B. Block and asynchronous two-stage methods for mildly nonlinear systems. *Numer. Math.* **1999**, *82*, 1–20. [CrossRef]
12. Zhu, M.-Z.; Zhang, G.-F. On CSCS-based iteration methods for Toeplitz system of weakly nonlinear equations. *J. Comput. Appl. Math.* **2011**, *235*, 5095–5104. [CrossRef]
13. Bai, Z.-Z.; Guo, X.-P. On Newton-HSS methods for systems of nonlinear equations with positive-definite Jacobian matrices. *J. Comput. Math.* **2010**, *28*, 235–260.
14. Li, X.; Wu, Y.-J. Accelerated Newton-GPSS methods for systems of nonlinear equations. *J. Comput. Anal. Appl.* **2014** , *17*, 245–254.
15. Edwards, C.H. *Advanced Calculus of Several Variables*; Academic Press: New York, NY, USA, 1973.
16. Cao, Y.; Tan, W.-W.; Jiang, M.-Q. A generalization of the positive-definite and skew-Hermitian splitting iteration. *Numer. Algebra Control Optim.* **2012**, *2*, 811–821.
17. Bai, Z.-Z.; Golub, G.H.; Ng, M.K. On inexact Hermitian and skew-Hermitian splitting methods for non-Hermitian positive definite linear systems. *Linear Algebra Appl.* **2008**, *428*, 413–440. [CrossRef]
18. Bai, Z.-Z.; Golub, G.H.; Lu, L.-Z.; Yin, J.-F. Block triangular and skew-Hermitian splitting methods for positive-definite linear systems. *SIAM J. Sci. Comput.* **2005**, *26*, 844–863.

4

Nonlinear Operators as Concerns Convex Programming and Applied to Signal Processing

Anantachai Padcharoen [1] and Pakeeta Sukprasert [2,*]

[1] Department of Mathematics, Faculty of Science and Technology, Rambhai Barni Rajabhat University, Chanthaburi 22000, Thailand; anantachai.p@rbru.ac.th

[2] Department of Mathematics and Computer Science, Faculty of Science and Technology, Rajamangala University of Technology Thanyaburi (RMUTT), Thanyaburi, Pathumthani 12110, Thailand

* Correspondence: pakeeta_s@rmutt.ac.th

Abstract: Splitting methods have received a lot of attention lately because many nonlinear problems that arise in the areas used, such as signal processing and image restoration, are modeled in mathematics as a nonlinear equation, and this operator is decomposed as the sum of two nonlinear operators. Most investigations about the methods of separation are carried out in the Hilbert spaces. This work develops an iterative scheme in Banach spaces. We prove the convergence theorem of our iterative scheme, applications in common zeros of accretive operators, convexly constrained least square problem, convex minimization problem and signal processing.

Keywords: convexity; least square problem; accretive operators; signal processing

---

1. Introduction

Let $\mathcal{E}$ be a real Banach space. The zero point problem is as follows:

$$\text{find} \quad x \in \mathcal{E} \quad \text{such that} \quad 0 \in \mathcal{A}u + \mathcal{B}u, \tag{1}$$

where $\mathcal{A} : \mathcal{E} \rightarrow \mathcal{E}$ is an operator and $\mathcal{B} : \mathcal{E} \rightarrow 2^{\mathcal{E}}$ is a set-valued operator. This problem includes, as special cases, convex programming, variational inequalities, split feasibility problem and minimization problem [1–7]. To be more precise, some concrete problems in machine learning, image processing [4,5], signal processing and linear inverse problem can be modeled mathematically as the form in Equation (1).

Signal processing and numerical optimization are independent scientific fields that have always been mutually influencing each other. Perhaps the most convincing example where the two fields have met is compressed sensing (CS) [2]. Several surveys dedicated to these algorithms and their applications in signal processing have appeared [3,6–8]

Fixed point iterations is an important tool for solving various problems and is known in a Banach space $\mathcal{E}$. Let $\mathcal{K}$ be a nonempty closed convex subset of $\mathcal{E}$ and $\mathcal{S} : \mathcal{K} \rightarrow \mathcal{K}$ is the operator with at least one fixed point. Then, for $u_1 \in \mathcal{K}$:

1. The Picard iterative scheme [9] is defined by:

$$u_{n+1} = \mathcal{S}u_n, \quad \forall\, n \in \mathbb{N}.$$

2. The Mann iterative scheme [10] is defined by:

$$u_{n+1} = (1 - \eta_n)u_n + \eta_n \mathcal{S}u_n, \quad \forall\, n \in \mathbb{N},$$

where $\{\eta_n\}$ is a sequence in $(0, 1)$.

3. The Ishikawa iterative scheme [11] is defined by:

$$u_{n+1} = (1-\eta_n)u_n + \eta_n \mathcal{S}[(1-\vartheta_n)u_n + \vartheta_n \mathcal{S}u_n], \quad \forall n \in \mathbb{N},$$

where $\{\eta_n\}$ and $\{\vartheta_n\}$ are sequences in $(0,1)$.

4. The $\mathcal{S}$-iterative scheme [12] is defined by:

$$u_{n+1} = (1-\eta_n)\mathcal{S}u_n + \eta_n \mathcal{S}[(1-\vartheta_n)u_n + \vartheta_n \mathcal{S}u_n], \quad \forall n \in \mathbb{N},$$

where $\{\eta_n\}$ and $\{\vartheta_n\}$ are sequences in $(0,1)$.

Recently, Sahu et al. [13] and Thakur et al. [14] introduced the following same iterative scheme for nonexpansive mappings in uniformly convex Banach space:

$$\begin{cases} w_n = (1-\xi_n)u_n + \xi_n \mathcal{S}u_n, \\ z_n = (1-\vartheta_n)w_n + \vartheta_n \mathcal{S}w_n, \\ u_{n+1} = (1-\eta_n)\mathcal{S}w_n + \eta_n \mathcal{S}z_n, \quad \forall n \in \mathbb{N}, \end{cases}$$

where $\{\eta_n\}$, $\{\vartheta_n\}$ and $\{\xi_n\}$ are sequences in $(0,1)$. The authors proved that this scheme converges to a fixed point of contraction mapping, faster than all known iterative schemes. In addition, the authors provided an example to support their claim.

In this paper, we first develop an iterative scheme for calculating common solutions and using our results to solve the problem in Equation (1). Secondly, we find common solutions of convexly constrained least square problems, convex minimization problems and applied to signal processing.

2. Preliminaries

Let $\mathcal{E}$ be a real Banach space with norm $\|\cdot\|$ and $\mathcal{E}^*$ be its dual. The value of $f \in \mathcal{E}^*$ at $u \in \mathcal{E}$ ia denoted by $\langle u, f\rangle$. A Banach space $\mathcal{E}$ is called strictly convex if $\frac{\|u+v\|}{2} < 1$, for all $u, v \in \mathcal{E}$ with $\|u\| = \|v\| = 1$. It is called uniformly convex if $\lim_{n\to\infty} \|u_n - v_n\| = 0$ for any two sequences $\{u_n\}, \{v_n\}$ in $\mathcal{E}$ such that $\|u\| = \|v\| = 1$ and $\lim_{n\to\infty} \frac{\|u+v\|}{2} = 1$.

The (normalized) duality mapping $\mathcal{J}$ from $\mathcal{E}$ into the family of nonempty (by Hahn Banach theorem) weak-star compact subsets of its dual $\mathcal{E}$ is defined by

$$\mathcal{J}(u) = \{f \in \mathcal{E}^* : \langle u, f\rangle = \|u\|^2 = \|f\|^2\}$$

for each $u \in \mathcal{E}$, where $\langle\cdot,\cdot\rangle$ denotes the generalized duality pairing.

For an operator $\mathcal{A} : \mathcal{E} \to 2^{\mathcal{E}}$, we denote its domain, range and graph as follows:

$$\mathcal{D}(\mathcal{A}) = \{u \in \mathcal{E} : \mathcal{A}u \neq \emptyset\}$$
$$\mathcal{R}(\mathcal{A}) = \cup\{\mathcal{A}p : p \in \mathcal{D}(\mathcal{A})\},$$

and

$$\mathcal{G}(\mathcal{A}) = \{(u, v) \in \mathcal{E} \times \mathcal{E} : u \in \mathcal{D}(\mathcal{A}), v \in \mathcal{A}u\},$$

respectively. The inverse $\mathcal{A}^{-1}$ of $\mathcal{A}$ is defined by $u \in \mathcal{A}^{-1}v$, if and only if $v \in \mathcal{A}u$. If $\forall u_i \in \mathcal{D}(\mathcal{A})$ and $v_i \in \mathcal{A}u_i$ $(i = 1, 2)$, and there is $j \in \mathcal{J}(u_1 - u_2)$ such that $\langle v_1 - v_2, j\rangle \geq 0$, then $\mathcal{A}$ is called accretive.

An accretive operator $\mathcal{A}$ in a Banach space $\mathcal{E}$ is said to satisfy the range condition if $\overline{\mathcal{D}(\mathcal{A})} \subset \mathcal{R}(I + \mu\mathcal{A})$ for all $\mu > 0$, where $\overline{\mathcal{D}(\mathcal{A})}$ denotes the closure of the domain of $\mathcal{A}$. We know that for an accretive operator $\mathcal{A}$ which satisfies the range condition, $\mathcal{A}^{-1}0 = \mathcal{F}ix(\mathcal{J}_\mu^{\mathcal{A}})$ for all $\mu > 0$.

A point $u \in \mathcal{K}$ is a fixed point of $\mathcal{S}$ provided $\mathcal{S}u = u$. Denote by $\mathcal{F}ix(\mathcal{S})$ the set of fixed points of $\mathcal{S}$, i.e., $\mathcal{F}ix(\mathcal{S}) = \{u \in \mathcal{K} : \mathcal{S}u = u\}$.

1. The mapping $\mathcal{S}$ is called $\mathcal{L}-$Lipschitz, $\mathcal{L} > 0$, if

$$||\mathcal{S}u - \mathcal{S}v|| \leq \mathcal{L}||u - v||, \quad \forall\, u, v \in \mathcal{K}.$$

2. The mapping $\mathcal{S}$ is called nonexpansive if

$$||\mathcal{S}u - \mathcal{S}v|| \leq ||u - v||, \quad \forall\, u, v \in \mathcal{K}.$$

3. The mapping $\mathcal{S}$ is called quasi-nonexpansive if $\mathcal{F}ix(\mathcal{S}) \neq \emptyset$ and

$$||\mathcal{S}u - v|| \leq ||u - v||, \quad \forall\, u \in \mathcal{K},\, v \in \mathcal{F}ix(\mathcal{S}).$$

In this case, $\mathcal{H}$ is a real Hilbert space. If $\mathcal{A} : \mathcal{E} \to 2^{\mathcal{E}}$ is an $m-$accretive operator (see [15–17]), then $\mathcal{A}$ is called maximal accretive operator [18], and for all $\mu > 0$, $\mathcal{R}(I + \mu\mathcal{A}) = \mathcal{H}$ if and only if $\mathcal{A}$ is called maximal monotone [19]. Denote by $\mathrm{dom}(h)$ the domain of a function $h : \mathcal{H} \to (-\infty, \infty]$, i.e.,

$$\mathrm{dom}(h) = \{u \in \mathcal{H} : h(u) < \infty\}.$$

The subdifferential of $h \in \Gamma_0(\mathcal{H})$ at $u \in \mathcal{H}$ is the set

$$\partial h(u) = \{z \in \mathcal{H} : h(u) \leq h(v) + \langle z, u - v\rangle, \quad \forall\, v \in \mathcal{H}\},$$

where $\Gamma_0(\mathcal{H})$ denotes the class of all *l.s.c.* functions from $\mathcal{H}$ to $(-\infty, \infty]$ with nonempty domains.

Lemma 1 ([20]). *Let $h \in \Gamma_0(\mathcal{H})$. Then, ∂h is maximal monotone.*

We denote by $B_\lambda[v]$ the closed ball with the center at v and radius λ :

$$B_\lambda[v] = \{u \in \mathcal{E} : ||v - u|| \leq \lambda\}.$$

Lemma 2 ([21]). *Let $\mathcal{E}$ be a Banach space, and $p > 1$ and $R > 0$ be two fixed numbers. Then, $\mathcal{E}$ is uniformly convex if and only if there exists a continuous, strictly increasing, and convex function $\varphi : [0, \infty) \to [0, \infty)$ with $\varphi(0) = 0$ such that*

$$||\alpha u + (1 - \alpha)v||^p \leq ||u||^p + (1 - \alpha)||v||^p - \alpha(1 - \alpha)\varphi(||u - v||),$$

for all $u, v \in B_R[0]$ and $\alpha \in [0, 1]$.

Definition 1 ([22]). *A vector space $\mathcal{H}$ is said to satisfy Opial's condition, if for each sequence $\{u_n\}$ in $\mathcal{H}$ which converges weakly to point $u \in \mathcal{H}$,*

$$\liminf_{n\to\infty} ||u_n - u|| < \liminf_{n\to\infty} ||u_n - v||, \quad v \in \mathcal{H},\, v \neq u.$$

Lemma 3 ([23]). *Let $\mathcal{K}$ be a nonempty subset of a Banach space $\mathcal{E}$, let $\mathcal{S} : \mathcal{K} \to \mathcal{E}$ be a uniformly continuous mapping, and let $\{u_n\} \subset \mathcal{K}$ an approximating fixed point sequence of $\mathcal{S}$. Then, $\{v_n\}$ is an approximating fixed point sequence of $\mathcal{S}$ whenever $\{v_n\}$ is in $\mathcal{K}$ such that $\lim_{n\to\infty} ||u_n - v_n|| = 0$.*

Lemma 4 ([16]). *Let $\mathcal{K}$ be a nonempty closed convex subset of a uniformly convex Banach space $\mathcal{E}$. If $\mathcal{S} : \mathcal{K} \to \mathcal{E}$ is a nonexpansive mapping, then $I - \mathcal{S}$ has the demiclosed property with respect to 0.*

A subset $\mathcal{K}$ of Banach space $\mathcal{E}$ is called a retract of $\mathcal{E}$ if there is a continuous mapping $\mathcal{Q}$ from $\mathcal{E}$ onto $\mathcal{K}$ such that $\mathcal{Q}u = u$ for all $u \in \mathcal{K}$. We call such $\mathcal{Q}$ a retraction of $\mathcal{E}$ onto $\mathcal{K}$. It follows that, if a mapping $\mathcal{Q}$ is a retraction, then $\mathcal{Q}v = v$ for all v in the range of $\mathcal{Q}$. A retraction $\mathcal{Q}$ is called a sunny if

$\mathcal{Q}(\mathcal{Q}u + \lambda(u - \mathcal{Q}u)) = \mathcal{Q}u$ for all $u \in \mathcal{E}$ and $\lambda \geq 0$. If a sunny retraction $\mathcal{Q}$ is also nonexpansive, then $\mathcal{K}$ is called a sunny nonexpansive retract of $\mathcal{E}$ [24].

Let $\mathcal{E}$ be a strictly convex reflexive Banach space and $\mathcal{K}$ be a nonempty closed convex subset of $\mathcal{E}$. Denote by $\mathcal{P}_{\mathcal{K}}$ the (metric) projection from $\mathcal{E}$ onto $\mathcal{K}$, namely, for $u \in \mathcal{E}$, $\mathcal{P}_{\mathcal{K}}(u)$ is the unique point in $\mathcal{K}$ with the property

$$\inf\{\|u - v\| : v \in \mathcal{K}\} = \|u - \mathcal{P}_{\mathcal{K}}(u)\|.$$

Let an inner product $\langle \cdot, \cdot \rangle$ and the induced norm $\|\cdot\|$ are specified with a real Hilbert space $\mathcal{H}$. Let $\mathcal{K}$ is a nonempty subset of $\mathcal{H}$, we have the nearest point projection $\mathcal{P}_{\mathcal{K}} : \mathcal{H} \to \mathcal{K}$ is the unique sunny nonexpansive retraction of $\mathcal{H}$ onto $\mathcal{K}$. It is also known that $\mathcal{P}_{\mathcal{K}}(u) \in \mathcal{K}$ and

$$\langle u - \mathcal{P}_{\mathcal{K}}(u), \mathcal{P}_{\mathcal{K}}(u) - v \rangle \geq 0, \quad \forall\, u \in \mathcal{H},\, v \in \mathcal{K}.$$

3. Main Results

Let $\mathcal{K}$ be a nonempty closed convex subset of a Banach space $\mathcal{E}$ with $\mathcal{Q}_{\mathcal{K}}$ as a sunny nonexpansive retraction. We denote by $\Psi := \mathcal{F}ix(\mathcal{S}) \cap \mathcal{F}ix(\mathcal{T})$.

Lemma 5. *Let $\mathcal{K}$ be a nonempty closed convex subset of a Banach space $\mathcal{E}$ with $\mathcal{Q}_{\mathcal{K}}$ as the sunny nonexpansive retraction, let $\mathcal{S}, \mathcal{T} : \mathcal{K} \to \mathcal{E}$ be quasi-nonexpansive mappings which $\Psi \neq \emptyset$, and let $\{\eta_n\}, \{\vartheta_n\}$ and $\{\xi_n\}$ be sequences in $(0,1)$ for all $n \in \mathbb{N}$. Let $\{u_n\}$ be defined by Algorithm 1. Then, for each $\bar{u} \in \Psi$, $\lim_{n\to\infty} \|u_n - \bar{u}\|$ exists and*

$$\|w_n - \bar{u}\| \leq \|u_n - \bar{u}\|, \quad \textit{and} \quad \|z_n - \bar{u}\| \leq \|u_n - \bar{u}\|, \quad \forall\, n \in \mathbb{N}. \tag{2}$$

Algorithm 1: Three-step sunny nonexpansive retraction

initialization: $\eta_n, \vartheta_n, \xi_n \in (0,1)$, $u_1 \in \mathcal{K}$ and $n = 1$.
while *stopping criterion not met* **do**
$w_n = \mathcal{Q}_{\mathcal{K}}[(1 - \xi_n)u_n + \xi_n \mathcal{S} u_n]$,
$z_n = \mathcal{Q}_{\mathcal{K}}[(1 - \vartheta_n)w_n + \vartheta_n \mathcal{T} w_n]$,
$u_{n+1} = \mathcal{Q}_{\mathcal{K}}[(1 - \eta_n)\mathcal{S} w_n + \eta_n \mathcal{T} z_n]$.
end

Proof. Let $\bar{u} \in \Psi$. Then, we have

$$\begin{aligned} \|w_n - \bar{u}\| &= \|\mathcal{Q}_{\mathcal{K}}[(1 - \xi_n)u_n + \xi_n \mathcal{S} u_n] - \bar{u}\| \\ &\leq \|(1 - \xi_n)(u_n - \bar{u}) + \xi_n(\mathcal{S} u_n - \bar{u})\| \\ &\leq (1 - \xi_n)\|u_n - \bar{u}\| + \xi_n \|\mathcal{S} u_n - \bar{u}\| \\ &\leq (1 - \xi_n)\|u_n - \bar{u}\| + \xi_n \|u_n - \bar{u}\| \\ &= \|u_n - \bar{u}\|, \end{aligned} \tag{3}$$

$$\begin{aligned} \|z_n - \bar{u}\| &= \|\mathcal{Q}_{\mathcal{K}}[(1 - \vartheta_n)w_n + \vartheta_n \mathcal{T} w_n] - \bar{u}\| \\ &\leq \|(1 - \vartheta_n)(w_n - \bar{u}) + \vartheta_n(\mathcal{T} w_n - \bar{u})\| \\ &\leq (1 - \vartheta_n)\|w_n - \bar{u}\| + \vartheta_n \|\mathcal{T} w_n - \bar{u}\| \\ &\leq (1 - \vartheta_n)\|w_n - \bar{u}\| + \vartheta_n \|w_n - \bar{u}\| \\ &= \|w_n - \bar{u}\| \\ &\leq \|u_n - \bar{u}\|, \end{aligned} \tag{4}$$

and

$$\begin{aligned}\|u_{n+1}-\bar{u}\| &= \|\mathcal{Q}_{\mathcal{K}}[(1-\eta_n)\mathcal{S}w_n+\eta_n\mathcal{T}z_n]-\bar{u}\| \\ &\le \|(1-\eta_n)(\mathcal{S}w_n-\bar{u})+\eta_n(\mathcal{T}z_n-\bar{u})\| \\ &\le (1-\eta_n)\|\mathcal{S}w_n-\bar{u}\|+\eta_n\|\mathcal{T}z_n-\bar{u}\| \\ &\le (1-\eta_n)\|w_n-\bar{u}\|+\eta_n\|z_n-\bar{u}\| \\ &\le (1-\eta_n)\|u_n-\bar{u}\|+\eta_n\|u_n-\bar{u}\| \\ &= \|u_n-\bar{u}\|.\end{aligned} \tag{5}$$

Therefore,

$$\|u_{n+1}-\bar{u}\| \le \|u_n-\bar{u}\| \le \cdots \le \|u_1-\bar{u}\|, \quad \forall n\in\mathbb{N}. \tag{6}$$

Since $\{\|u_n-\bar{u}\|\}$ is monotonically decreasing, we have that the sequence $\{\|u_n-\bar{u}\|\}$ is convergent. □

From Lemma 5, we have results:

Theorem 1. *Let $\mathcal{K}$ be a nonempty closed convex subset of a Banach space $\mathcal{E}$ with $\mathcal{Q}_{\mathcal{K}}$ as the sunny nonexpansive retraction, let $\mathcal{S},\mathcal{T}:\mathcal{K}\to\mathcal{E}$ be quasi-nonexpansive mappings with $\Psi\neq\emptyset$, and let $\{\eta_n\},\{\vartheta_n\}$ and $\{\xi_n\}$ be sequences of real numbers, for which $0<c_1\le\eta_n\le\hat{c}_1<1$, $0<c_2\le\vartheta_n\le\hat{c}_2<1$, $0<c_3\le\xi_n\le\hat{c}_3<1$ for all $n\in\mathbb{N}$. Let $u_1\in\mathcal{K}$, $\mathcal{P}_{\Psi}(u_1)=u_*$ and $\{u_n\}$ is defined by Algorithm 1. Then, we have the following:*

(i) *$\{u_n\}$ is in a closed convex bounded set $B_\lambda[u_*]\cap\mathcal{K}$, where λ is a constant in $(0,\infty)$ such that $\|u_1-u_*\|\le\lambda$.*
(ii) *If $\mathcal{S}$ is uniformly continuous, then $\lim_{n\to\infty}\|u_n-\mathcal{S}u_n\|=0$ and $\lim_{n\to\infty}\|u_n-\mathcal{T}u_n\|=0$.*
(iii) *If $\mathcal{E}$ fulfills the Opial's condition and $I-\mathcal{S}$ and $I-\mathcal{T}$ are demiclosed at 0, then $\{u_n\}$ converges weakly to an element of $\Psi\cap B_\lambda[u_*]$.*

Proof. (i) Since $u_*\in\Psi$, from Equation (6), we obtain

$$\|u_{n+1}-u_*\| \le \|u_n-u_*\| \le \cdots \le \|u_1-u_*\| \le \lambda, \quad \forall n\in\mathbb{N}. \tag{7}$$

Therefore, $\{u_n\}$ is in the closed convex bounded set $B_\lambda[u_*]\cap\mathcal{K}$.

(ii) Suppose that $\mathcal{S}$ is uniformly continuous. Using Lemma 5, we get that $\{u_n\}$, $\{z_n\}$ and $\{w_n\}$ are in $B_\lambda[u_*]\cap\mathcal{K}$, and hence, from Equation (2), we obtain

$$\|\mathcal{T}w_n-u_*\|\le\lambda, \quad \|\mathcal{S}w_n-u_*\|\le\lambda \quad \text{and} \quad \|\mathcal{S}u_n-u_*\|\le\lambda, \quad \forall n\in\mathbb{N}.$$

Using Lemma 2 for $p=2$ and $R=\lambda$, from Equation (5), we obtain

$$\begin{aligned}\|u_{n+1}-u_*\|^2 &\le \|(1-\eta_n)(\mathcal{S}w_n-u_*)+\eta_n(\mathcal{T}z_n-u_*)\|^2 \\ &\le (1-\eta_n)\|\mathcal{S}w_n-u_*\|^2+\eta_n\|\mathcal{T}z_n-u_*\|^2 \\ &\quad -\eta_n(1-\eta_n)\varphi(\|\mathcal{S}w_n-\mathcal{T}z_n\|) \\ &\le (1-\eta_n)\|w_n-u_*\|^2+\eta_n\|z_n-u_*\|^2 \\ &\quad -\eta_n(1-\eta_n)\varphi(\|\mathcal{S}w_n-\mathcal{T}z_n\|) \\ &\le (1-\eta_n)\|u_n-u_*\|^2+\eta_n\|u_n-u_*\|^2 \\ &\quad -\eta_n(1-\eta_n)\varphi(\|\mathcal{S}w_n-\mathcal{T}z_n\|) \\ &= \|u_n-u_*\|^2-\eta_n(1-\eta_n)\varphi(\|\mathcal{S}w_n-\mathcal{T}z_n\|),\end{aligned} \tag{8}$$

which implies that

$$\eta_n(1-\eta_n)\varphi(\|\mathcal{S}w_n-\mathcal{T}z_n\|) = \|u_n-u_*\| - \|u_{n+1}-u_*\|^2. \tag{9}$$

Note that: $c_1(1-\hat{c}_1) \leq \eta_n(1-\eta_n)$. Thus,

$$c_1(1-\hat{c}_1)\sum_{i=1}^{n}\varphi(\|\mathcal{S}w_i - \mathcal{T}z_i\|) = \|u_1 - u_*\| - \|u_{n+1} - u_*\|^2, \quad \forall n \in \mathbb{N}. \tag{10}$$

In the same way, we obtain

$$c_1(1-\hat{c}_1)\sum_{n=1}^{\infty}\varphi(\|\mathcal{S}w_n - \mathcal{T}z_n\|) \leq \|u_1 - u_*\| < \infty. \tag{11}$$

Therefore, we have $\lim_{n\to\infty}\|\mathcal{S}w_n - \mathcal{T}z_n\| = 0$. From the relations in Algorithm 1, we obtain

$$\begin{aligned}\|w_n - u_*\|^2 &\leq (1-\xi_n)\|u_n - u_*\|^2 + \xi_n\|\mathcal{S}u_n - u_*\|^2 \\ &\quad - \xi_n(1-\xi_n)\varphi(\|u_n - \mathcal{S}u_n\|) \\ &\leq (1-\xi_n)\|u_n - u_*\|^2 + \xi_n\|u_n - u_*\|^2 \\ &\quad - \xi_n(1-\xi_n)\varphi(\|u_n - \mathcal{S}u_n\|) \\ &= \|u_n - u_*\|^2 - \xi_n(1-\xi_n)\varphi(\|u_n - \mathcal{S}u_n\|)\end{aligned} \tag{12}$$

and

$$\begin{aligned}\|z_n - u_*\|^2 &\leq \|(1-\vartheta_n)(w_n - u_*) + \vartheta_n(\mathcal{T}w_n - u_*)\|^2 \\ &\leq (1-\vartheta_n)\|w_n - u_*\|^2 + \vartheta_n\|\mathcal{T}w_n - u_*\|^2 \\ &\quad - \vartheta_n(1-\vartheta_n)\varphi(\|w_n - \mathcal{T}w_n\|) \\ &\leq (1-\vartheta_n)\|w_n - u_*\|^2 + \vartheta_n\|w_n - u_*\|^2 \\ &= \|w_n - u_*\|^2 - \vartheta_n(1-\vartheta_n)\varphi(\|w_n - \mathcal{T}w_n\|) \\ &\leq \|u_n - u_*\|^2 - \vartheta_n(1-\vartheta_n)\varphi(\|w_n - \mathcal{T}w_n\|).\end{aligned} \tag{13}$$

From Equations (8), (13) and (12), we obtain

$$\begin{aligned}\|u_{n+1} - u_*\|^2 &\leq \|(1-\eta_n)(\mathcal{S}w_n - u_*) + \eta_n(\mathcal{T}z_n - u_*)\|^2 \\ &\leq (1-\eta_n)\|\mathcal{S}w_n - u_*\|^2 + \eta_n\|\mathcal{T}z_n - u_*\|^2 \\ &\quad - \eta_n(1-\eta_n)\varphi(\|\mathcal{S}w_n - \mathcal{T}z_n\|) \\ &\leq (1-\eta_n)\|w_n - u_*\|^2 + \eta_n\|z_n - u_*\|^2 \\ &\quad - \eta_n(1-\eta_n)\varphi(\|\mathcal{S}w_n - \mathcal{T}z_n\|) \\ &\leq (1-\eta_n)[\|u_n - u_*\|^2 - \xi_n(1-\xi_n)\varphi(\|u_n - \mathcal{S}u_n\|)] \\ &\quad + \eta_n[\|u_n - u_*\|^2 - \vartheta_n(1-\vartheta_n)\varphi(\|w_n - \mathcal{T}w_n\|)] \\ &\quad - \eta_n(1-\eta_n)\varphi(\|\mathcal{S}w_n - \mathcal{T}z_n\|) \\ &= \|u_n - u_*\|^2 - (1-\eta_n)\xi_n(1-\xi_n)\varphi(\|u_n - \mathcal{S}u_n\|) \\ &\quad - \eta_n\vartheta_n(1-\vartheta_n)\varphi(\|w_n - \mathcal{T}w_n\|) \\ &\quad - \eta_n(1-\eta_n)\varphi(\|\mathcal{S}w_n - \mathcal{T}z_n\|).\end{aligned} \tag{14}$$

Note that: $(1-\hat{c}_1)c_3(1-\hat{c}_3) \leq (1-\eta_n)\xi_n(1-\xi_n)$ and $c_1c_2(1-\hat{c}_2) \leq \eta_n\vartheta_n(1-\vartheta_n)$. Thus,

$$(1-\hat{c}_1)c_3(1-\hat{c}_3)\sum_{i=1}^{n}\varphi(\|u_i - \mathcal{S}u_i\|) \leq \|u_1 - u_*\|^2 - \|u_{n+1} - u_*\|^2, \quad \forall n \in \mathbb{N}.$$

It follows that $\lim_{n\to\infty}\|u_n - \mathcal{S}u_n\| = 0$. Note that:

$$\begin{aligned}\|w_n - u_n\| &= \|\mathcal{Q}_{\mathcal{K}}[(1-\xi_n)u_n + \xi_n\mathcal{S}u_n] - \mathcal{Q}_{\mathcal{K}}[u_n]\| \\ &\leq \|\mathcal{S}u_n - u_n\| \to 0 \quad \text{as} \quad n \to \infty.\end{aligned}$$

Since $\mathcal{S}$ is uniformly continuous, it follows from Lemma 3 that $\lim_{n\to\infty} \|w_n - \mathcal{S}w_n\| = 0$. Thus, from $\lim_{n\to\infty} \|\mathcal{S}w_n - \mathcal{T}z_n\| = 0$, we obtain $\lim_{n\to\infty} \|u_n - \mathcal{T}u_n\| = 0$.

(iii) By assumption, $\mathcal{E}$ satisfies the Opial's condition. Let $w^* \in \Psi$ such that $w^* \in B_\lambda[u_*] \cap \mathcal{K}$. From Lemma 5, we have $\lim_{n\to\infty} \|u_n - w^*\|$ exists. Suppose there are two subsequences $\{u_{n_q}\}$ and $\{u_{m_l}\}$ which converge to two distinct points u^* and v^* in $B_\lambda[u_*] \cap \mathcal{K}$, respectively. Then, since both $I - \mathcal{S}$ and $I - \mathcal{T}$ have the demiclosed property at 0, we have $\mathcal{S}u^* = \mathcal{T}u^* = u^*$ and $\mathcal{S}v^* = \mathcal{T}v^* = v^*$. Moreover, using the Opial's condition:

$$\lim_{n\to\infty} \|u_n - u^*\| = \lim_{q\to\infty} \|u_{n_q} - u^*\| < \lim_{l\to\infty} \|u_{m_l} - v^*\| = \lim_{n\to\infty} \|u_n - v^*\|.$$

Similarly, we obtain

$$\lim_{n\to\infty} \|u_n - v^*\| < \lim_{n\to\infty} \|u_n - u^*\|,$$

which is a contradiction. Therefore, $u^* = v^*$. Hence, the sequence $\{u_n\}$ converges weakly to an element of $\Psi \cap B_\lambda[u_*] \cap \mathcal{K}$. □

Theorem 2. *Let $\mathcal{K}$ be a nonempty closed convex subset of a Banach space $\mathcal{E}$ with $\mathcal{Q}_{\mathcal{K}}$ as the sunny nonexpansive retraction, let $\mathcal{S}, \mathcal{T} : \mathcal{K} \to \mathcal{E}$ be nonexpansive mappings with $\Psi \neq \emptyset$, and let $\{\eta_n\}, \{\vartheta_n\}$ and $\{\xi_n\}$ be sequences of real numbers, for which $0 < c_1 \leq \eta_n \leq \hat{c}_1 < 1$, $0 < c_2 \leq \vartheta_n \leq \hat{c}_2 < 1$, $0 < c_3 \leq \xi_n \leq \hat{c}_3 < 1$ for all $n \in \mathbb{N}$. Let $u_1 \in \mathcal{K}$, $\mathcal{P}_\Psi(u_1) = u_*$ and $\{u_n\}$ is defined by Algorithm 1. Then, we have the following:*

(i) *$\{u_n\}$ is in a closed convex bounded set $B_\lambda[u_*] \cap \mathcal{K}$, where λ is a constant in $(0, \infty)$ such that $\|u_1 - u_*\| \leq \lambda$.*
(ii) *$\lim_{n\to\infty} \|u_n - \mathcal{S}u_n\| = 0$ and $\lim_{n\to\infty} \|u_n - \mathcal{T}u_n\| = 0$.*
(iii) *If $\mathcal{E}$ fulfills the Opial's condition, then $\{u_n\}$ converges weakly to an element of $\Psi \cap B_\lambda[u_*]$.*

Proof. It follows from Theorem 1. □

Corollary 1. *Let $\mathcal{K}$ be a nonempty closed convex subset of a real Hilbert space $\mathcal{H}$, let $\mathcal{S}, \mathcal{T} : \mathcal{K} \to \mathcal{E}$ be nonexpansive mappings with $\Psi \neq \emptyset$, and let $\{\eta_n\}$, $\{\vartheta_n\}$ and $\{\xi_n\}$ be sequences of real numbers, for which $0 < c_1 \leq \eta_n \leq \hat{c}_1 < 1$, $0 < c_2 \leq \vartheta_n \leq \hat{c}_2 < 1$, $0 < c_3 \leq \xi_n \leq \hat{c}_3 < 1$ for all $n \in \mathbb{N}$. Let $\{u_n\}$ be defined by*

$$\begin{cases} w_n = (1 - \xi_n)u_n + \xi_n \mathcal{S}u_n, \\ z_n = (1 - \vartheta_n)w_n + \vartheta_n \mathcal{T}w_n, \\ u_{n+1} = (1 - \eta_n)\mathcal{S}w_n + \eta_n \mathcal{T}z_n, \quad \forall n \in \mathbb{N}. \end{cases} \tag{15}$$

Then, $\{u_n\}$ converges weakly to an element of Ψ.

Proof. It follows from Theorem 1. □

4. Applications

4.1. Common Zeros of Accretive Operators

From Equation (15), we set $\mathcal{S} = J_\mu^{\mathcal{A}}$ and $\mathcal{T} = J_\mu^{\mathcal{B}}$, and inherit the convergence analysis for solving Equation (1).

Theorem 3. *Let $\mathcal{K}$ be a nonempty closed convex subset of a r.u.c. Banach space $\mathcal{E}$ satisfying the Opial's condition. Let $\mathcal{A} : \mathcal{D}(\mathcal{A}) \subseteq \mathcal{K} \to 2^{\mathcal{E}}$, $\mathcal{B} : \mathcal{D}(\mathcal{B}) \subseteq \mathcal{K} \to 2^{\mathcal{E}}$ be accretive operators, for which $\mathcal{D}(\mathcal{A}) \subseteq \mathcal{K} \subseteq \cap_{\mu>0}\mathcal{R}(I + \mu\mathcal{A})$, $\mathcal{D}(\mathcal{B}) \subseteq \mathcal{K} \subseteq \cap_{\mu>0}\mathcal{R}(I + \mu\mathcal{B})$ and $\mathcal{A}^{-1}(0) \cap \mathcal{B}^{-1}(0) \neq \emptyset$. Let $\{\eta_n\}$, $\{\vartheta_n\}$ and $\{\xi_n\}$ be sequences of real numbers, for which $0 < c_1 \leq \eta_n \leq \hat{c}_1 < 1$, $0 < c_2 \leq$*

$\vartheta_n \leq \hat{c}_2 < 1$, $0 < c_3 \leq \xi_n \leq \hat{c}_3 < 1$ *for all* $n \in \mathbb{N}$. *Let* $\mu > 0$, $u_1 \in \mathcal{K}$ *and* $\mathcal{P}_{\mathcal{A}^{-1}(0)\cap\mathcal{B}^{-1}(0)}(u_1) = u_*$. *Let* $\{u_n\}$ *be defined by*

$$\begin{cases} w_n = (1-\xi_n)u_n + \xi_n J_\mu^{\mathcal{A}} u_n, \\ z_n = (1-\vartheta_n)w_n + \vartheta_n J_\mu^{\mathcal{B}} w_n, \\ u_{n+1} = (1-\eta_n)J_\mu^{\mathcal{A}} w_n + \eta_n J_\mu^{\mathcal{B}} z_n, \quad \forall n \in \mathbb{N}. \end{cases} \tag{16}$$

Then, we have the following:

(i) $\{u_n\}$ *is in a closed convex bounded set* $B_\lambda[u_*] \cap \mathcal{K}$, *where* λ *is a constant in* $(0, \infty)$ *such that* $\|u_1 - u_*\| \leq \lambda$.
(ii) $\lim_{n\to\infty} \|u_n - J_\mu^{\mathcal{A}} u_n\| = 0$ *and* $\lim_{n\to\infty} \|u_n - J_\mu^{\mathcal{B}} u_n\| = 0$.
(iii) $\{u_n\}$ *converges weakly to an element of* $\mathcal{A}^{-1}(0) \cap \mathcal{B}^{-1}(0) \cap B_\lambda[u_*]$.

Proof. By assumption $\mathcal{D}(\mathcal{A}) \subseteq \mathcal{K} \subseteq \cap_{\mu>0}\mathcal{R}(I + \mu\mathcal{A})$, we known that $J_\mu^{\mathcal{A}}$, $J_\mu^{\mathcal{B}} : \mathcal{K} \to \mathcal{K}$ be nonexpansive. Note that $\mathcal{D}(\mathcal{A}) \cap \mathcal{D}(\mathcal{B}) \subseteq \mathcal{K}$ and hence

$$\begin{aligned} u_* \in \mathcal{A}^{-1}(0) \cap \mathcal{B}^{-1}(0) &\Rightarrow u_* \in \mathcal{D}(\mathcal{A}) \cap \mathcal{D}(\mathcal{B}) \text{ with } 0 \in \mathcal{A}u_* \text{ and } 0 \in \mathcal{B}u_* \\ &\Rightarrow u_* \in \mathcal{K} \text{ with } J_\mu^{\mathcal{A}} u_* = u_* \text{ and } J_\mu^{\mathcal{B}} u_* = u_* \\ &\Rightarrow u_* \in \mathcal{F}ix(J_\mu^{\mathcal{A}}, J_\mu^{\mathcal{B}}) \cap \mathcal{K}. \end{aligned}$$

Next, set $\mathcal{S} = J_\mu^{\mathcal{A}}$ and $\mathcal{T} = J_\mu^{\mathcal{B}}$. Hence, Theorem 3 is the same way as Theorem 2. □

4.2. Convexly Constrained Least Square Problem

We provide applications of Theorem 2 for finding solutions to common problems with two convexly constrained least square problems. We consider the following problem:

Let $\mathcal{A}, \mathcal{B} \in \mathcal{B}(\mathcal{H})$, and $y, z \in \mathcal{H}$. Define $\varphi, \psi : \mathcal{H} \to \mathbb{R}$ by

$$\varphi = \|\mathcal{A}u - y\|^2 \quad \text{and} \quad \psi = \|\mathcal{B}u - z\|^2, \quad \forall u \in \mathcal{H},$$

where $\mathcal{H}$ is a real Hilbert space.

Let $\mathcal{K}$ be a nonempty closed convex subset of $\mathcal{H}$. The objective is to find $b \in \mathcal{K}$ such that

$$b \in \arg\min_{u\in\mathcal{K}} \varphi(u) \cap \arg\min_{u\in\mathcal{K}} \psi(u), \tag{17}$$

where

$$\arg\min_{u\in\mathcal{K}} \varphi(u) := \{\bar{u} \in \mathcal{K} : \varphi(u_*) = \inf_{u\in\mathcal{K}} \varphi(u)\}.$$

Proposition 1 ([8]). *Let* $\mathcal{H}$ *be a real Hilbert space,* $\mathcal{A} \in \mathcal{B}(\mathcal{H})$ *with the adjoint* $\mathcal{A}^*$ *and* $y \in \mathcal{H}$. *Let* $\mathcal{K}$ *be a nonempty closed convex subset of* $\mathcal{H}$. *Let* $b \in \mathcal{H}$ *and* $\delta \in (0, \infty)$. *Then, the following statements are equivalent:*

(i) b *solves the following problem:*

$$\min_{u\in\mathcal{K}} \|\mathcal{A}u - y\|^2.$$

(ii) $b = \mathcal{P}_{\mathcal{K}}(b - \delta\mathcal{A}^*(\mathcal{A}b - y))$.
(iii) $\langle \mathcal{A}v - \mathcal{A}b, y - \mathcal{A}b \rangle \leq 0$, *for all* $v \in \mathcal{K}$.

Theorem 4. *Let* $\mathcal{K}$ *be a nonempty closed convex subset of a real Hilbert space* $\mathcal{H}$, $y, z \in \mathcal{H}$ *and* $\mathcal{A}, \mathcal{B} \in \mathcal{B}(\mathcal{H})$, *for which the solution set of the problem in Equation* (17) *is nonempty. Let* $\{\eta_n\}$, $\{\vartheta_n\}$ *and* $\{\xi_n\}$ *be sequences of real numbers, for which* $0 < c_1 \leq \eta_n \leq \hat{c}_1 < 1$, $0 < c_2 \leq \vartheta_n \leq \hat{c}_2 < 1$, $0 < c_3 \leq \xi_n \leq \hat{c}_3 < 1$ *for all*

$n \in \mathbb{N}$. Let $u_1 \in \mathcal{H}$, $\mathcal{P}_{\arg\min_{u\in\mathcal{K}} \varphi(u) \cap \arg\min_{u\in\mathcal{K}} \psi(u)}(u_1) = u_*$, $\delta \in (0, 2\min\{\frac{1}{\|\mathcal{A}\|^2}, \frac{1}{\|\mathcal{B}\|^2}\})$, $u_1 \in \mathcal{K}$ and $\{u_n\}$ is defined by

$$\begin{cases} w_n = (1-\xi_n)u_n + \xi_n \mathcal{S}u_n, \\ z_n = (1-\vartheta_n)w_n + \vartheta_n \mathcal{T}w_n, \\ u_{n+1} = (1-\eta_n)\mathcal{S}w_n + \eta_n \mathcal{T}z_n, \quad \forall n \in \mathbb{N}. \end{cases} \tag{18}$$

where $\mathcal{S}, \mathcal{T} : \mathcal{K} \to \mathcal{K}$ defined by $\mathcal{S}u = \mathcal{P}_{\mathcal{K}}(u - \delta\mathcal{A}^*(\mathcal{A}u - y))$ and $\mathcal{T}u = \mathcal{P}_{\mathcal{K}}(u - \delta\mathcal{B}^*(\mathcal{B}u - z))$ for all $u \in \mathcal{K}$. Then, we have the following:

(i) *$\{u_n\}$ is in the closed ball $B_\lambda[u_*]$, where λ is a constant in $(0, \infty)$ such that $\|u_1 - u_*\| \leq \lambda$.*
(ii) *$\lim_{n\to\infty} \|u_n - \mathcal{S}u_n\| = 0$ and $\lim_{n\to\infty} \|u_n - \mathcal{T}u_n\| = 0$.*
(iii) *$\{u_n\}$ converges weakly to an element of $\arg\min_{u\in\mathcal{K}} \varphi(u) \cap \arg\min_{u\in\mathcal{K}} \psi(u) \cap B_\lambda[u_*]$.*

Proof. Note that: $\nabla\varphi(u) = \mathcal{A}^*(\mathcal{A}u - y)$, for all $u \in \mathcal{H}$; we obtain that $\|\nabla\varphi(u) - \nabla\varphi(v)\| = \|\mathcal{A}^*(\mathcal{A}u - y) - \mathcal{A}^*(\mathcal{A}v - y)\| \leq \|\mathcal{A}\|^2\|u - v\|$, for all $u, v \in \mathcal{H}$. Thus, $\nabla\varphi$ is $\frac{1}{\|\mathcal{A}\|^2}$-ism and hence $(I - \delta\nabla\varphi)$ is nonexpansive from $\mathcal{K}$ into $\mathcal{H}$ for $\sigma \in (0, \frac{2}{\|\mathcal{A}\|^2})$. Therefore, $\mathcal{S} = \mathcal{P}_{\mathcal{K}}(I - \sigma\nabla\varphi)$ and $\mathcal{T} = \mathcal{P}_{\mathcal{K}}(I - \tau\nabla\varphi)$ are nonexpansive mappings from $\mathcal{K}$ into itself for $\sigma \in (0, \frac{2}{\|\mathcal{A}\|^2})$ and $\tau \in (0, \frac{2}{\|\mathcal{B}\|^2})$, respectively. Hence, Theorem 4 is the same way as Theorem 2. □

4.3. Convex Minimization Problem

We give an application to common solutions to convex programming problems in a Hilbert space $\mathcal{H}$. We consider the following problem:

Let $g_1, g_2 : \mathcal{H} \to (-\infty, \infty]$ be proper *l.s.c.* functions. The objective is to find $x \in \mathcal{H}$ such that:

$$x \in \partial g_1^{-1}(0) \cap g_2^{-1}(0). \tag{19}$$

Note that: $J_\mu^{\partial g_1} = prox_{\mu g_1}$.

Theorem 5. *Let $\mathcal{K}$ be a nonempty closed convex subset of a real Hilbert space $\mathcal{H}$. Let $g_1, g_2 \in \Gamma_0(\mathcal{H})$, for which the solution set of the problem in Equation (19) is nonempty. Let $\{\eta_n\}$, $\{\vartheta_n\}$ and $\{\xi_n\}$ be sequences of real numbers, for which $0 < c_1 \leq \eta_n \leq \hat{c}_1 < 1$, $0 < c_2 \leq \vartheta_n \leq \hat{c}_2 < 1$, $0 < c_3 \leq \xi_n \leq \hat{c}_3 < 1$ for all $n \in \mathbb{N}$. Let $\mu > 0$, $u_1 \in \mathcal{H}$ and $\mathcal{P}_{\partial g_1^{-1}(0)\cap g_2^{-1}(0)}(u_1) = u_*$. Let $u_1 \in \mathcal{K}$ and $\{u_n\}$ is defined by*

$$\begin{cases} w_n = (1-\xi_n)u_n + \xi_n prox_{\mu g_1}(u_n), \\ z_n = (1-\vartheta_n)w_n + \vartheta_n prox_{\mu g_2}(w_n), \\ u_{n+1} = (1-\eta_n)prox_{\mu g_1}(w_n) + \eta_n prox_{\mu g_2}(z_n), \quad \forall n \in \mathbb{N}. \end{cases} \tag{20}$$

Then, we have the following:

(i) *$\{u_n\}$ is in the closed ball $B_\lambda[u_*]$, where λ is a constant in $(0, \infty)$ such that $\|u_1 - u_*\| \leq \lambda$.*
(ii) *$\lim_{n\to\infty} \|u_n - prox_{\mu g_1}(u_n)\| = 0$ and $\lim_{n\to\infty} \|u_n - prox_{\mu g_2}(u_n)\| = 0$.*
(iii) *$\{u_n\}$ converges weakly to an element of $\partial g_1^{-1}(0) \cap g_2^{-1}(0) \cap B_\lambda[u_*]$.*

Proof. Using Lemma 1, we have that ∂g_1 is maximal monotone. We know that $\mathcal{R}(I + \mu\partial f) = \mathcal{H}$ and using the maximal monotonicity of ∂g_1. Thus, $J_\mu^{\partial g_1} = prox_{\mu g_1} : \mathcal{H} \to \mathcal{H}$ is nonexpansive. Similarly, $J_\mu^{\partial g_2} = prox_{\mu g_2} : \mathcal{H} \to \mathcal{H}$ is nonexpansive. Hence, Theorem 5 is the same way as Theorem 2. □

4.4. Signal Processing

We consider some applications of our algorithm to inverse problems occurring from signal processing. For example, we consider the following underdeterminated linear equation system:

$$y = \mathcal{A}u + e, \tag{21}$$

where $u \in \mathbb{R}^N$ is recovered, $y \in \mathbb{R}^M$ is observations or measured data with noisy e, and $\mathcal{A} : \mathbb{R}^N \to \mathbb{R}^M$ is a bounded linear observation operator. It determines a process with loss of information. For finding solutions of the linear inverse problems in Equation (21), a successful one of some models is the convex unconstrained minimization problem:

$$\min_{u \in \mathcal{R}^N} \frac{1}{2}\|\mathcal{A}u - y\|^2 + d\|u\|_1, \tag{22}$$

where $d > 0$ and $\|\cdot\|_1$ is the l_1-norm. Thus, we can find solution to Equation (22) by applying our method in the case $g_1(u) = \frac{1}{2}\|\mathcal{A}u - y\|^2$ and $g_2(u) = d\|u\|_1$. For any $\alpha \in (0, \frac{2}{\mathcal{L}}]$, the corresponding forward-backward operator $J_\alpha^{g_1, d\|\cdot\|_1}$ as follows:

$$J_\alpha^{g_1, d\|\cdot\|_1}(u) = prox_{\alpha d\|\cdot\|_1}(u - \alpha\nabla g_1(u)), \tag{23}$$

where g_1 is the squared loss function of the Lasso problem in Equation (22). The proximity operator for l_1-norm is defined as the shrinkage operator as follows:

$$prox_{\alpha d\|\cdot\|_1}(u) = \max(|u_i| - \alpha d, 0) \cdot \mathrm{sgn}(u_i), \tag{24}$$

where $\mathrm{sgn}(\cdot)$ is the signum function. We apply the algorithm to the problem in Equation (22) follow as Algorithm 2:

Algorithm 2: Three-step forward-backward operator

---

initialization: $\eta_n, \vartheta_n, \xi_n \in (0,1)$, α, $d \in (0,1)$ $u_1 \in \mathcal{K}$ and $n = 1$.
while *stopping criterion not met* **do**

$w_n = (1 - \xi_n)u_n + \xi_n J_\alpha^{g_1, d\|\cdot\|_1}(u_n)$,

$z_n = (1 - \vartheta_n)w_n + \vartheta_n J_\alpha^{g_1, d\|\cdot\|_1}(w_n)$,

$u_{n+1} = (1 - \eta_n)J_\alpha^{g_1, d\|\cdot\|_1}(w_n) + \eta_n J_\alpha^{g_1, d\|\cdot\|_1}(z_n)$.

end

In our experiment, we set the hits of a signal $u \in \mathbb{R}^N$. The matrix $\mathcal{A} \in \mathbb{R}^{M \times N}$ was generated from a normal distribution with mean zero and one invariance. The observation y is generated by Gaussian noise distributed normally with mean 0 and variance 10^{-4}. We compared our Algorithm 2 with SPGA [12]. Let $\eta_n = \vartheta_n = \xi_n = 0.5, \alpha = 0.1$ and $d = 0.01$ in both Algorithm 2 and SPGA. The experiment was initialized by $u_1 = A^*y$ and terminated when $\frac{\|u_{n+1} - u_n\|}{\|u_n\|} < 10^{-4}$. The restoration accuracy was measured by means of the mean squared error: MSE = $\frac{\|u_* - u\|^2}{N}$, where u_* is an estimated signal of u. All codes were written in Matlab 2016b and run on Dell i-5 Core laptop. We present the numerical comparison of the results in Figures 1–6.

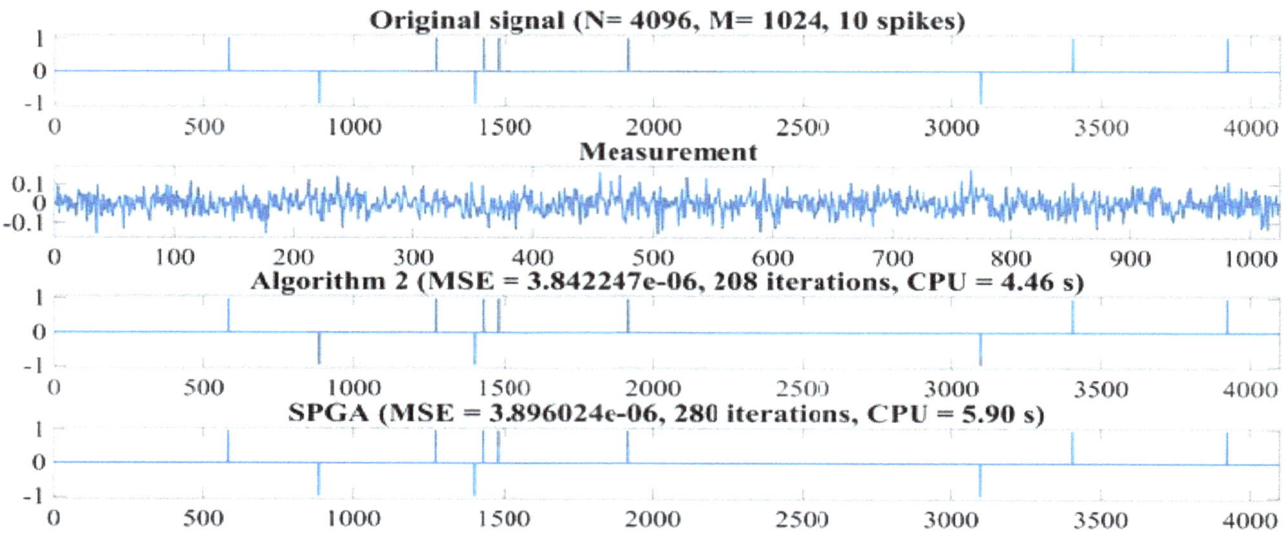

Figure 1. From top to bottom: Original signal, observation data, recovered signal by Algorithm 2 and SPGA with $N = 4096$, $M = 1024$ and 10 spikes, respectively.

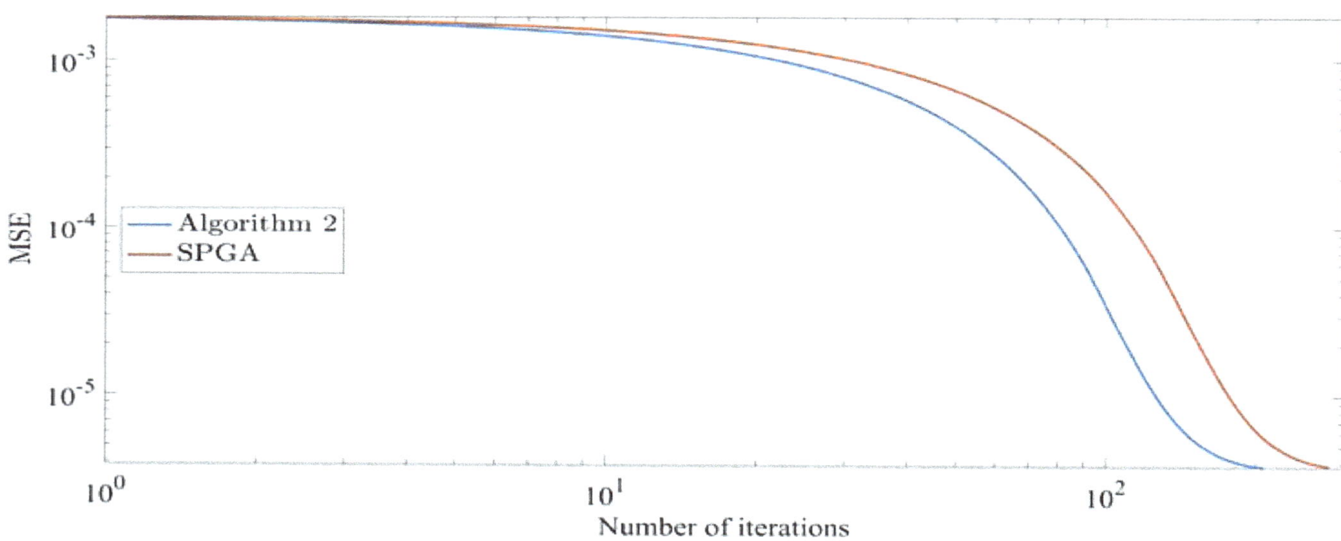

Figure 2. Comparison MSE of two algorithms for recovered signal with $N = 4096$, $M = 1024$ and 10 spikes, respectively.

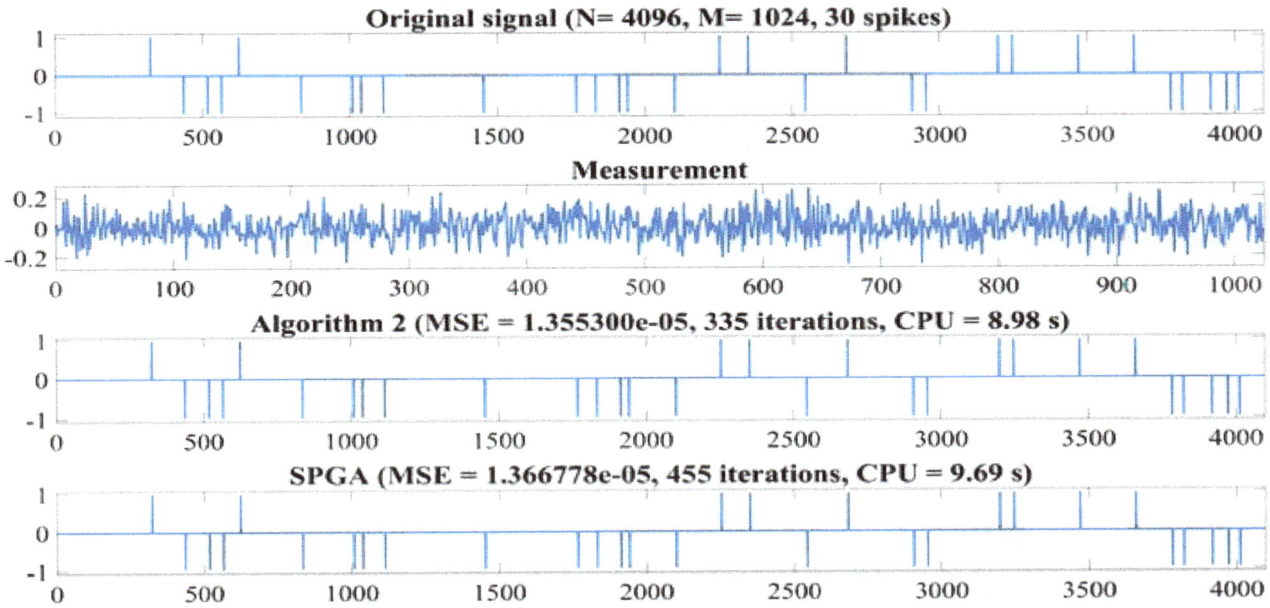

Figure 3. From top to bottom: Original signal, observation data, recovered signal by Algorithm 2 and SPGA with $N = 4096$, $M = 1024$ and 30 spikes, respectively.

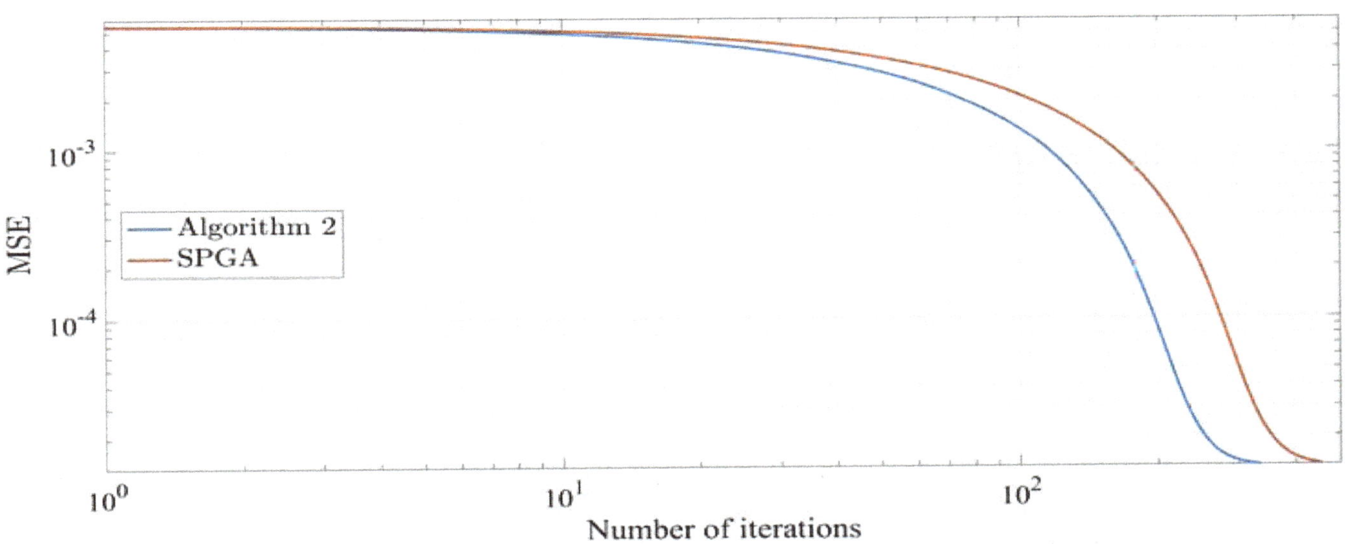

Figure 4. Comparison MSE of two algorithms for recovered signal with $N = 4096$, $M = 1024$ and 30 spikes, respectively.

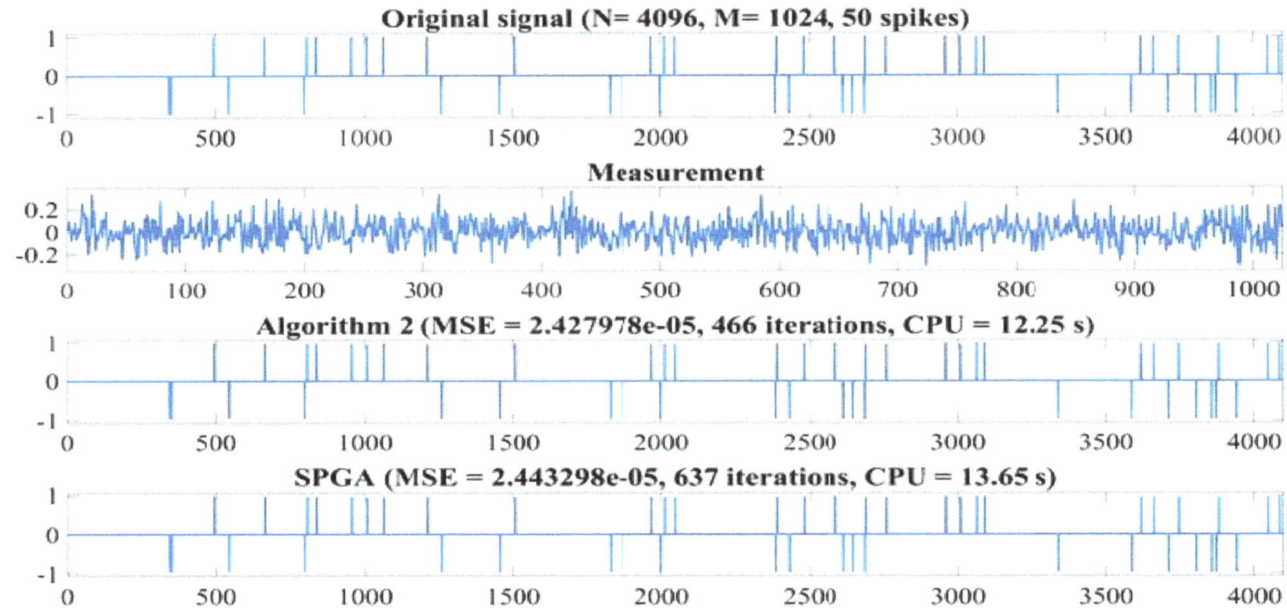

Figure 5. From top to bottom: Original signal, observation data, recovered signal by Algorithm 2 and SPGA with $N = 4096$, $M = 1024$ and 50 spikes, respectively.

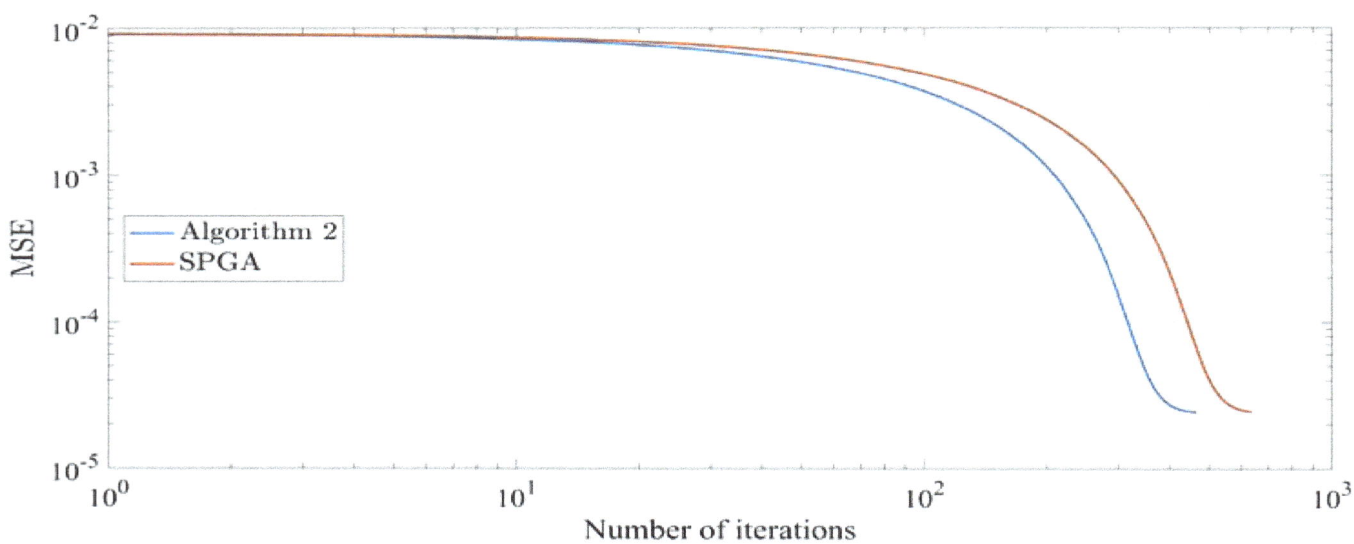

Figure 6. Comparison MSE of two algorithms for recovered signal with $N = 4096$, $M = 1024$ and 50 spikes, respectively.

5. Conclusions

In this work, we introduce a modified iterative scheme in Banach spaces and solve common zeros of accretive operators, convexly constrained least square problem, convex minimization problem and signal processing. In the case of signal processing, all results are compared with the forward-backward method in Algorithm 2 and SPGA, as proposed in [12]. The numerical results show that Algorithm 2 has a better convergence behavior than SPGA when using the same step sizes for both.

Author Contributions: A.P. and P.S.; writing original draft, A.P. and P.S.; data analysis, A.P. and P.S.; formal analysis and methodology

Acknowledgments: The first author thanks Rambhai Barni Rajabhat University for the support. Pakeeta Sukprasert was financially supported by Rajamangala University of Technology Thanyaburi (RMUTT).

Abbreviations

The following abbreviations are used in this manuscript:

Symbols	Display
$l.s.c.$	lower semicontinuous, convex
$\mathcal{B}(\mathcal{H})$	the set of all bounded and linear operators from $\mathcal{H}$ into itself
$r.u.c.$	real uniformly convex

References

1. Kankam, K.; Pholasa, N.; Cholamjiak, P. On convergence and complexity of the modified forward-backward method involving new linesearches for convex minimization. *Math. Meth. Appl. Sci.* **2019**, *42*, 1352–1362. [CrossRef]
2. Candès, E.J.; Wakin, M.B. An introduction to compressive sampling. *IEEE Signal Process. Mag.* **2008**, *25*, 21–30. [CrossRef]
3. Suantai, S.; Kesornprom, S.; Cholamjiak, P. A new hybrid CQ algorithm for the split feasibility problem in Hilbert spaces and Its applications to compressed Sensing. *Mathematics* **2019**, *7*, 789; doi:10.3390/math7090789. [CrossRef]
4. Kitkuan, D.; Kumam, P.; Padcharoen, A.; Kumam, W.; Thounthong, P. Algorithms for zeros of two accretive operators for solving convex minimization problems and its application to image restoration problems. *J. Comput. Appl. Math.* **2019**, *354*, 471–495. [CrossRef]
5. Padcharoen, A.; Kumam, P.; Cho, Y.J. Split common fixed point problems for demicontractive operators. *Numer. Algorithms* **2019**, *82*, 297–320. [CrossRef]
6. Cholamjiak, P.; Shehu, Y. Inertial forward-backward splitting method in Banach spaces with application to compressed sensing. *Appl. Math.* **2019**, *64*, 409–435. [CrossRef]
7. Jirakitpuwapat, W.; Kumam, P.; Cho, Y.J.; Sitthithakerngkiet, K. A general algorithm for the split common fixed point problem with its applications to signal processing. *Mathematics* **2019**, *7*, 226. [CrossRef]
8. Combettes, P.L.; Wajs, V.R. Signal recovery by proximal forward-backward splitting. *Multiscale Model Simul.* **2005**, *4*, 1168–1200. [CrossRef]
9. Picard, E. Memoire sur la theorie des equations aux d'erives partielles et la methode des approximations successives. *J. Math Pures Appl.* **1890**, *231*, 145–210.
10. Mann, W.R. Mean value methods in iteration. *Proc. Am. Math. Soc.* **1953**, *4*, 506–510. [CrossRef]
11. Ishikawa, S. Fixed points by a new iteration method. *Proc. Am. Math. Soc.* **1974**, *44*, 147–150. [CrossRef]
12. Agarwal, R.P.; O'Regan, D.; Sahu, D.R. Iterative construction of fixed points of nearly asymptotically nonexpansive mappings. *J. Nonlinear Convex Anal.* **2007**, *8*, 61–79.
13. Sahu, V.K.; Pathak, H.K.; Tiwari, R. Convergence theorems for new iteration scheme and comparison results. *Aligarh Bull. Math.* **2016**, *35*, 19–42.
14. Thakur, B.S.; Thakur, D.; Postolache, M. New iteration scheme for approximating fixed point of non-expansive mappings. *Filomat* **2016**, **30**, 2711–2720. [CrossRef]
15. Chang, S.S.; Wen, C.F.; Yao, J.C. Zero point problem of accretive operators in Banach spaces. *Bull. Malays. Math. Sci. Soc.* **2019**, *42*, 105–118. [CrossRef]
16. Browder, F.E. Nonlinear mappings of nonexpansive and accretive type in Banach spaces. *Bull. Am. Math. Soc.* **1967**, *73*, 875–882. [CrossRef]
17. Browder, F.E. Semicontractive and semiaccretive nonlinear mappings in Banach spaces. *Bull. Am. Math. Soc.* **1968**, *7*, 660–665. [CrossRef]
18. Cioranescu, I. *Geometry of Banach Spaces, Duality Mapping and Nonlinear Problems*; Kluwer: Amsterdam, The Netherlands, 1990.
19. Takahashi, W. *Nonlinear Functional Analysis, Fixed Point Theory and Its Applications*; Yokohama Publishers: Yokohama, Japan, 2000.
20. Rockafellar, R.T. On the maximal monotonicity of subdifferential mappings. *Pac. J. Math.* **1970**, *33*, 209–216. [CrossRef]

21. Xu, H.K. Inequalities in Banach spaces with applications. *Nonlinear Anal.* **1991**, *16*, 1127–1138. [CrossRef]
22. Opial, Z. Weak convergence of the sequence of successive approximations for nonexpansive mappings. *Bull. Am. Math. Soc.* **1967**, *73*, 591–597. [CrossRef]
23. Sahu, D.R.; Pitea, A.; Verma, M. A new iteration technique for nonlinear operators as concerns convex programming and feasibility problems. *Numer. Algorithms* **2019**. [CrossRef]
24. Goebel, K.; Reich, S. *Uniform Convexity, Hyperbolic Geometry and Non Expansive Mappings*; Marcel Dekker: New York, NY, USA; Basel, Switzerland, 1984 .

5

Improving the Computational Efficiency of a Variant of Steffensen's Method for Nonlinear Equations

Fuad W. Khdhr [1], Rostam K. Saeed [1] and Fazlollah Soleymani [2,*]

[1] Department of Mathematics, College of Science, Salahaddin University, Erbil, Iraq; fuad.khdhr@su.edu.krd (F.W.K.); rostam.saeed@su.edu.krd (R.K.S.)

[2] Department of Mathematics, Institute for Advanced Studies in Basic Sciences (IASBS), Zanjan 45137-66731, Iran

* Correspondence: fazlollah.soleymani@gmail.com

Abstract: Steffensen-type methods with memory were originally designed to solve nonlinear equations without the use of additional functional evaluations per computing step. In this paper, a variant of Steffensen's method is proposed which is derivative-free and with memory. In fact, using an acceleration technique via interpolation polynomials of appropriate degrees, the computational efficiency index of this scheme is improved. It is discussed that the new scheme is quite fast and has a high efficiency index. Finally, numerical investigations are brought forward to uphold the theoretical discussions.

Keywords: iterative methods; Steffensen's method; R-order; with memory; computational efficiency

---

1. Introduction

One of the commonly encountered topics in computational mathematics is to tackle solving a nonlinear algebraic equation. The equation can be presented as in the scalar case $f(x) = 0$, or more complicated as a system of nonlinear algebraic equations. The procedure of finding the solutions (if it exists) cannot be done analytically. In some cases, the analytic techniques only give the real result while its complex zeros should be found and reported. As such, numerical techniques are a viable choice for solving such nonlinear problems. Each of the existing computational procedures has their own domain of validity with some pros and cons [1,2].

Two classes of methods with the use of derivatives and without the use of derivatives are known to be useful depending on the application dealing with [3]. In the derivative-involved methods, a larger attraction basin along with a simple coding effort for higher dimensional problems is at hand which, in derivative-free methods, the area of choosing the initial approximations is smaller and extending to higher dimensional problems is via the application of a divided difference operator matrix, which is basically a dense matrix. However, the ease in not computing the derivative and, subsequently, the Jacobians, make the application of derivative-free methods more practical in several problems [4–7].

Here, an attempt is made at developing a computational method which is not only efficient in terms of the computational efficiency index, but also in terms of larger domains for the choice of the initial guesses/approximations for starting the proposed numerical method.

The Steffensen's method [8] for solving nonlinear scalar equations has quadratic convergence for simple zeros and given by:

$$\begin{cases} x_{k+1} = x_k - \dfrac{f(x_k)}{f[x_k, w_k]}, \\ w_k = x_k + \beta f(x_k),\ \beta \in R\backslash\{0\},\ k \geq 0, \end{cases} \tag{1}$$

where the two-point divided difference is defined by:

$$f[x_k, w_k] = \frac{f(x_k) - f(w_x)}{x_k - w_k},$$

This scheme needs two function evaluations per cycle. Scheme (1) shows an excellent tool for constructing efficient iterative methods for nonlinear equations. This is because it is derivative-free with a free parameter. This parameter can, first of all, enlarge the attraction basins of Equation (1) or any of its subsequent methods and, second, can directly affect the improvement of the R-order of convergence and the efficiency index.

Recalling that Kung and Traub conjectured that the iterative method without memory based on m functions evaluation per iteration attain the optimal convergence of order 2^{m-1} [9,10].

The term "with memory" means that the values of the function associated with the computed approximations of the roots are used in subsequent iterations. This is unlike the term "without memory" in which the method only uses the current values to find the next estimate. As such, in a method with memory, the calculated results up to the desired numbers of iterations should be stored and then called to proceed.

Before proceeding the given idea to improve the speed of convergence, efficiency index, and the attraction basins, we provide a short literature by reviewing some of the existing methods with accelerated convergence order. Traub [11] proposed the following two-point method with memory of order 2.414:

$$\begin{cases} x_{k+1} = x_k - \dfrac{f(x_k)}{f[x_k, x_k + \beta_k f(x_k)]}, \\ \beta_k = \dfrac{-1}{f[x_k, z_{k-1}]}, \end{cases} \tag{2}$$

where $z_{k-1} = x_{k-1} + \beta_{k-1} f(x_{k-1})$, and $\beta_0 = -\text{sign}(f'(x_0))$ or $-\dfrac{1}{f[x_0, x_0 + f(x_0)]}$. This is one of the pioneering and fundamental methods with memory for solving nonlinear equations.

Džunić in [12] suggested an effective bi-parametric iterative method with memory of $\frac{1}{2}\left(3 + \sqrt{17}\right)$ R-order of convergence as follows:

$$\begin{cases} w_k = x_k + \beta_k f(x_k), \\ \beta_k = -\dfrac{1}{N_2'(x_k)}, \ \zeta_k = -\dfrac{N_3''(w_k)}{2N_3'(w_k)}, \ k \geq 1, \\ x_{k+1} = x_k - \dfrac{f(x_k)}{f[x_k, w_k] + \zeta_k f(w_k)} \ k \geq 0. \end{cases} \tag{3}$$

Moreover, Džunić and Petković [13] derived the following cubically convergent Steffensen-like method with memory:

$$\begin{cases} x_{k+1} = x_k - \dfrac{f(x_k)}{f[x_k, x_k + \beta_k f(x_k)]}, \\ \beta_k = \dfrac{-1}{f[x_k, z_{k-1}] + f[x_k, x_{k-1}] + f[x_{k-1}, z_{k-1}]}, \end{cases} \tag{4}$$

where $z_{k-1} = x_{k-1} + \beta_{k-1} f(x_{k-1})$ depending on the second-order Newton interpolation polynomial.

Various Steffensen-type methods are proposed in [14–17].

In fact, it is possible to improve the performance of the aforementioned method by considering several more sub-steps and improve the computational efficiency index via multi-step iterative methods. However, this procedure is more computational burdensome. Thus, the motivation here is to know that is it possible to improve the performance of numerical methods in terms of the computational efficiency index, basins of attraction, and the rate of convergence without adding more sub-steps and propose a numerical method as a one-step solver.

Hence, the aim of this paper is to design a one-step method with memory which is quite fast and has an improved efficiency index, based on the modification of the one-step method of Steffensen (Equation (1)) and increase the convergence order to 3.90057 without any additional functional evaluations.

The rest of this paper is ordered as follows: In Section 2, we develop the one-point Steffensen-type iterative scheme (Equation (1)) with memory which was proposed by [18]. We present the main goal in Section 3 by approximating the acceleration parameters involved in our contributed scheme by Newton's interpolating polynomial and, thus, improve the convergence R-order. The numerical reports are suggested in Section 4 to confirm the theoretical results. Some discussions are given in Section 5.

2. An Iterative Method

The following iterative method without memory was proposed by [18]:

$$\begin{cases} w_k = x_k - \beta f(x_k), \\ x_{k+1} = x_k - \dfrac{f(x_k)}{f[x_k, w_k]}\left(1 + \zeta \dfrac{f(w_k)}{f[x_k, w_k]}\right), \ \zeta \in \mathbb{R}, \end{cases} \tag{5}$$

with the following error equation to improve the performance of (1) in terms of having more free parameters:

$$e_{k+1} = -(-1 + \beta f'(\alpha))(c_2 - \zeta)e_k^2 + O\left(e_k^3\right), \tag{6}$$

where $c_i = \frac{1}{i!}\frac{f^{(i)}(\alpha)}{f'(\alpha)}$. Using the error Equation (6), to derive Steffensen-type iterative methods with memory, we calculate the following parameters: $\beta = \beta_k$, $\zeta = \zeta_k$, by the formula:

$$\begin{cases} \beta_k = \dfrac{1}{\overline{f'}(x_\alpha)}, \\ \zeta_k = \overline{c}_2, \end{cases} \tag{7}$$

for $k = 1, 2, 3, \cdots$, while $\overline{f'}(x_\alpha)$, $\overline{c}_2$ are approximations to $f(\alpha)$ and c_2, respectively; where α is a simple zero of $f(x)$. In fact, Equation (7) shows a way to minimize the asymptotic error constant of Equation (6) by making this coefficient closer and closer to zero when the iterative method is converging to the true solution.

The initial estimates β_0 and ζ_0 must be chosen before starting the process of iterations. We state the Newton's interpolating polynomial of fourth and fifth-degree passing through the saved points as follows:

$$\begin{cases} N_4(t) = N_4(t; x_k, w_{k-1}, x_{k-1}, w_{k-2}, x_{k-2}), \\ N_5(t) = N_5(t; w_k, x_k, w_{k-1}, x_{k-1}, w_{k-2}, x_{k-2}). \end{cases} \tag{8}$$

Recalling that $N(t)$ is an interpolation polynomial for a given set of data points also known as the Newton's divided differences interpolation polynomial because the coefficients of the polynomial are calculated using Newton's divided differences method. For instance, here the set of data points for $N_4(t)$ are $\{\{x_k, f(x_k)\}, \{w_{k-1}, f(w_{k-1})\}, \{x_{k-1}, f(x_{k-1})\}, \{w_{k-2}, f(w_{k-2})\}, \{x_{k-2}, f(x_{k-2})\}\}$.

Now, using some modification on Equation (5) we present the following scheme:

$$\begin{cases} w_k = x_k - \beta_k f(x_k), \\ \beta_k = \dfrac{1}{N_4'(x_k)}, \ \zeta_k = \dfrac{N_5''(w_k)}{2N_5'(w_k)}, \ k \geq 2, \\ x_{k+1} = x_k - \dfrac{f(x_k)}{f[x_k, w_k]}\left(1 + \zeta_k \dfrac{f(w_k)}{f[x_k, w_k]}\right), \ k \geq 0. \end{cases} \tag{9}$$

Noting that the accelerator parameters β_k, ζ_k are getting updated and then used in the iterative method right after the second iterations, viz, $k \geq 2$. This means that the third line of Equation (9) is imposed at the beginning and after that the computed values are stored and used in the subsequent iterates. For $k = 1$, the degree of Newton interpolation polynomials would be two and three. However, for $k \geq 2$, interpolations of degrees four and five as given in Equation (8) can be used to increase the convergence order.

Additionally speaking, this acceleration of convergence would be attained without the use any more functional evaluations as well as imposing more steps. Thus, the proposed scheme with memory (Equation (9)) can be attractive for solving nonlinear equations.

3. Convergence Analysis

In this section, we show the convergence criteria of Equation (9) using Taylor's series expansion and several extensive symbolic computations.

Theorem 1. *Let the function $f(x)$ be sufficiently differentiable in a neighborhood of its simple zero α. If an initial approximation x_0 is necessarily close to α. Then, R-order of convergence for the one-step method (Equation (9)) with memory is 3.90057.*

Proof. The proof is done using the definition of the error equation as the difference between the k-estimate and the exact zero along with symbolic computations. Let the sequence $\{x_k\}$ and $\{w_k\}$ have convergence orders r and p, respectively. Namely,

$$e_{k+1} \sim e_k^r, \tag{10}$$

and:

$$e_{w,k} \sim e_k^p, \tag{11}$$

Therefore, using Equations (10) and (11), we have:

$$e_{k+1} \sim e_k^r \sim e_{k-1}^{r^2} \sim e_{k-2}^{r^3}, \tag{12}$$

and:

$$e_{w,k} \sim e_k^p \sim (e_{k-1}^r)^p \sim e_{k-2}^{pr^2}. \tag{13}$$

The associated error equations to the accelerating parameters β_k and ζ_k for Equation (9) can now be written as follows:

$$e_{w,k} \sim (-1 + \beta_k f'(\alpha))e_k \ , \tag{14}$$

and:

$$e_{k+1} \sim -(-1 + \beta_k f'(\alpha))(c_2 - \zeta_k)e_k^2. \tag{15}$$

On the other hand, by using a symbolic language and extensive computations one can find the following error terms for the involved terms existing in the fundamental error Equation (6):

$$-1 + \beta_k f'(\alpha) \sim c_5 e_{k-2} e_{k-1} e_{w,k-1} e_{w,k-2}, \tag{16}$$

$$c_2 - \zeta_k \sim c_6 e_{k-2} e_{k-1} e_{w,k-1} e_{w,k-2} \tag{17}$$

Combining Equations (14)–(17), we get that:

$$e_{w,k} \sim e_{k-2}^{r^2+pr+r+p+1}, \tag{18}$$

$$e_{k+1} \sim e_{k-2}^{2(r^2+pr+r+p+1)}. \tag{19}$$

We now compare the left and right hand side of Equations (12)–(19) and Equations (13)–(18), respectively. Thus, we have the following nonlinear system of equations in order to find the final R-orders:

$$\begin{cases} r^2p-(r^2+pr+r+p+1)=0, \\ r^3-2(r^2+pr+r+p+1)=0. \end{cases} \tag{20}$$

The positive real solution of (20) is $r=3.90057$ and $p=1.9502$. Therefore, the convergence R-order for Equation (9) is 3.90057. □

Since improving the convergence R-order is useless if the whole computational method is expensive, basically researcher judge on a new scheme based upon its computational efficiency index which is a tool in order to provide a trade-off between the whole computational cost and the attained R-order. Assuming the cost of calculating each functional evaluation is one, we can use the definition of efficiency index as $EI=p^{1/\theta}$, θ is the whole computational cost [19].

The computational efficiency index of Equation (9) is $3.90057^{\frac{1}{2}}\approx 1.97499\approx 2$, which is clearly higher than efficiency index $2^{\frac{1}{2}}\approx 1.4142$ of Newton's and Steffensen's methods, $3.56155^{\frac{1}{2}}\approx 1.8872$ of (3) $3^{1/2}\approx 1.73205$ of Equation (4).

However, this improved computational efficiency is reported by ignoring the number of multiplication and division per computing cycle. By imposing a slight weight for such calculations one may once again obtain the improved computational efficiency of (9) in contrast to the existing schemes of the same type.

4. Numerical Computations

In this section, we compare the convergence performance of Equation (9), with three well-known iterative methods for solving four test problems numerically carried out in Mathematica 11.1. [20].

We denote Equations (1), (3), (5) and (9) with SM, DZ, PM, M4, respectively. We compare the our method with different methods, using $\beta_0=0.1$ and $\zeta_0=0.1$. Here, the computational order of convergence (coc) has been computed by the following formula [21]:

$$coc=\frac{\ln|(f(x_k)/f(x_{k-1})|}{\ln|(f(x_{k-1})/f(x_{k-2})|} \tag{21}$$

Recalling that using a complex initial approximation, one is able to find the complex roots of the nonlinear equations using (9).

Experiment 1. *Let us consider the following nonlinear test function:*

$$f_1(x)=(x-2\tan(x))(x^3-8), \tag{22}$$

where $\alpha=2$ *and* $x_0=1.7$.

Experiment 2. *We take into account the following nonlinear test function:*

$$f_2(x)=(x-1)(x^{10}+x^3+1)\sin(x), \tag{23}$$

where $\alpha=1$ *and* $x_0=0.7$.

Experiment 3. *We consider the following test problem now:*

$$f_3(x)=\frac{-x^3}{2}+2\tan^{-1}(x)+1, \tag{24}$$

where $\alpha\approx 1.8467200$ *and* $x_0=4$.

Experiment 4. *The last test problem is taken into consideration as follows:*

$$f_4(x) = \tan^{-1}(\exp(x+2)+1) + \tanh(\exp(-x\cos(x))) - \sin(\pi x), \tag{25}$$

where $\alpha \approx -3.6323572\cdots$ *and* $x_0 = -4.1$.

Tables 1–4 show that the proposed Equation (9) is of order 3.90057 and it is obviously believed to be of more advantageous than the other methods listed due to its fast speed and better accuracy.

For better comparisons, we present absolute residual errors $|f(x)|$, for each test function which are displayed in Tables 1–4. Additionally, we compute the computational order of convergence. Noting that we have used multiple precision arithmetic considering 2000 significant digits to observe and the asymptotic error constant and the coc as obviously as possible.

The results obtained by our proposed Equation (M4) are efficient and show better performance than other existing methods.

A significant challenge of executing high-order nonlinear solvers is in finding initial approximation to start the iterations when high accuracy calculating is needed.

Table 1. Result of comparisons for the function f_1.

Methods	$\|f_1(x_3)\|$	$\|f_1(x_4)\|$	$\|f_1(x_5)\|$	$\|f_1(x_6)\|$	coc
SM	4.1583	3.0743	1.4436	0.25430	2.00
DZ	0.13132	2.0026×10^{-7}	1.0181×10^{-27}	7.1731×10^{-99}	3.57
PM	1.8921×10^{-6}	4.5864×10^{-24}	1.0569×10^{-88}	7.5269×10^{-318}	3.55
M4	9.1741×10^{-6}	3.3242×10^{-26}	4.4181×10^{-103}	1.1147×10^{-404}	3.92

Table 2. Result of comparisons for the function f_2.

Methods	$\|f_2(x_5)\|$	$\|f_2(x_6)\|$	$\|f_2(x_7)\|$	$\|f_2(x_8)\|$	coc
SM	—	—	—	—	—
DZ	0.14774	0.0016019.	1.3204×10^{-10}	1.5335×10^{-35}	3.56
PM	2.1191×10^{-10}	8.0792×10^{-35}	1.9037×10^{-121}	3.7062×10^{-430}	3.56
M4	5.9738×10^{-15}	4.1615×10^{-57}	1.7309×10^{-220}	1.8231×10^{-857}	3.90

Table 3. Result of comparisons for the function f_3.

Methods	$\|f_3(x_3)\|$	$\|f_3(x_4)\|$	$\|f_3(x_5)\|$	$\|f_3(x_6)\|$	coc
SM	0.042162	0.00012627	1.1589×10^{-9}	9.7638×10^{-20}	2.00
DZ	1.0219×10^{-11}	4.4086×10^{-44}	1.6412×10^{-157}	1.5347×10^{-562}	3.57
PM	7.9792×10^{-8}	3.712×10^{-30}	4.9556×10^{-108}	2.9954×10^{-386}	3.57
M4	4.4718×10^{-6}	2.9187×10^{-25}	4.7057×10^{-101}	1.0495×10^{-395}	3.89

To discuss further, mostly based on interval mathematics, one can find a close enough guess to start the process. There are some other ways to determine the real initial approximation to start the process. An idea of finding such initial guesses given in [22] is based on the useful commands in Mathematica 11.1 NDSolve [] for the nonlinear function on the interval $D = [a,b]$.

Following this the following piece of Mathematica code could give a list of initial approximations in the working interval for Experiment 4:

```
ClearAll["Global`*"]

(*Defining the nonlinear function.*)
f[x_]:=ArcTan[Exp[x+2]+1]+Tanh[Exp[-x Cos[x]]]-Sin[Pi x];

(*Defining the interval.*)
a=-4.; b=4.;

(*Find the list of initial estimates.*)
Zeros = Quiet@Reap[soln=y[x]/.First[NDSolve[{y'[x]
==Evaluate[D[f[x],x]],y[b]==(f[b])},y[x],{x,a,b},
Method->{"EventLocator","Event"->y[x], "EventAction":>Sow[{x,y[x]}]}]]][[2,1]];
initialPoints = Sort[Flatten[Take[zeros,Length[zeros],1]]]
```

To check the position of the zero and the graph of the function, we can use the following code to obtain Figure 1.

```
Length[initialPoints]
Plot[f[x],{x,a,b}, Epilog->{PointSize[Medium], Red, Point[zeros]},PlotRange->All, PerformanceGoal-
>"Quality", PlotStyle->{Thick, Blue}]
```

Table 4. Result of comparisons for the function f_4.

Methods	$\lvert f_4(x_3)\rvert$	$\lvert f_4(x_4)\rvert$	$\lvert f_4(x_5)\rvert$	$\lvert f_4(x_6)\rvert$	coc
SM	0.00001166	3.7123×10^{-10}	3.7616×10^{-19}	3.8622×10^{-37}	2.00
DZ	1.6×10^{-13}	6.9981×10^{-47}	1.0583×10^{-164}	7.0664×10^{-585}	3.57
PM	3.0531×10^{-11}	3.2196×10^{-38}	3.7357×10^{-134}	6.5771×10^{-476}	3.56
M4	2.5268×10^{-13}	1.5972×10^{-49}	2.8738×10^{-191}	1.6018×10^{-744}	3.90

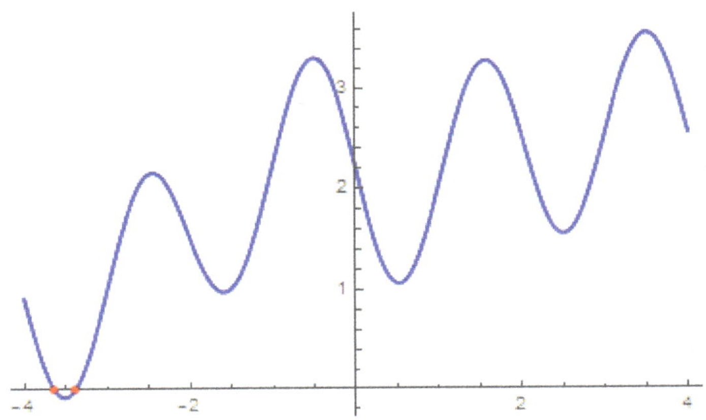

Figure 1. The plot of the nonlinear function in Experiment 4 along with its roots colored in red.

As a harder test problem, for the nonlinear function $g(x) = 2x + 0.5\sin(20\pi\, x) - x^2$, we can simply find a list of estimates as initial guesses using the above piece of codes as follows: $\{-0.185014, -0.162392, -0.0935912, -0.0535277, 6.73675 \times 10^{-9}, 0.0533287, 0.0941576, 0.160021, 0.188066, 0.269075, 0.279428, 1.76552, 1.78616, 1.8588, 1.89339, 1.95294, 2., 2.04692, 2.10748, 2.13979, 2.2228, 2.22471\}$. The plot of the function in this case is brought forward in Figure 2.

We observe that the two self-accelerating parameters β_0 and ζ_0 have to be selected before the iterative procedure is started. That is, they are calculated by using information existing from the present and previous iterations (see, e.g., [23]). The initial estimates β_0 and ζ_0 should be preserved as precise small positive values. We use $\beta_0 = \zeta_0 = 0.1$ whenever required.

After a number of iterates, the (nonzero) free parameters start converging to a particular value which makes the coefficient of Equation (6) zero as well as make the numerical scheme to converge with high R-order.

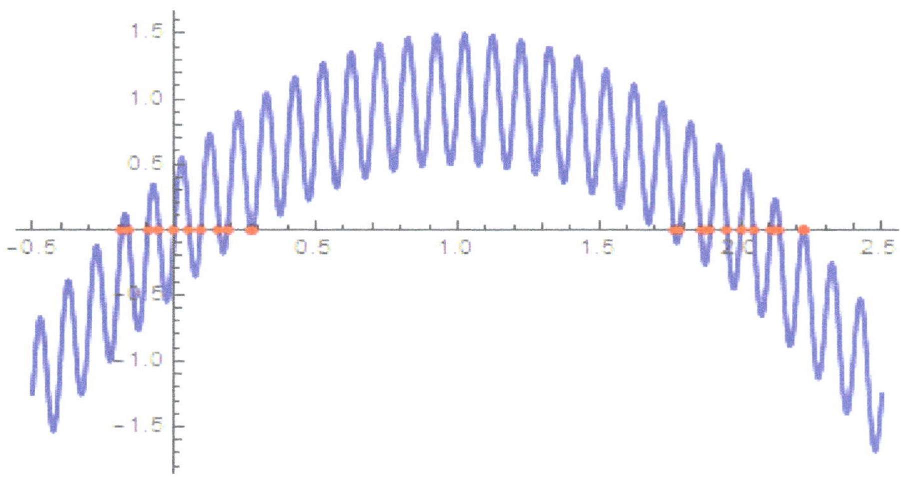

Figure 2. The behavior of the function g and the position of its roots (the red dots show the location of the zeros of the nonlinear functions).

5. Ending Comments

In this paper, we have constructed a one-step method with memory to solve nonlinear equations. By using two self-accelerator parameters our scheme equipped with Newton's interpolation polynomial without any additional functional calculation possesses the high computational efficiency index 1.97499, which is higher than many of the existing methods.

The efficacy of our scheme is confirmed by some of numerical examples. The results in Tables 1–4 shows that our method (Equation (M4)) is valuable to find an adequate estimate of the exact solution of nonlinear equations.

Author Contributions: The authors contributed equally to this paper.

Acknowledgments: The authors are thankful to two anonymous referees for careful reading and valuable comments which improved the quality of this manuscript.

References

1. Cordero, A.; Hueso, J.L.; Martinez, E.; Torregrosa, J.R. Steffensen type methods for solving nonlinear equations. *J. Comput. Appl. Math.* **2012**, *236*, 3058–3064. [CrossRef]
2. Soleymani, F. Some optimal iterative methods and their with memory variants. *J. Egypt. Math. Soc.* **2013**, *21*, 133–141. [CrossRef]
3. Praks, P.; Brkić, D. Choosing the optimal multi-point iterative method for the Colebrook Flow friction equation. *Processes* **2018**, *6*, 130. [CrossRef]
4. Zafar, F.; Cordero, A.; Torregrosa, J.R. An efficient family of optimal eighth-order multiple root finders. *Mathematics* **2018**, *6*, 310. [CrossRef]
5. Saeed, R.K.; Aziz, K.M. An iterative method with quartic convergence for solving nonlinear equations. *Appl. Math. Comput.* **2008**, *202*, 435–440. [CrossRef]
6. Saeed, R.K. Six order iterative method for solving nonlinear equations. *World Appl. Sci. J.* **2010**, *11*, 1393–1397.
7. Torkashvand, V.; Lotfi, T.; Araghi, M.A.F. A new family of adaptive methods with memory for solving nonlinear equations. *Math. Sci.* **2019**, 1–20.
8. Noda, T. The Steffensen iteration method for systems of nonlinear equations. *Proc. Jpn. Acad.* **1987**, *63*, 186–189.
9. Kung, H.T.; Traub, J.F. Optimal order of one-point and multipoint iteration. *J. Assoc. Comput. Math.* **1974**, *21*, 634–651. [CrossRef]

10. Ahmad, F. Comment on: On the Kung-Traub conjecture for iterative methods for solving quadratic equations. *Algorithms* **2016**, *9*, 30. [CrossRef]
11. Traub, J.F. *Iterative Methods for the Solution of Equations*; Prentice-Hall: Englewood Cliffs, NJ, USA, 1964.
12. Džunić, J. On efficient two-parameter methods for solving nonlinear equations. *Numer. Algorithms* **2013**, *63*, 549–569. [CrossRef]
13. Džunić, J.; Petković, M.S. On generalized biparametric multipoint root finding methods with memory. *J. Comput. Appl. Math.* **2014**, *255*, 362–375. [CrossRef]
14. Zheng, O.; Wang, J.; Zhang, P.L. A Steffensen-like method and its higher-order variants. *Appl. Math. Comput.* **2009**, *214*, 10–16. [CrossRef]
15. Lotfi, T.; Tavakoli, E. On a new efficient Steffensen-like iterative class by applying a suitable self-accelerator parameter. *Sci. World J.* **2014**, *2014*, 769758. [CrossRef]
16. Zheng, O.P.; Zhao, L.; Ma, W. Variants of Steffensen-Secant method and applications. *Appl. Math. Comput.* **2010**, *216*, 3486–3496. [CrossRef]
17. Petković, M.S.; Ilić, S.; Džunić, J. Derivative free two-point methods with and without memory for solving nonlinear equations. *Appl. Math. Comput.* **2010**, *217*, 1887–1895.
18. Khaksar Haghani, F. A modiffied Steffensen's method with memory for nonlinear equations. *Int. J. Math. Model. Comput.* **2015**, *5*, 41–48.
19. Howk, C.L.; Hueso, J.L.; Martinez, E.; Teruel, C. A class of efficient high-order iterative methods with memory for nonlinear equations and their dynamics. *Math. Meth. Appl. Sci.* **2018**, 1–20. [CrossRef]
20. Cliff, H.; Kelvin, M.; Michael, M. *Hands-on Start to Wolfram Mathematica and Programming with the Wolfram Language*, 2nd ed.; Wolfram Media, Inc.: Champaign, IL, USA, 2016; ISBN 9781579550127.
21. Weerakoon, S.; Fernando, T.G.I. A variant of Newton's method with accelerated third-order convergence. *Appl. Math. Lett.* **2000**, *13*, 87–93. [CrossRef]
22. Soleymani, F.; Shateyi, S. Two optimal eighth-order derivative-free classes of iterative methods. *Abstr. Appl. Anal.* **2012**, *2012*, 318165. [CrossRef]
23. Zaka, M.U.; Kosari, S.; Soleymani, F.; Khaksar, F.H.; Al-Fhaid, A.S. A super-fast tri-parametric iterative method with memory. *Appl. Math. Comput.* **2016**, *289*, 486–491.

6

A Third Order Newton-Like Method and its Applications

D. R. Sahu [1,*], Ravi P. Agarwal [2] and Vipin Kumar Singh [3]

1 Department of Mathematics, Banaras Hindu University, Varanasi-221005, India
2 Department of Mathematics, Texas A&M University-Kingsville, Kingsville, TX 78363-8202, USA; Ravi.Agarwal@tamuk.edu
3 Department of Mathematics, Banaras Hindu University, Varanasi-221005, India; vipinkumarsingh666@gmail.com
* Correspondence: drsahudr@gmail.com

Abstract: In this paper, we design a new third order Newton-like method and establish its convergence theory for finding the approximate solutions of nonlinear operator equations in the setting of Banach spaces. First, we discuss the convergence analysis of our third order Newton-like method under the ω-continuity condition. Then we apply our approach to solve nonlinear fixed point problems and Fredholm integral equations, where the first derivative of an involved operator does not necessarily satisfy the Hölder and Lipschitz continuity conditions. Several numerical examples are given, which compare the applicability of our convergence theory with the ones in the literature.

Keywords: nonlinear operator equation; Fréchet derivative; ω-continuity condition; Newton-like method; Frédholm integral equation

---

1. Introduction

Our purpose of this paper is to compute solution of nonlinear operator equation of the form

$$F(x) = 0, \tag{1}$$

where $F : D \subset X \to Y$ is a nonlinear operator defined on an open convex subset D of a Banach space X with values into a Banach space Y.

A lot of challenging problems in physics, numerical analysis, engineering, and applied mathematics are formulated in terms of finding roots of the equation of the form Equation (1). In order to solve such problems, we often use iterative methods. There are many iterative methods available in literature. One of the central method for solving such problems is the Newton method [1,2] defined by

$$x_{n+1} = x_n - (F'_{x_n})^{-1}F(x_n) \tag{2}$$

for each $n \geq 0$, where F'_x denotes the Fréchet derivative of F at point $x \in D$.

The Newton method and the Newton-like method are attractive because it converges rapidly from any sufficient initial guess. A number of researchers [3–20] have generalized and established local as well as semilocal convergence analysis of the Newton method Equation (2) under the following conditions:

(a) *Lipschitz condition*: $\|F'_x - F'_y\| \leq K\|x - y\|$ for all $x, y \in D$ and for some $K > 0$;
(b) *Hölder Lipschitz condition*: $\|F'_x - F'_y\| \leq K\|x - y\|^p$ for all $x, y \in D$ and for some $p \in (0, 1]$ and $K > 0$;

(c) *ω-continuity condition*: $\|F'_x - F'_y\| \leq \omega(\|x-y\|)$ for all $x, y \in D$,where $\omega : [0,\infty) \to [0,\infty)$ is a nondecreasing and continuous function.

One can observe that the condition (c) is more general than the conditions (a) and (b). One can find numerical examples where the Lipschitz condition (a) and the Hölder continuity condition (b) on the first Fréchet derivative do not hold, but the ω-continuity condition (c) on first Fréchet derivative holds (see Example 1, [21]).

On the other hand, many mathematical problems such as differential equations, integral equations, economics theory, game theory, variational inequalities, and optimization theory ([22,23]) can be formulated into the fixed point problem:

$$\text{Find } x \in C \text{ such that } x = G(x), \tag{3}$$

where $G : C \to X$ is an operator defined on a nonempty subset C of a metric space X. The easiest iterative method for constructing a sequence is Picard iterative method [24] which is given by

$$x_{n+1} = G(x_n) \tag{4}$$

for each $n \geq 0$. The Banach contraction principle (see [1,22,23,25]) provides sufficient conditions for the convergence of the iterative method Equation (4) to the fixed point of G. Banach spaces have more geometrical stricture with respect to metric spaces. For study fixed points of nonlinear smooth operators, Banach space structure is required. More details of Banach space theory and fixed point theory of nonlinear operators can be found in [1,22,23,26–28].

The Newton method and its variant [29,30] are also used to solve the fixed point problem of the form:

$$(I-G)(x) = 0, \tag{5}$$

where I is the identity operator defined on X and $G : D \subset X \to X$ is a nonlinear Fréchet differentiable operator defined on an open convex subset D of a Banach space X. For finding approximate solution of the Equation (5), Bartle [31] used the Newton-like iterative method of the form

$$x_{n+1} = x_n - (I - G'_{y_n})^{-1}(I - G(x_n)) \tag{6}$$

for each $n \geq 0$, where G'_x is Fréchet derivative of G at point $x \in D$ and $\{y_n\}$ is the sequence of arbitrary points in D which are sufficiently closed to the desired solution of the Equation (5). Bartle [31] has discussed the convergence analysis of the form Equation (6) under the assumption that G is Fréchet differentiable at least at desired points and a modulus of continuity is known for G' as a function of x. The Newton method Equation (2) and the modified Newton method are the special cases of the form Equation (6).

Following the idea of Bartle [31], Rall [32] introduced the following Stirling method for finding a solution of the fixed point problem Equation (5):

$$\begin{cases} y_n = G(x_n), \\ x_{n+1} = x_n - (I - G'_{y_n})^{-1}(x_n - G(x_n)) \end{cases} \tag{7}$$

for each $n \geq 0$. Many researchers [33–35] have studied the Stirling method Equation (7) and established local as well as semilocal convergence analysis of the Stirling-like method.

Recently, Parhi and Gupta [21,36] have discussed the semilocal convergence analysis of the following Stirling-like iterative method for computing a solution of operator Equation (5):

$$\begin{cases} z_n = G(x_n), \\ y_n = x_n - (I - G'_{z_n})^{-1}(x_n - G(x_n)), \\ x_{n+1} = y_n - (I - G'_{z_n})^{-1}(y_n - G(y_n)) \end{cases} \tag{8}$$

for each $n \geq 0$. More precisely, Parhi and Gupta [21] have studied the semilocal convergence analysis of Equation (8) for computing a solution of the operator Equation (5), where $G : D \subset X \to X$ is a nonlinear Fréchet differentiable operator defined on an open convex subset D under the condition:

$$(\Omega)\ \|G'_x\| \leq k \text{ for all } x \in D \text{ and for some } k \in (0, \tfrac{1}{3}].$$

There are some nonlinear Fréchet differentiable operators $G : D \subset X \to X$ defined on an open convex subset D which fail to satisfy the condition (Ω) (see Example 1). Therefore, ref. [21] (Theorem 1) is not applicable for such operators. So, there is the following natural question:

Problem 1. *Is it possible to develop the Stirling-like iterative method for computing a solution of the operator Equation (5), where the condition (Ω) does not hold?*

The main purpose of this paper is to design a new Newton-like method for solving the operator Equation (1) and provide an affirmative answer of the Problem 1. We prove our proposed Newton-like method has R-order of convergence at least $2p+1$ under the ω-continuity condition and it covers a wide variety of iterative methods. We derive the Stirling-like iterative method for computing a solution of the fixed point problem Equation (5), where (Ω) does not hold and hence it gives an affirmative answer to Question 1 and generalizes the results of Parhi and Gupta [21,36] in the context of the condition (Ω).

In Section 2, we summarize some known concepts and results. In Section 3, we introduce a new Newton-like method for solving the operator Equation (1) and establish convergence theory of the proposed Newton-like method. In Section 4, we derive the Stirling-like iterative method from the proposed Newton-like method and establish a convergence theorem for computing a solution of the fixed point problem. Applications to Fredholm integral equations are also presented in Section 5, together with several numerical examples, which compare the applicability of our iterative technique with the ones in the literature.

2. Preliminary

In this section, we discuss some technical results. Throughout the paper, we denote $B(X,Y)$ a collection of bounded linear operators from a Banach space X into a Banach space Y and $B(X) = B(X,X)$. For some $r > 0$, $B_r[x]$ and $B_r(x)$ are the closed and open balls with center x and radius r, respectively, $\mathbb{N}_0 = \mathbb{N} \cup \{0\}$ and Φ denote the collection of nonnegative, nondecreasing, continuous real valued functions defined on $[0,\infty)$.

Lemma 1. *(Rall [37] (p. 50)) Let $\mathcal{L}$ be a bounded linear operator on a Banach space X. Then $\mathcal{L}^{-1}$ exists if and only if there is a bounded linear operator M in X such that M^{-1} exists and*

$$\|M - \mathcal{L}\| \quad < \quad \frac{1}{\|M^{-1}\|}.$$

If $\mathcal{L}^{-1}$ exists, then we have

$$\left\|\mathcal{L}^{-1}\right\| \leq \frac{\|M^{-1}\|}{1 - \|1 - M^{-1}\mathcal{L}\|} \leq \frac{\|M^{-1}\|}{1 - \|M^{-1}\|\,\|M - \mathcal{L}\|}.$$

Lemma 2. *Let $0 < k \leq \frac{1}{3}$ be a real number. Assume that $q = \frac{1}{p+1} + k^p$ for any $p \in (0,1]$ and the scalar equation*

$$(1 - k^p(1+qt)^p t)^{p+1} - \left(\frac{q^p t^p}{p+1} + k^p\right)^p q^p t^{2p} = 0$$

has a minimum positive root α. Then we have the following:

(1) $q > k$ *for all* $p \in (0,1]$.
(2) $\alpha \in (0,1)$.

Proof. (1) This part is obvious. Indeed, we have

$$\frac{1}{p+1} + k^p - \frac{1}{3} = \frac{2-p}{3(1+p)} + k^p > 0$$

for all $p \in (0,1]$ and $0 < k \leq \frac{1}{3}$.

(2) Set

$$g(t) = (1 - k^p(1+qt)^p t)^{p+1} - \left(\frac{q^p t^p}{p+1} + k^p\right)^p q^p t^{2p}. \tag{9}$$

It is clear from the definition of $g(t)$ that $g(0) > 0, g(1) < 0$ and $g'(t) < 0$ in $(0,1)$. Therefore, $g(t)$ is decreasing in $(0,1)$ and hence the Equation (9) has a minimum positive root $\alpha \in (0,1)$. This completes the proof. □

Lemma 3. *Let* $b_0 \in (0,\alpha)$ *be a number such that* $k^p(1+qb_0)^p b_0 < 1$, *where* k, p, α *and* q *are same as in Lemma 2. Define the real sequences* $\{b_n\}$, $\{\theta_n\}$ *and* $\{\gamma_n\}$ *by*

$$b_{n+1} = \frac{\left(\frac{q^p b_n^p}{p+1} + k^p\right)^p q^p b_n^{2p}}{(1 - k^p(1+qb_n)^p b_n)^{p+1}} b_n, \tag{10}$$

$$\theta_n = \frac{\left(\frac{q^p b_n^p}{p+1} + k^p\right) q b_n^2}{1 - k^p(1+qb_n)^p b_n}, \quad \gamma_n = \frac{1}{1 - k^p(1+qb_n)^p b_n} \tag{11}$$

for each $n \in \mathbb{N}_0$. *Then we have the following:*

(1) $\frac{\left(\frac{q^p b_0^p}{p+1} + k^p\right)^p q^p b_0^{2p}}{(1-k^p(1+qb_0)^p b_0)^{p+1}} < 1.$
(2) *The sequence* $\{b_n\}$ *is decreasing, that is* $b_{n+1} \leq b_n$ *for all* $n \in \mathbb{N}_0$.
(3) $k^p(1+qb_n)^p b_n < 1$ *for all* $n \in \mathbb{N}_0$.
(4) $b_{n+1} \leq \xi^{(2p+1)^n} b_n$ *for all* $n \in \mathbb{N}_0$.
(5) $\theta_n \leq \xi^{\frac{(2p+1)^n - 1}{p}} \theta$ *for all* $n \in \mathbb{N}_0$, *where* $\theta_0 = \theta$ *and* $\xi = \gamma_0 \theta^p$.

Proof. (1) Since the scalar equation $g(t) = 0$ defined by Equation (9) has a minimum positive root $\alpha \in (0,1)$ and $g(t)$ is decreasing in $(0,1)$ with $g(0) > 0$ and $g(1) < 0$. Therefore, $g(t) > 0$ in the interval $(0,\alpha)$ and hence

$$\frac{\left(\frac{q^p b_0^p}{p+1} + k^p\right)^p q^p b_0^{2p}}{(1 - k^p(1+qb_0)^p b_0)^{p+1}} < 1.$$

(2) From (1) and Equation (10), we have $b_1 \leq b_0$. This shows that (2) is true for $n = 0$. Let $j \geq 0$ be a fixed positive integer. Assume that (2) is true for $n = 0, 1, 2, \cdots, j$. Now, using Equation (10), we have

$$b_{j+2} = \frac{\left(\frac{q^p b_{j+1}^p}{p+1} + k^p\right)^p q^p b_{j+1}^{2p}}{(1 - k^p(1+qb_{j+1})^p b_{j+1})^{p+1}} b_{j+1} \leq \frac{\left(\frac{q^p b_j^p}{p+1} + k^p\right)^p q^p b_j^{2p}}{(1 - k^p(1+qb_j)^p b_n)^{p+1}} b_j = b_{j+1}.$$

Thus (2) holds for $n = j+1$. Therefore, by induction, (2) holds for all $n \in \mathbb{N}_0$.

(3) Since $b_n < b_{n-1}$ for each $n = 1, 2, 3, \cdots$ and $k^p(1 + qb_0)^p b_0 < 1$ for all $p \in (0, 1]$, it follows that

$$k^p(1 + qb_n)^p b_n < k^p(1 + qb_0)^p b_0 < 1.$$

(4) From (3), one can easily prove that the sequences $\{\gamma_n\}$ and $\{\theta_n\}$ are well defined. Using Equations (10) and (11), one can easily observe that

$$b_{n+1} = \gamma_n \theta_n^p b_n \tag{12}$$

for each $n \in \mathbb{N}_0$. Put $n = 0$ and $n = 1$ in Equation (12), we have

$$b_1 = \gamma_0 \theta^p b_0 = \xi^{(2p+1)^0} b_0$$

and

$$\begin{aligned} b_2 &= \frac{\left(\frac{q^p b_1^p}{p+1} + k^p\right)^p q^p b_1^{2p}}{(1 - k^p(1 + qb_1)^p b_1)^{p+1}} b_1 \\ &\leq \frac{\left(\frac{q^p b_0^p}{p+1} + k^p\right)^p q^p (\xi b_0)^{2p}}{(1 - k^p(1 + qb_0)^p b_0)^{p+1}} b_1 \\ &\leq \xi^{2p} \frac{\left(\frac{q^p b_0^p}{p+1} + k^p\right)^p q^p b_0^{2p}}{(1 - k^p(1 + qb_0)^p b_0)^{p+1}} b_1 \\ &= \xi^{2p} \gamma_0 \theta^p b_1 = \xi^{2p+1} b_1. \end{aligned}$$

Hence (4) holds for $n = 0$ and $n = 1$. Let $j > 1$ be a fixed integer. Assume that (4) holds for each $n = 0, 1, 2 \cdots, j$. From Equations (11) and (12), we have

$$\begin{aligned} b_{j+2} &= \frac{\left(\frac{q^p b_{j+1}^p}{p+1} + k^p\right)^p q^p b_{j+1}^{2p}}{(1 - k^p(1 + qb_{j+1})^p b_{j+1})^{p+1}} b_{j+1} \\ &\leq \frac{\left(\frac{q^p b_{j+1}^p}{p+1} + k^p\right)^p q^p (\xi^{(2p+1)^j} b_j)^{2p}}{(1 - k^p(1 + qb_{j+1})^p b_{j+1})^{p+1}} b_{j+1} \\ &\leq (\xi^{2p(2p+1)^j}) \frac{\left(\frac{q^p b_j^p}{p+1} + k^p\right)^p q^p b_j^{2p}}{(1 - k^p(1 + qb_j)^p b_j)^{p+1}} b_{j+1} \\ &\leq \xi^{2p(2p+1)^j} \xi^{2p(2p+1)^{j-1}} \cdots \xi^{2p(2p+1)} \xi^{(2p+1)} b_{j+1} \\ &= \xi^{(2p+1)^{j+1}} b_{j+1}. \end{aligned}$$

Thus (4) holds for $n = j + 1$. Therefore, by induction, (4) holds for all $n \in \mathbb{N}_0$.

(5) From Equation (11) and (4), one can easily observe that

$$\theta_1 = \frac{\left(\frac{q^p b_1^p}{p+1} + k^p\right) q b_1^2}{1 - k^p(1 + qb_1)^p b_1} \leq \frac{\left(\frac{q^p b_0^p}{p+1} + k^p\right) q (\xi b_0)^2}{1 - k^p(1 + qb_0)^p b_0} \leq \xi^{\frac{(2p+1)^1 - 1}{p}} \theta.$$

Hence (5) holds for $n = 1$. Let $j > 1$ be a fixed integer. Assume that (5) holds for each $n = 0, 1, 2 \cdots, j$. From Equation (11), we have

$$\begin{aligned}
\theta_{j+1} &= \frac{\left(\frac{q^p b_{j+1}^p}{p+1} + k^p\right) q b_{j+1}^2}{1 - k^p(1 + q b_{j+1})^p b_{j+1}} \\
&\leq \frac{\left(\frac{q^p b_0^p}{p+1} + k^p\right) q \left(\xi^{\frac{(2p+1)^{j+1}-1}{2p}} b_0\right)^2}{1 - k^p(1 + q b_0)^p b_0} \\
&= \xi^{\frac{(2p+1)^{j+1}-1}{p}} \theta.
\end{aligned}$$

Thus (5) holds for $n = j+1$. Therefore, by induction, (v) holds for all $n \in \mathbb{N}_0$. This completes the proof. □

3. Computation of a Solution of the Operator Equation (1)

Let X and Y be Banach spaces and D be a nonempty open convex subset of X. Let $F : D \subset X \to Y$ be a nonlinear operator such that F is Fréchet differentiable at each point of D and let $L \in B(Y, X)$ such that $(I - LF)(D) \subseteq D$. To solve the operator Equation (1), we introduce the Newton-like algorithm as follows:

Starting with $x_0 \in D$ and after $x_n \in D$ is defined, we define the next iterate x_{n+1} as follows:

$$\begin{cases} z_n = (I - LF)(x_n), \\ y_n = (I - F_{z_n}'^{-1} F)(x_n), \\ x_{n+1} = (I - F_{z_n}'^{-1} F)(y_n) \end{cases} \tag{13}$$

for each $n \in \mathbb{N}_0$.

If we take $X = Y, F = I - G$ and $L = I \in B(X)$ in Equation (13), then the iteration process Equation (13) reduces to the Stirling-like iteration process Equation (8).

Before proving the main result of the paper, we establish the following:

Proposition 1. *Let D be a nonempty open convex subset of a Banach space X, $F : D \subset X \to Y$ be a Fréchet differentiable at each point of D with values in a Banach space Y and $L \in B(Y, X)$ such that $(I - LF)(D) \subseteq D$. Let $\omega : [0, \infty) \to [0, \infty)$ be a nondecreasing and continuous real-valued function. Assume that F satisfies the following conditions:*

(1) $\|F_x' - F_y'\| \leq \omega(\|x - y\|)$ *for all* $x, y \in D$;
(2) $\|I - LF_x'\| \leq c$ *for all* $x \in D$ *and for some* $c \in (0, \infty)$.

Define a mapping $T : D \to D$ by

$$T(x) = (I - LF)(x) \tag{14}$$

for all $x \in D$. Then we have

$$\|I - F_{Tx}'^{-1} F_{Ty}'\| \leq \|F_{Tx}'^{-1}\| \omega(c\|x - y\|)$$

for all $x, y \in D$.

Proof. For any $x, y \in D$, we have

$$\begin{aligned}
\|I - F_{Tx}'^{-1}F_{Ty}'\| &\leq \|F_{Tx}'^{-1}\|\|F_{Tx}' - F_{Ty}'\| \\
&\leq \|F_{Tx}'^{-1}\|\omega(\|Tx - Ty\|) \\
&= \|F_{Tx}'^{-1}\|\omega(\|x - y - L(F(x) - F(y))\|) \\
&= \|F_{Tx}'^{-1}\|\omega\left(\|x - y - L\int_0^1 F_{y+t(x-y)}'(x-y)dt\|\right) \\
&\leq \|F_{Tx}'^{-1}\|\omega\left(\int_0^1 \|I - LF_{y+t(x-y)}'\|dt\|x-y\|\right) \\
&\leq \|F_{Tx}'^{-1}\|\omega(c\|x-y\|).
\end{aligned}$$

This completes the proof. □

Now, we are ready to prove our main results for solving the problem Equation (1) in the framework of Banach spaces.

Theorem 1. *Let D be a nonempty open convex subset of a Banach space X, $F : D \subset X \to Y$ a Fréchet differentiable at each point of D with values in a Banach space Y and $L \in B(Y,X)$ such that $(I - LF)(D) \subseteq D$. Let $x_0 \in D$ be such that $z_0 = x_0 - LF(x_0)$ and $F_{z_0}'^{-1} \in B(Y,X)$ exist. Let $\omega \in \Phi$ and let α be the solution of the Equation (9). Assume that the following conditions hold:*

(C1) $\|F_x' - F_y'\| \leq \omega(\|x-y\|)$ for all $x, y \in D$;
(C2) $\|I - LF_x'\| \leq k$ for all $x \in D$ and for some $k \in (0, \frac{1}{3}]$;
(C3) $\|F_{z_0}'^{-1}\| \leq \beta$ for some $\beta > 0$;
(C4) $\|F_{z_0}'^{-1}F(x_0)\| \leq \eta$ for some $\eta > 0$;
(C5) $\omega(ts) \leq t^p\omega(s)$, $s \in [0,\infty)$, $t \in [0,1]$ and $p \in (0,1]$;
(C6) $b_0 = \beta\omega(\eta) < \alpha$, $q = \frac{1}{p+1} + k^p$, $\theta = \frac{\left(\frac{q^p b_0^p}{p+1} + k^p\right)qb_0^2}{1 - k^p(1+qb_0)^p b_0}$ and $B_r[x_0] \subset D$, where $r = \frac{1+q}{1-\theta}\eta$.

Then we have the following:

(1) The sequence $\{x_n\}$ generated by Equation (13) is well defined, remains in $B_r[x_0]$ and satisfies the following estimates:

$$\begin{cases}
\|y_{n-1} - z_{n-1}\| \leq k\|y_{n-1} - x_{n-1}\|, \\
\|x_n - y_{n-1}\| \leq qb_{n-1}\|y_{n-1} - x_{n-1}\|, \\
\|x_n - x_{n-1}\| \leq (1 + qb_{n-1})\|y_{n-1} - x_{n-1}\|, \\
F_{z_n}'^{-1} \text{ exists and } \|F_{z_n}'^{-1}\| \leq \gamma_{n-1}\|F_{z_{n-1}}'^{-1}\|, \\
\|y_n - x_n\| \leq \theta_{n-1}\|y_{n-1} - x_{n-1}\| \leq \theta^n\|y_0 - x_0\|, \\
\|F_{z_n}'^{-1}\|\omega(\|y_n - x_n\|) \leq b_n
\end{cases} \tag{15}$$

for all $n \in \mathbb{N}$, where $z_n, y_n \in B_r[x_0]$, the sequences $\{b_n\}$, $\{\theta_n\}$, and $\{\gamma_n\}$ are defined by Equations (10) and (11), respectively.

(2) The sequence $\{x_n\}$ converges to the solution $x^* \in B_r[x_0]$ of the Equation (1).

(3) The priory error bounds on x^* is given by:

$$\|x_n - x^*\| \leq \frac{(1+qb_0)\eta}{\xi^{1/2p^2}\left(1 - \xi^{\frac{(2p+1)^n}{p}}\gamma_0^{-\frac{1}{p}}\right)\gamma_0^{n/p}}\left(\xi^{1/2p^2}\right)^{(2p+1)^n}$$

for each $n \in \mathbb{N}_0$.

(4) The sequence $\{x_n\}$ has R-order of convergence at least $2p+1$.

Proof. (1) First, we show that Equation (15) is true for $n = 1$. Since $x_0 \in D$, $y_0 = x_0 - F_{z_0}'^{-1}F(x_0)$ is well defined. Note that

$$\|y_0 - x_0\| = \|F_{z_0}'^{-1}F(x_0)\| \le \eta < r.$$

Hence $y_0 \in B_r[x_0]$. Using Equation (13), we have

$$\begin{aligned}\|y_0 - z_0\| &= \|-F_{z_0}'^{-1}F(x_0) + LF(x_0)\| \\ &= \|y_0 - x_0 - LF_{z_0}'(y_0 - x_0)\| \\ &\le \|I - LF_{z_0}'\|\|y_0 - x_0\| \\ &\le k\|y_0 - x_0\|.\end{aligned}$$

By Proposition 1 and (C2), we have

$$\begin{aligned}\|x_1 - y_0\| &= \|F_{z_0}'^{-1}(F(y_0) - F(x_0) - F_{z_0}'(y_0 - x_0))\| \\ &\le \int_0^1 \|F_{z_0}'^{-1}(F'_{x_0+t(y_0-x_0)} - F'_{y_0} + F'_{y_0} - F'_{z_0})\|\|y_0 - x_0\|dt \\ &\le \beta\left[\int_0^1 \|(F'_{x_0+t(y_0-x_0)} - F'_{y_0})\|dt + \|F'_{y_0} - F'_{z_0}\|\right]\|y_0 - x_0\| \\ &= \beta\left[\int_0^1 \omega((1-t)\|y_0 - x_0\|)dt + \omega(\|y_0 - z_0\|)\right]\|y_0 - x_0\| \\ &= \beta\left[\int_0^1 (1-t)^p\omega(\|y_0 - x_0\|)dt + \omega(k\|y_0 - x_0\|)\right]\|y_0 - x_0\| \\ &= \beta\left[\int_0^1 (1-t)^p\omega(\|y_0 - x_0\|)dt + k^p\omega(\|y_0 - x_0\|)\right]\|y_0 - x_0\| \\ &\le \beta\left[\frac{1}{p+1} + k^p\right]\omega(\|y_0 - x_0\|)\|y_0 - x_0\| \\ &\le q\beta\omega(\eta)\|y_0 - x_0\| \le qb_0\|y_0 - x_0\|.\end{aligned}$$

Thus we have

$$\begin{aligned}\|x_1 - x_0\| &\le \|x_1 - y_0\| + \|y_0 - x_0\| \le qb_0\|y_0 - x_0\| + \|y_0 - x_0\| \\ &\le (1 + qb_0)\|y_0 - x_0\| < r,\end{aligned} \tag{16}$$

which shows that $x_1 \in B_r[x_0]$. Note that $z_1 = (I - LF)(x_1) \in D$. Using Proposition 1 and (C3)–(C5), we have

$$\begin{aligned}\|I - F_{z_0}'^{-1}F_{z_1}'\| &\le \|F_{z_0}'^{-1}\|\omega(k\|x_1 - x_0\|) \\ &\le \beta\omega(k(1 + qb_0)\|y_0 - x_0\|) \\ &\le \beta k^p(1 + qb_0)^p\omega(\|y_0 - x_0\|) \\ &\le k^p(1 + qb_0)^p\beta\omega(\eta) \\ &\le (k(1 + qb_0))^p b_0 < 1.\end{aligned}$$

Therefore, by Lemma 1, $F_{z_1}'^{-1}$ exists and

$$\|F_{z_1}'^{-1}\| \le \frac{\|F_{z_0}'^{-1}\|}{1 - (k(1 + qb_0))^p b_0} = \gamma_0\|F_{z_0}'^{-1}\|. \tag{17}$$

Subsequently, we have

$$\begin{aligned}
\|y_1 - x_1\| &= \|F_{z_1}'^{-1}F(x_1)\| \\
&= \|F_{z_1}'^{-1}(F(x_1) - F(y_0) - F_{z_0}'(x_1 - y_0))\| \\
&\leq \|F_{z_1}'^{-1}\| \left[\int_0^1 \|(F'_{y_0+t(x_1-y_0)} - F'_{y_0})\|dt + \|F'_{y_0} - F'_{z_0}\|\right] \|x_1 - y_0\| \\
&\leq \|F_{z_1}'^{-1}\| \left[\tfrac{1}{p+1}\omega(\|x_1 - y_0\|) + \omega(k\|y_0 - x_0\|)\right] \|x_1 - y_0\| \\
&\leq \|F_{z_1}'^{-1}\| \left[\tfrac{1}{p+1}\omega(qb_0\|x_0 - y_0\|) + k^p\omega(\|y_0 - x_0\|)\right] qb_0\|x_0 - y_0\| \\
&\leq \|F_{z_1}'^{-1}\| \left[\tfrac{q^p b_0^p}{p+1}\omega(\|y_0 - x_0\|) + k^p\omega(\|y_0 - x_0\|)\right] qb_0\|y_0 - x_0\| \\
&\leq \gamma_0 \left[\tfrac{q^p b_0^p}{p+1} + k^p\right] \beta\omega(\eta) qb_0\|y_0 - x_0\| \\
&\leq \frac{\left(\frac{q^p b_0^p}{p+1} + k^p\right) qb_0^2}{1-(k(1+qb_0))^p b_0}\|y_0 - x_0\| \\
&\leq \theta\|y_0 - x_0\|.
\end{aligned} \tag{18}$$

From Equations (16) and (18), we have

$$\begin{aligned}
\|y_1 - x_0\| &\leq \|y_1 - x_1\| + \|x_1 - x_0\| \\
&\leq \theta\|y_0 - x_0\| + (1 + qb_0)\|y_0 - x_0\| \\
&\leq (1 + qb_0)\theta\|y_0 - x_0\| + (1 + qb_0)\|y_0 - x_0\| \\
&\leq (1 + qb_0)(1 + \theta)\eta < r
\end{aligned}$$

and

$$\begin{aligned}
\|z_1 - x_0| &\leq \|z_1 - y_1\| + \|y_1 - x_1\| + \|x_1 - x_0\| \\
&\leq (1 + k)\|y_1 - x_1\| + (1 + qb_0)\|y_0 - x_0\| \\
&\leq (1 + q)\theta\eta + (1 + q)\eta \\
&= (1 + q)(1 + \theta)\eta < r.
\end{aligned}$$

This shows that $z_1, y_1 \in B_r[x_0]$. From Equations (17) and (18), we get

$$\begin{aligned}
\|F_{z_1}'^{-1}\|\omega(\|y_1 - x_1\|) &\leq \gamma_0\|F_{z_0}'^{-1}\|\omega(\theta\|y_0 - x_0\|) \\
&\leq \gamma_0\theta^p\beta\omega(\eta) \\
&\leq \gamma_0\theta^p b_0 = b_1.
\end{aligned}$$

Thus we see that Equation (15) holds for $n = 1$.

Let $j > 1$ be a fixed integer. Assume that Equation (15) is true for $n = 1, 2, \cdots, j$. Since $x_j \in B_r[x_0]$, it follows $z_j = (I - LF)(x_j) \in D$. Using (C3), (C4), Equations (13) and (15), we have

$$\begin{aligned}
\|y_j - z_j\| &= \|LF(x_j) - F_{z_j}'^{-1}F(x_j)\| = \|(L - F_{z_j}'^{-1})F(x_j)\| \\
&= \|(L - F_{z_j}'^{-1})F'_{z_j}(x_j - y_j)\| \\
&\leq \|I - LF'_{z_j}\|\|y_j - x_j\| \\
&\leq k\|y_j - x_j\|.
\end{aligned} \tag{19}$$

Using Equations (13) and (19), we have

$$\begin{aligned}
\|x_{j+1} - y_j\| &= \|F_{z_j}'^{-1} F(y_j)\| \\
&\leq \|F_{z_j}'^{-1}\| \|F(y_j) - F(x_j) - F_{z_j}'(y_j - x_j)\| \\
&\leq \|F_{z_j}'^{-1}\| \left[\int_0^1 \|F'_{x_j+t(y_j-x_j)} - F'_{z_j}\|dt\right] \|y_j - x_j\| \\
&\leq \|F_{z_j}'^{-1}\| \left[\int_0^1 \|F'_{x_j+t(y_j-x_j)} - F'_{y_j}\|dt + \|F'_{y_j} - F'_{z_j}\|\right] \|y_j - x_j\| \\
&= \|F_{z_j}'^{-1}\| \left[\int_0^1 \omega(x_j + t(y_j - x_j) - y_j)dt + \omega(\|y_j - z_j\|)\right] \|y_j - x_j\| \\
&\leq \|F_{z_j}'^{-1}\| \left[\int_0^1 \omega((1-t)\|y_j - x_j\|)dt + \omega(k\|y_j - x_j\|)\right] \|y_j - x_j\| \\
&\leq \|F_{z_j}'^{-1}\| \left[\int_0^1 ((1-t)^p + k^p)\omega(\|y_j - x_j\|)dt\right] \|y_j - x_j\| \\
&\leq \|F_{z_j}'^{-1}\| \left[\frac{1}{p+1} + k^p\right] \omega(\|y_j - x_j\|)\|y_j - x_j\| \\
&= q\|F_{z_j}'^{-1}\|\omega(\|y_j - x_j\|)\|y_j - x_j\| \\
&= qb_j\|y_j - x_j\|.
\end{aligned} \tag{20}$$

From Equation (20), we have

$$\begin{aligned}
\|x_{j+1} - x_j\| &\leq \|x_{j+1} - y_j\| + \|y_j - x_j\| \\
&\leq qb_j\|y_j - x_j\| + \|y_j - x_j\| \\
&\leq (1 + qb_j)\|y_j - x_j\|.
\end{aligned} \tag{21}$$

Using Equations (20) and (21) and the triangular inequality, we have

$$\begin{aligned}
\|x_{j+1} - x_0\| &\leq \sum_{s=0}^{j} \|x_{s+1} - x_s\| \\
&\leq \sum_{s=0}^{j} (1 + qb_s)\|y_s - x_s\| \\
&\leq \sum_{s=0}^{j} (1 + qb_0)\theta^s\|y_0 - x_0\| \\
&\leq (1 + qb_0)\frac{1 - \theta^{j+1}}{1 - \theta}\eta \\
&\leq \frac{(1+q)\eta}{1 - \theta} = r,
\end{aligned}$$

which implies that $x_{k+1} \in B_r[x_0]$. Again, by using Proposition 1, (C2), (C5), and Equation (21), we have

$$\begin{aligned}
\|I - F_{z_j}'^{-1}F'_{z_{j+1}}\| &\leq \|F_{z_j}'^{-1}\|\omega(k\|x_{j+1} - x_j\|) \\
&\leq \|F_{z_j}'^{-1}\|k^p(1 + qb_j)^p\omega(\|y_j - x_j\|) \\
&\leq k^p(1 + qb_j)^p b_j < 1.
\end{aligned}$$

Therefore, by Lemma 1, $F_{z_{j+1}}'^{-1}$ exists and

$$\|F_{z_{j+1}}'^{-1}\| \leq \frac{\|F_{z_j}'^{-1}\|}{1 - k^p(1 + qb_j)^p b_j} = \gamma_j\|F_{z_j}'^{-1}\|.$$

Using Equations (13), (C2), and (21), we have

$$\begin{aligned}
\|y_{j+1}-x_{j+1}\| &= \|F'^{-1}_{z_{j+1}}F(x_{j+1})\| \\
&= \|F'^{-1}_{z_{j+1}}(F(x_{j+1})-F(y_j)-F'_{z_j}(x_{j+1}-y_j))\| \\
&\leq \|F'^{-1}_{z_{j+1}}\|\left[\int_0^1 \|F'_{y_j+t(x_{j+1}-y_j)}-F'_{y_j}\|dt+\|F'_{y_j}-F'_{z_j}\|\right]\|x_{j+1}-y_j\| \\
&\leq \|F'^{-1}_{z_{j+1}}\|\left[\int_0^1 \omega(t\|x_{j+1}-y_j\|)dt+\omega(\|y_j-z_j\|)\right]\|x_{j+1}-y_j\| \\
&\leq \|F'^{-1}_{z_{j+1}}\|\left[\int_0^1 \omega(tqb_j\|y_j-x_j\|)dt+\omega(k\|y_j-x_j\|)\right]qb_j\|y_j-x_j\| \\
&\leq \gamma_j\|F'^{-1}_{z_j}\|\left[\frac{q^pb_j^p}{p+1}\omega(\|y_j-x_j\|)+k^p\omega(\|y_j-x_j\|)\right]qb_j\|y_j-x_j\| \\
&= \gamma_j\left[\frac{q^pb_j^p}{p+1}+k^p\right]\|F'^{-1}_{z_j}\|\omega(\|y_j-x_j\|)qb_j\|y_j-x_j\| \\
&\leq \gamma_j\left[\frac{q^pb_j^p}{p+1}+k^p\right]qb_j^2\|y_j-x_j\| \\
&\leq \theta_j\|y_j-x_j\|\leq\theta^{j+1}\|y_0-x_0\|,
\end{aligned}$$

$$\begin{aligned}
\|y_{j+1}-x_0\| &\leq \|y_{j+1}-x_{j+1}\|+\|x_{j+1}-x_0\| \\
&\leq \theta^{j+1}\|y_0-x_0\|+\sum_{s=0}^{j}\|x_{s+1}-x_s\| \\
&\leq \theta^{j+1}\|y_0-x_0\|+\sum_{s=0}^{j}(1+qb_0)\theta^s\|y_0-x_0\| \\
&\leq (1+qb_0)\sum_{s=0}^{j+1}\theta^s\eta \\
&\leq \frac{(1+q)\eta}{1-\theta}=r
\end{aligned}$$

and

$$\begin{aligned}
\|z_{j+1}-x_0\| &\leq \|z_{j+1}-y_{j+1}\|+\|y_{j+1}-x_{j+1}\|+\|x_{j+1}-x_0\| \\
&\leq (1+k)\|y_{j+1}-x_{j+1}\|+\sum_{s=0}^{j}(1+qb_0)\theta^s\eta \\
&\leq (1+q)\theta^{j+1}\eta+\sum_{s=0}^{j}(1+q)\theta^s\eta \\
&\leq \sum_{s=0}^{j+1}(1+q)\theta^s\eta<r
\end{aligned}$$

which implies that $z_{j+1}, y_{j+1}\in B_r(x_0)$. Also, we have

$$\begin{aligned}
\|F'^{-1}_{z_{j+1}}\|\omega(\|y_{j+1}-x_{j+1}\|) &\leq \gamma_j\|F'^{-1}_{z_j}\|\omega(\theta_j\|y_j-x_j\|) \\
&\leq \gamma_j\theta_j^p\|F'^{-1}_{z_j}\|\omega(\|y_j-x_j\|) \\
&\leq \gamma_j\theta_j^pb_j=b_{j+1}.
\end{aligned}$$

Hence we conclude that Equation (15) is true for $n=j+1$. Therefore, by induction, Equation (15) is true for all $n\in\mathbb{N}_0$.

(2) First, we show that the sequence $\{x_n\}$ is a Cauchy sequence. For this, letting $m, n \in \mathbb{N}_0$ and using Lemma 3, we have

$$\begin{aligned}
\|x_{m+n} - x_n\| &\leq \sum_{j=n}^{m+n-1} \|x_{j+1} - x_j\| \\
&\leq \sum_{j=n}^{m+n-1} (1 + qb_j)\|y_j - x_j\| \\
&\leq (1 + qb_0) \sum_{j=n}^{m+n-1} \prod_{i=0}^{j-1} \theta_i \|y_0 - x_0\| \\
&\leq (1 + qb_0) \sum_{j=n}^{m+n-1} \prod_{i=0}^{j-1} \xi^{\frac{(2p+1)^i - 1}{p}} \theta \|y_0 - x_0\| \\
&\leq (1 + qb_0) \sum_{j=n}^{m+n-1} \prod_{i=0}^{j-1} \xi^{\frac{(2p+1)^i}{p}} \gamma_0^{-\frac{1}{p}} \|y_0 - x_0\| \\
&= (1 + qb_0) \sum_{j=n}^{m+n-1} \xi^{\sum_{i=0}^{j-1} \frac{(2p+1)^i}{p}} \gamma_0^{-\frac{1}{p}} \|y_0 - x_0\| \\
&\leq (1 + qb_0) \left(\sum_{j=n}^{m+n-1} \xi^{\frac{(2p+1)^j - 1}{2p^2}} \gamma_0^{-\frac{j}{p}} \right) \|y_0 - x_0\|.
\end{aligned}$$

By Bernoulli's inequality, for each $j \geq 0$ and $y > -1$, we have $(1 + y)^j \geq 1 + jy$. Hence we have

$$\begin{aligned}
&\|x_{m+n} - x_n\| \\
\leq\ & (1 + qb_0)\xi^{-\frac{1}{2p^2}} \gamma_0^{-\frac{n}{p}} \left(\xi^{\frac{(2p+1)^n}{2p^2}} + \xi^{\frac{(2p+1)^n (2p+1)}{2p^2}} \gamma_0^{-\frac{1}{p}} + \cdots + \xi^{\frac{(2p+1)^n (2p+1)^{m-1}}{2p^2}} \gamma_0^{-\frac{(m-1)}{p}} \right) \eta \\
\leq\ & (1 + qb_0)\xi^{-\frac{1}{2p^2}} \gamma_0^{-\frac{n}{p}} \left(\xi^{\frac{(2p+1)^n}{2p^2}} + \xi^{\frac{(2p+1)^n (1+2p)}{2p^2}} \gamma_0^{-\frac{1}{p}} + \cdots + \xi^{\frac{(2p+1)^n (1+2(m-1)p)}{2p^2}} \gamma_0^{-\frac{(m-1)}{p}} \right) \eta \\
=\ & (1 + qb_0)\xi^{-\frac{1}{2p^2}} \gamma_0^{-\frac{n}{p}} \left(\xi^{\frac{(2p+1)^n}{2p^2}} + \xi^{(2p+1)^n \left(\frac{1}{2p^2} + \frac{1}{p}\right)} \gamma_0^{-\frac{1}{p}} + \cdots + \xi^{(2p+1)^n \left(\frac{1}{2p^2} + \frac{m-1}{p}\right)} \gamma_0^{-\frac{(m-1)}{p}} \right) \eta \qquad (22) \\
=\ & (1 + qb_0)\xi^{\frac{(2p+1)^n - 1}{2p^2}} \gamma_0^{-\frac{n}{p}} \left(1 + \left(\xi^{(2p+1)^n} \gamma_0^{-1}\right)^{\frac{1}{p}} + \cdots + \left(\xi^{(2p+1)^n} \gamma_0^{-1}\right)^{\frac{m-1}{p}} \right) \eta \\
=\ & (1 + qb_0)\xi^{\frac{(2p+1)^n - 1}{2p^2}} \gamma_0^{-\frac{n}{p}} \left(\frac{1 - \left(\xi^{(2p+1)^n} \gamma_0^{-1}\right)^{\frac{m}{p}}}{1 - \left(\xi^{(2p+1)^n} \gamma_0^{-1}\right)^{\frac{1}{p}}} \right) \eta.
\end{aligned}$$

Since the sequence $\{x_n\}$ is a Cauchy sequence and hence it converges to some point $x^* \in B_r[x_0]$. From Equations (13), (C2), and (15), we have

$$\begin{aligned}
\|LF(x_n)\| &\leq \|z_n - y_n\| + \|y_n - x_n\| \\
&\leq k\|y_n - x_n\| + \|y_n - x_n\| \\
&\leq (1 + k)\theta^n \eta.
\end{aligned}$$

Taking the limit as $n \to \infty$ and using the continuity of F and the linearity of L, we have

$$F(x^*) = 0.$$

(3) Taking the limit as $m \to \infty$ in Equation (22), we have

$$\|x^* - x_n\| \leq \frac{(1+qb_0)\eta}{\xi^{1/2p^2}\left(1 - \xi^{\frac{(2p+1)^n}{p}} \gamma_0^{-\frac{1}{p}}\right)\gamma_0^{n/p}} \left(\xi^{1/2p^2}\right)^{(2p+1)^n} \tag{23}$$

for each $n \in \mathbb{N}_0$.

(4) Here we prove

$$\frac{\|x_{n+1} - x^*\|}{\|x_n - x^*\|^{2p+1}} \leq K$$

for all $n \in \mathbb{N}_0$ and for some $K > 0$. One can easily observe that there exists $n_0 > 0$ such that

$$\|x_n - x^*\| < 1 \tag{24}$$

whenever $n \geq n_0$. Using Equations (13) and (24), we have

$$\begin{aligned}
\|z_n - x^*\| &= \|x_n - x^* - LF(x_n)\| \\
&= \|x_n - x^* - L(F(x_n) - F(x^*))\| \\
&= \|x_n - x^* - L\int_0^1 F'_{x^*+t(x_n-x^*)}(x_n - x^*)dt\| \\
&\leq \int_0^1 \|I - LF'_{x^*+t(x_n-x^*)}\|\|x_n - x^*\|dt \\
&\leq k\|x_n - x^*\|
\end{aligned}$$

and

$$\begin{aligned}
\|y_n - x^*\| &= \|x_n - x^* - F'^{-1}_{z_n}F(x_n)\| \\
&= \|F'^{-1}_{z_n}[F'_{z_n}(x_n - x^*) - F(x_n)]\| \\
&\leq \|F'^{-1}_{z_n}\|\|F(x_n) - x^* - F'_{z_n}(x_n - x^*)\| \\
&= \|F'^{-1}_{z_n}\|\|\textstyle\int_0^1 (F'_{x^*+t(x_n-x^*)} - F'_{z_n})(x_n - x^*)\|dt \\
&\leq \|F'^{-1}_{z_n}\| \textstyle\int_0^1 \|F'_{x^*+t(x_n-x^*)} - F'_{z_n}\|\|x_n - x^*\|dt \\
&\leq \|F'^{-1}_{z_n}\| \textstyle\int_0^1 (\|F'_{x^*+t(x_n-x^*)} - F'_{x^*}\| + \|F'_{x^*} - F'_{z_n}\|)\|x_n - x^*\|dt \\
&\leq \|F'^{-1}_{z_n}\| \textstyle\int_0^1 (\omega(t\|x_n - x^*\|) + \omega(\|z_n - x^*\|))\|x_n - x^*\|dt \\
&\leq \|F'^{-1}_{z_n}\| \textstyle\int_0^1 (t^p\|x_n - x^*\|^p\omega(1) + \omega(k\|x_n - x^*\|))\|x_n - x^*\|dt \\
&\leq \|F'^{-1}_{z_n}\| \left(\tfrac{1}{p+1} + k^p\right)\omega(1)\|x_n - x^*\|^{p+1} \\
&= \|F'^{-1}_{z_n}\|q\omega(1)\|x_n - x^*\|^{p+1}.
\end{aligned} \tag{25}$$

Using Equations (13), (24) and (25), we have

$$
\begin{aligned}
& \|x_{n+1} - x^*\| \\
= \;& \|y_n - F_{z_n}'^{-1}F(y_n) - x^*\| \\
\leq \;& \|F_{z_n}'^{-1}\|\|F(y_n) - F(x^*) - F_{z_n}'(y_n - x^*)\| \\
= \;& \|F_{z_n}'^{-1}\|\|\int_0^1 (F'_{x^*+t(y_n-x^*)} - F'_{z_n})(y_n - x^*)\|dt \\
\leq \;& \|F_{z_n}'^{-1}\|\int_0^1 \|F'_{x^*+t(y_n-x^*)} - F'_{z_n}\|\|y_n - x^*\|dt \\
= \;& \|F_{z_n}'^{-1}\|\int_0^1 (\|F'_{x^*+t(y_n-x^*)} - F'_{x^*}\| + \|F'_{x^*} - F'_{z_n}\|)\|y_n - x^*\|dt \\
\leq \;& \|F_{z_n}'^{-1}\|\int_0^1 (\omega(t\|y_n - x^*\|) + \omega(k\|x_n - x^*\|))\|y_n - x^*\|dt \\
= \;& \|F_{z_n}'^{-1}\|\int_0^1 \Big(t^p\omega\left(\|F_{z_n}'^{-1}\|q\omega(1)\|x_n - x^*\|^{p+1}\right) + \omega(k\|x_n - x^*\|)\Big)dt \\
& \times\|F_{z_n}'^{-1}\|q\omega(1)\|x_n - x^*\|^{p+1} \\
\leq \;& \|F_{z_n}'^{-1}\|^2\int_0^1 \Big(t^p\|x_n - x^*\|^{p(p+1)}\omega\left(\|F_{z_n}'^{-1}\|q\omega(1)\right) + k^p\|x_n - x^*\|^p\omega(1)\Big)dt \\
& \times q\omega(1)\|x_n - x^*\|^{p+1} \\
= \;& \|F_{z_n}'^{-1}\|^2\Big(\frac{\|x_n - x^*\|^{p^2}\omega\left(\|F_{z_n}'^{-1}\|q\omega(1)\right)}{p+1} + k^p\Big)q\omega(1)\|x_n - x^*\|^{2p+1} \\
= \;& K_n\|x_n - x^*\|^{2p+1},
\end{aligned}
$$

where

$$K_n = \|F_{z_n}'^{-1}\|^2\Big(\frac{\|x_n - x^*\|^{p^2}\omega\left(\|F_{z_n}'^{-1}\|q\omega(1)\right)}{p+1} + k^p\Big)q\omega(1).$$

Let $\|F_{x^*}'^{-1}\| \leq d$ and $0 < d < \omega(\sigma)^{-1}$, where $\sigma > 0$. Then, for all $x \in B_\sigma(x^*)$, we have

$$\|I - F_{x^*}'^{-1}F_x'\| \leq \|F_{x^*}'^{-1}\|\|F'_{x^*} - F'_x\| \leq d\omega(\sigma) < 1$$

and so, by Lemma 1, we have

$$\|F_x'^{-1}\| \leq \frac{d}{1 - d\omega(\sigma)} := \lambda.$$

Since $x_n \to x^*$ and $z_n \to x^*$ as $n \to \infty$, there exists a positive integer N_0 such that

$$\|F_{z_n}'^{-1}\| \leq \frac{d}{1 - d\omega(\sigma)}$$

for all $n \geq N_0$. Thus, for all $n \geq N_0$, one can easily observe that

$$K_n \leq \lambda^2\Big(\frac{\sigma^{p^2}\omega(\lambda q\omega(1))}{p+1} + k^p\Big)q\omega(1) = K.$$

This shows that the R-order of convergence at least $(2p+1)$. This completes the proof. □

4. Applications

4.1. Fixed Points of Smooth Operators

Let X be a Banach spaces and D be a nonempty open convex subset of X. Let $G : D \subset X \to X$ be a nonlinear operator such that D is Fréchet differentiable at each point of D and let $L \in B(X, X)$ such that $(I - L(I - G))(D) \subseteq D$. For $F = I - G$, the Newton-like algorithm Equation (13) reduces to the following Stirling-like method for computing fixed point of the operator G:

Starting with $x_0 \in D$ and after $x_n \in D$ is defined, we define the next iterate x_{n+1} as follows:

$$\begin{cases} z_n = (I - L(I - G))(x_n), \\ y_n = (I - (I - G'_{z_n})^{-1}(I - G))(x_n), \\ x_{n+1} = (I - (I - G'_{z_n})^{-1}(I - G))(y_n) \end{cases} \tag{26}$$

for each $n \in \mathbb{N}_0$.

For the choice of $X = Y$ and $F = I - G$, Theorem 1 reduces to the following:

Theorem 2. *Let D be a nonempty open convex subset of a Banach space X, $G : D \to X$ be a Fréchet differentiable at each point of D with values into itself. Let $L \in B(X)$ be such that $(I - L(I - G))(D) \subseteq D$. Let $x_0 \in D$ be such that $z_0 = x_0 - L(x_0 - G(x_0))$ and let $(I - G'_{z_0})^{-1} \in B(X)$ exist. Let $\omega \in \Phi$ and α be a solution of the Equation (9). Assume that the conditions (C5)–(C6) and the following conditions hold:*

(C7) $\|(I - G'_{z_0})^{-1}\| \le \beta$ *for some* $\beta > 0$;
(C8) $\|(I - G'_{z_0})^{-1}(x_0 - G(x_0))\| \le \eta$ *for some* $\eta > 0$;
(C9) $\|G'_x - G'_y\| \le \omega(\|x - y\|)$ *for all* $x, y \in D$;
(C10) $\|I - L(I - G'_x)\| \le k$ *for all* $x \in D$ *and for some* $k \in (0, \frac{1}{3}]$.

Then the sequence $\{x_n\}$ generated by Equation (26) is well defined, remains in $B_r[x_0]$ and converges to the fixed point $x^* \in B_r[x_0]$ of the operator G and the sequence $\{x_n\}$ has R-order of convergence at least $2p + 1$.

We give an example to illustrate Theorem 2.

Example 1. *Let $X = Y = \mathbb{R}$ and $D = (-1, 1) \subset X$. Define a mapping $G : D \to \mathbb{R}$ by*

$$G(x) = \frac{1.1x^3 - x}{6} \tag{27}$$

for all $x \in D$. Define $L : \mathbb{R} \to \mathbb{R}$ by $L(x) = \frac{7.9}{7}x$ for all $x \in \mathbb{R}$. One can easily observe that

$$(I - L(I - G))(x) \in D$$

for all $x \in D$. Clearly, G is differentiable on D and its derivative at $x \in D$ is $G'_x = \frac{3.3x^2 - 1}{6}$ and G'_x is bounded with $\|G'_x\| \le 0.3833$ for all $x \in D$ and G' satisfies the Lipschitz condition

$$\|G'_x - G'_y\| \le K\|x - y\|$$

for all $x, y \in D$, where $K = 1.1$. For $x_0 = 0.3$, we have

$$z_0 = (I - L(I - G))(x_0) = -0.0894135714, \quad \|(I - G'_{z_0})^{-1}\| \le 0.860385626 = \beta,$$

$$\|(I - G'_{z_0})^{-1}(x_0 - G(x_0))\| \le 0.29687606 = \eta.$$

For $p = 1$, $q = \frac{5}{6}$ and $\omega(t) = Kt$ for all $t \ge 0$, we have

$$b_0 = \beta\omega(\eta) = 0.280970684 < 1,$$

$$\theta = \frac{\left(\frac{q^p b_0^p}{p+1} + k^p\right) q b_0^2}{1 - k(1 + q b_0) b_0} = 0.0335033167 < 1$$

and $r = 0.563139828$. Hence all the conditions of Theorem 2 are satisfied. Therefore, the sequence $\{x_n\}$ generated by Equation (26) is in $B_r[x_0]$ and it converges to the fixed point $x^* = 0 \in B_r[x_0]$ of G.

Remark 1. *In Example 1,* $||G'_x|| \leq 0.38333 > \frac{1}{3}$. *Thus the condition* ($\Omega$) *does not hold and so we can not apply Parhi and Gupta [21] (Theorem 1) for finding fixed points of the operators like G defined by (27). Thus the Stirling-like method defined by Equation (26) provides an affirmative answer of the Problem 1.*

If the condition (Ω) holds, then Theorem 2 with $L = I$ reduces to the main result of Parhi and Gupta [21] as follows:

Corollary 1. *[21] (Theorem 1) Let D be a nonempty open convex subset of a Banach space X and* $G : D \to D$ *be a Fréchet differentiable operator and let* $x_0 \in D$ *with* $z_0 = G(x_0)$. *Let* $(I - G'_{z_0})^{-1} \in B(X)$ *exists and* $\omega \in \Phi$. *Assume that the conditions (C5)–(C9) and the following condition holds:*

(C11) $||G'_x|| \leq k$ *for all* $x \in D$ *and for some* $k \in (0, \frac{1}{3}]$.

Then the sequence $\{x_n\}$ *generated by Equation (8) is well defined, remains in* $B_r[x_0]$ *and converges to the fixed point* $x^* \in B_r[x_0]$ *of the operator G with R−order of the convergence at least* $2p + 1$.

Example 2. *Let* $X = Y = \mathbb{R}$ *and* $D = (-6, 6) \subset X$. *Define a mapping* $G : D \to \mathbb{R}$ *by*

$$G(x) = 2 + e^{\frac{\sin x}{5}}$$

for all $x \in D$. *It is obvious that G is Fréchet differentiable on D and its Fréchet derivative at* $x \in D$ *is* $G'_x = \frac{\cos x}{5} e^{\frac{\sin x}{5}}$. *Clearly,* G'_x *is bounded with* $||G'_x|| \leq 0.22 < \frac{1}{3} = k$ *and*

$$||G'_x - G'_y|| \leq K||x - y||$$

for all $x, y \in D$, *where* $K = 0.245$. *For* $x_0 = 0$, *we have*

$$z_0 = G(x_0) = 3, \quad ||(I - G'_{z_0})^{-1}|| \leq 0.834725586524139 = \beta$$

and

$$||(I - G'_{z_0})^{-1}(x_0 - G(x_0))|| \leq 2.504176759572418 = \eta.$$

For $p = 1$, $q = \frac{5}{6}$ *and* $\omega(t) = Kt$ *for all* $t \geq 0$, *we have*

$$b_0 = \beta K \eta = 0.512123601526580 < 1,$$

$$\theta = \frac{\left(\frac{q^p b_0^p}{p+1} + k^p\right) q b_0^2}{1 - k(1 + q b_0) b_0} = 0.073280601270728 < 1$$

and $r = 5.147038576039456$.

Hence all the conditions of Theorem 2 with $L = I$ *are satisfied. Therefore, the sequence* $\{x_n\}$ *generated by Equation (26) is in* $B_r[x_0]$ *and it converges to the fixed point* $x^* = 3.023785446275295 \in B_r[x_0]$ *of G (Table 1).*

Table 1. A priori error bounds.

n	$\|\|x_n - x^*\|\|$
0	3.0237854462752
1	$1.7795738211156 \times 10^{-2}$
2	$6.216484249588206 \times 10^{-6}$
3	$2.335501569916687 \times 10^{-9}$
4	$8.775202786637237 \times 10^{-13}$
5	$4.440892098500626 \times 10^{-16}$

4.2. Fredholm Integral Equations

Let X be a Banach space over the field $\mathbb{F}$ ($\mathbb{R}$ or $\mathbb{C}$) with the norm $\|\cdot\|$ and D be an open convex subset of X. Further, let $B(X)$ be the Banach space of bounded linear operators from X into itself. Let $S \in B(X)$, $u \in X$ and $\lambda \in \mathbb{F}$. We investigate a solution $x \in X$ of the nonlinear Fredholm-type operator equation:

$$x - \lambda SQ(x) = u, \tag{28}$$

where $Q : D \to X$ is continuously Fréchet differentiable on D. The operator Equation (28) has been discussed in [10,38,39]. Define an operator $F : D \to X$ by

$$F(x) = x - \lambda SQ(x) - u \tag{29}$$

for all $x \in D$. Then solving the operator Equation (29) is equivalent to solving the operator Equation (1). From Equation (29), we have

$$F'_x(h) = h - \lambda SQ'_x(h) \tag{30}$$

for all $h \in X$. Now, we apply Theorem 1 to solve the operator Equation (28).

Theorem 3. *Let X be a Banach space and D an open convex subset of X. Let $Q : D \to X$ be a continuously Fréchet differentiable mapping at each point of D. Let $L, S \in B(X)$ and $u \in X$. Assume that, for any $x_0 \in D$, $z_0 = x_0 - L(x_0 - \lambda SQ(x_0) - u)$ and $(I - \lambda SQ'_{z_0})^{-1}$ exist. Assume that the condition (C6) and the following conditions hold:*

(C12) $(I - L(I - \lambda SQ))(x) - u \in D$ *for all* $x \in D$;
(C13) $\|(I - \lambda SQ'_{z_0})^{-1}\| \leq \beta$ *for some* $\beta > 0$;
(C14) $\|(I - \lambda SQ'_{z_0})^{-1}(x_0 - \lambda SQ(x_0) - u)\| \leq \eta$ *for some* $\eta > 0$;
(C15) $\|Q'_x - Q'_y\| \leq \omega_0(\|x - y\|)$ *for all* $x, y \in D$, *where* $\omega_0 \in \Phi$;
(C16) $\omega_0(st) \leq s^p \omega_0(t), s \in [0,1]$ *and* $t \in [0, \infty)$;
(C17) $\|I - L(I - \lambda SQ'_x)\| \leq k, k \leq \frac{1}{3}$ *for all* $x \in D$.

Then we have the following:

(1) The sequence $\{x_n\}$ generated by

$$\begin{cases} z_n = x_n - L(x_n - \lambda SQ(x_n) - u), \\ y_n = x_n - (I - \lambda SQ'_{z_n})^{-1}(x_n - \lambda SQ(x_n) - u), \\ x_{n+1} = y_n - (I - \lambda SQ'_{z_n})^{-1}(y_n - \lambda SQ(y_n) - u) \end{cases} \tag{31}$$

for each $n \in \mathbb{N}_0$ is well defined, remains in $B_r[x_0]$ and converges to a solution x^* of the Equation (28).

(2) The R-order convergence of sequence $\{x_n\}$ is at least $2p + 1$.

Proof. Let $F : D \to X$ be an operator defined by Equation (29). Clearly, F is Fréchet differentiable at each point of D and its Fréchet derivative at $x \in D$ is given by Equation (30). Now, from (C13) and Equation (30), we have $\|F'^{-1}_{z_0}\| \leq \beta$ and so it follows that (C3) holds. From (C14), Equations (29) and (30), we have $\|F'^{-1}_{z_0}(F(x_0))\| \leq \eta$. Hence (C4) is satisfied. For all $x, y \in D$, using (C15), we have

$$\begin{aligned} \|F'_x - F'_y\| &= \sup\{\|(F'_x - F'_y)z\| : z \in X, \|z\| = 1\} \\ &\leq |\lambda|\|S\| \sup\{\|Q'_x - Q'_y\|\|z\| : z \in X, \|z\| = 1\} \\ &\leq |\lambda|\|S\|\omega_0(\|x - y\|) \\ &= \omega(\|x - y\|), \end{aligned}$$

where $\omega(t) = |\lambda|\|S\|\omega_0(t)$. Clearly, $\omega \in \Phi$ and, from (C16), we have

$$\omega(st) \leq s^p \omega(t)$$

for all $s \in [0,1]$ and $t \in (0,\infty]$. Thus (C1) and (C5) hold. (C2) follows from (C17) for $c = k \in (0, \frac{1}{3}]$. Hence all the conditions of Theorem 1 are satisfied. Therefore, Theorem 3 follows from Theorem 1. This completes the proof. □

Let $D = X = Y = C[a,b]$ be the space of all continuous real valued functions defined on $[a,b] \subset \mathbb{R}$ with the norm $\|x\| = \sup_{t\in[a,b]} |x(t)|$. Consider, the following nonlinear integral equation:

$$x(s) = g(s) + \lambda \int_a^b K(s,t)(\mu(x(t))^{1+p} + \nu(x(t))^2)dt \tag{32}$$

for all $s \in [a,b]$ and $p \in (0,1]$, where $g, x \in C[a,b]$ with $g(s) \geq 0$ for all $s \in [a,b]$, $K : [a,b] \times [a,b] \to \mathbb{R}$ is a continuous nonnegative real-valued function and $\mu, \nu, \lambda \in \mathbb{R}$. Define two mappings $S, Q : D \to X$ by

$$Sx(s) = \int_a^b K(s,t)x(t)dt \tag{33}$$

for all $s \in [a,b]$ and

$$Qx(s) = \mu(x(s))^{1+p} + \nu(x(s))^2 \tag{34}$$

for all $\mu, \nu \in \mathbb{R}$ and $s \in [a,b]$.

One can easily observe that K is bounded on $[a,b] \times [a,b]$, that is, there exists a number $M \geq 0$ such that $|K(s,t)| \leq M$ for all $s,t \in [a,b]$. Clearly, S is bounded linear operator with $\|S\| \leq M(b-a)$ and Q is Fréchet differentiable and its Fréchet derivative at $x \in D$ is given by

$$Q'_x h(s) = (\mu(1+p)x^p + 2\nu x)h(s) \tag{35}$$

for all $h \in C[a,b]$. For all $x, y \in D$, we have

$$\begin{aligned} \|Q'_x - Q'_y\| &= \sup\{\|(Q'_x - Q'_y)h\| : h \in C[a,b], \|h\| = 1\} \\ &\leq \sup\{\|(\mu(1+p)(x^p - y^p) + 2\nu(x-y))h\| : h \in C[a,b], \|h\| = 1\} \\ &\leq \sup\{(|\mu|(1+p)\|x^p - y^p\| + 2|\nu|\|x-y\|)\|h\| : h \in C[a,b], \|h\| = 1\} \\ &\leq |\mu|(1+p)\|x-y\|^p + 2|\nu|\|x-y\| \\ &= \omega_0(\|x-y\|), \end{aligned} \tag{36}$$

where $\omega_0(t) = |\mu|(1+p)t^p + 2|\nu|t, t \geq 0$ with

$$\omega_0(st) \leq s^p \omega_0(t) \tag{37}$$

for all $s \in [0,1]$ and $t \in [0,\infty)$. For any $x \in D$, using Equations (33) and (35), we have

$$\begin{aligned} &\|SQ'_x\| \\ &= \sup\{\|SQ'_x h\| : h \in X, \|h\| = 1\} \\ &= \sup\left\{\sup_{s\in[a,b]} \left|\int_a^b K(s,t)(\mu(1+p)(x(t))^p + 2\nu x(t))h(t)dt\right| : h \in X, \|h\| = 1\right\} \\ &\leq \sup\left\{\int_a^b |K(s,t)|(|\mu|(1+p)|x(t)|^p + 2|\nu||x(t)|)|h(t)|dt : h \in X, \|h\| = 1\right\} \\ &\leq (|\mu|(1+p)\|x\|^p + 2|\nu|\|x\|)M(b-a) < 1. \end{aligned} \tag{38}$$

We now apply Theorem 3 to solve the Fredholm integral Equation (32).

Theorem 4. *Let $D = X = Y = C[a,b]$ and $\mu, \nu, \lambda, M \in \mathbb{R}$. Let $S, Q : D \to X$ be operators defined by Equations (33) and (34), respectively. Let $L \in B(X)$ and $x_0 \in D$ be such that $z_0 = x_0 - L(x_0 - \lambda SQ(x_0) - g) \in D$. Assume that the condition (C6) and the following conditions hold:*

(C18) $\frac{1}{1-|\lambda|(|\mu|(1+p)\|z_0\|^p+2|\nu|\|z_0\|)M(b-a)} = \beta$ *for some* $\beta > 0$;

(C19) $\frac{\|x_0-g\|+|\lambda|(|\mu|\|x_0\|^{p+1}+2|\nu|\|x_0\|^2)M(b-a)}{1-|\lambda|(|\mu|(1+p)\|z_0\|^p+2|\nu|\|z_0\|)M(b-a)} = \eta$ *for some* $\eta > 0$;

(C20) $\|I - L\| + |\lambda|\|L\|(|\mu|(1+p)\|x\|^p + 2|\nu|\|x\|)M(b-a) \leq \frac{1}{3}$ *for all* $x \in D$.

Then the sequence generated by Equation (31) with $u = g \in X$ is well defined, remains in $B_r[x_0]$ and converges to the solution $x^* \in B_r[x_0]$ of the Equation (32) with R-order convergence at least $(2p+1)$.

Proof. Note that $D = X = Y = C[a,b]$. Obviously, (C12) holds. Using Equations (C20), (33), (35) and (38), we have

$$\|I - (I - \lambda SQ'_{z_0})\| \leq |\lambda|(|\mu|(1+p)\|z_0\|^p + 2|\nu|\|z_0\|)M(b-a) < 1.$$

Therefore, by Lemma 1, $(I - \lambda SQ'_{z_0})^{-1}$ exists and

$$\|(I - \lambda SQ'_{z_0})^{-1}\| \leq \frac{1}{1 - |\lambda|(|\mu|(1+p)\|z_0\|^p + 2|\nu|\|z_0\|)M(b-a)}. \quad (39)$$

Hence Equations (C18) and (39) implies (C13) holds. Using Equations (C19), (38) and (39), we have

$$\begin{aligned} & \|(I - \lambda SQ'_{z_0})^{-1}(x_0 - \lambda SQ(x_0) - g)\| \\ \leq\ & \|(I - \lambda SQ'_{z_0})^{-1}\|(\|x_0 - g\| + \|\lambda SQ(x_0)\|) \\ \leq\ & \frac{\|x_0 - g\| + |\lambda|(|\mu|\|x_0\|^{p+1} + 2|\nu|\|x_0\|^2)M(b-a)}{1 - |\lambda|(|\mu|(1+p)\|z_0\|^p + 2|\nu|\|z_0\|)M(b-a)} \\ \leq\ & \eta. \end{aligned}$$

Thus the condition (C14) is satisfied. The conditions (C15) and (C16) follow from Equations (36) and (37), respectively. Now, from Equation (C20) and (38), we have

$$\begin{aligned} \|I - L(I - \lambda SQ'_x)\| &\leq \|I - L\| + \|L\|\|\lambda SQ'_x\| \\ &\leq \|I - L\| + \|L\||\lambda|(|\mu|(1+p)\|x\|^p + 2|\nu|\|x\|)M(b-a) \\ &\leq \frac{1}{3}. \end{aligned}$$

This implies that (C17) holds. Hence all the conditions of Theorem 3 are satisfied. Therefore, Theorem 4 follows from Theorem 3. This completes the proof. □

Now, we give one example to illustrate Theorem 3.

Example 3. *Let $X = Y = C[0,1]$ be the space of all continuous real valued functions defined on $[0,1]$. Let $D = \{x : x \in C[0,1], \|x\| < \frac{3}{2}\} \subset C[0,1]$. Consider the following nonlinear integral equation:*

$$x(s) = \sin(\pi s) + \frac{1}{10}\int_0^1 \cos(\pi s)\sin(\pi t)(x(t))^{p+1}dt, p \in (0,1]. \quad (40)$$

Define two mappings $S : X \to X$ and $Q : D \to Y$ by

$$S(x)(s) = \int_0^1 K(s,t)x(t)dt, \quad Q(x)(s) = (x(s))^{p+1},$$

where $K(s,t) = \cos(\pi s)\sin(\pi t)$. *For* $u = \sin(\pi s)$, *the problem Equation (40) is equivalent to the problem Equation (28). Here, one can easily observe that S is bounded linear operator with* $\|S\| \leq 1$ *and Q is Fréchet differentiable with* $Q'_x h(s) = (p+1)(x(s))^p h(s)$ *for all* $h \in X$ *and* $s \in [0,1]$. *For all* $x, y \in D$, *we have*

$$\|Q'_x - Q'_y\| \leq (p+1)\|x-y\|^p = \omega_0(\|x-y\|),$$

where $\omega_0(t) = (p+1)t^p$ *for any* $t \geq 0$. *Clearly,* $\omega_0 \in \Phi$. *Define a mapping* $F : D \to X$ *by*

$$F(x)(s) = x(s) - \frac{1}{10}SQ(x)(s) - \sin(\pi s).$$

Clearly, F is Fréchet differentiable on D. We now show that (C12) *holds for* $L = I \in B(X)$. *Note that*

$$\|(I - L(I - \lambda SQ))(x) - u\| = \left\|\frac{1}{10}SQ(x)(s) + \sin(\pi s)\right\| \leq \frac{1}{10}\left(\frac{3}{2}\right)^{p+1} + 1 < \frac{3}{2}$$

for all $x \in D$. *Thus* $(I - L(I - \lambda SQ))(x) - u \in D$ *for all* $x \in D$. *For all* $x \in D$, *we have*

$$\|I - F'_x\| \leq \frac{p+1}{10}\|x\|^p \leq \frac{p+1}{10}\left(\frac{3}{2}\right)^p = k.$$

Therefore, by Lemma 1, F'^{-1}_x *exists and*

$$F'^{-1}_x U(s) = U(s) + \frac{(p+1)\cos(\pi s)\int_0^1 \sin(\pi t)(x(t))^p U(t)dt}{10 - (p+1)\int_0^1 \sin(\pi t)\cos(\pi t)(x(t))^p dt} \tag{41}$$

for all $U \in Y$.

Let $x_0(s) = \sin(\pi s), \omega(t) = \frac{\omega_0(t)}{10} = \frac{p+1}{10}t^p$. Then we have the following:

(a) $x_0 \in X, F(x_0(s)) = -\frac{\cos(\pi s)}{10}\int_0^1 (\sin(\pi t))^{p+2}dt;$
(b) $z_0(s) = x_0(s) - F(x_0(s)) = \sin(\pi s) + \frac{\cos(\pi s)}{10}\int_0^1 (\sin(\pi t))^{p+2}dt;$
(c) $\|F'^{-1}_{z_0}\| \leq \frac{10^{p+1}}{10^{p+1}-(p+1)11^p} = \beta;$
(d) $\|F'^{-1}_{z_0}F(x_0)\| \leq \frac{10^p}{10^{p+1}-(p+1)11^p} = \eta;$
(e) $b_0 = \beta\omega(\eta) = \frac{(p+1)10^{p(p+1)}}{(10^{p+1}-(p+1)11^p)^{p+1}}$ and $q = \frac{1}{p+1} + \left(\frac{p+1}{10}\right)^p \left(\frac{3}{2}\right)^{p^2}$.

One can easily observe that $\theta = \frac{\left(\frac{qb_0}{2}+k\right)qb_0{}^2}{1-k(1+q)b_0} < 1$ for all $p \in (0,1]$ and $r = \frac{(1+qb_0)\eta}{1-\theta}$. Hence all the conditions of Theorem 3 are satisfied. Therefore, the sequence $\{x_n\}$ generated by Equation (31) is well defined, remains in $B_r[x_0]$ and converges to a solution of the integral Equation (40).

For $p = 1$, the convergence behavior of Newton-like method Equation (31) is given in Table 2.

Table 2. Iterates of Newton-like method Equation (31).

n	$x_n(s)$	$z_n(s)$	$y_n(s)$
0	$\sin(\pi s)$	$\sin(\pi s) + 0.0424413182\cos(\pi s)$	$\sin(\pi s) + 0.0425947671\cos(\pi s)$
1	$\sin(\pi s) + 0.0424794035\cos(\pi s)$	$\sin(\pi s) + 0.0424791962\cos(\pi s)$	$\sin(\pi s) + 0.0424796116\cos(\pi s)$
2	$\sin(\pi s) + 0.0424795616\cos(\pi s)$	$\sin(\pi s) + 0.042479611\cos(\pi s)$	$\sin(\pi s) + 0.0424796112\cos(\pi s)$
3	$\sin(\pi s) + 0.0424796111\cos(\pi s)$	$\sin(\pi s) + 0.0424796109\cos(\pi s)$	$\sin(\pi s) + 0.0424796113\cos(\pi s)$

5. Conclusions

The semilocal convergence of the third order Newton-like method for finding zeros of an operator from a Banach space to another Banach space and the corresponding Stirling-like method for finding fixed points of an operator on a Banach space are established under the ω-continuity condition. Our iterative technique is applied to nonlinear Fredholm-type operator equations. The R-order of our methods are clearly shown to be equal to at least $2p+1$ for any $p \in (0,1]$. Some numerical examples are given in support of our work, where earlier work cannot apply. In future, our iterative techniques can be applied in optimization problems.

Author Contributions: The authors have contributed equally to this paper.

Acknowledgments: We would like to thank the reviewers for their valuable suggestions for improvement of this paper.

Reference

1. Kantorovich, L.V.; Akilov, G.P. *Functional Analysis*; Aregamon Press: Oxford, UK, 1982.
2. Kantorovich, L.V. On Newton's method for functional equations. *Dokl. Akad. Nauk. SSSR* **1948**, *59*, 1237–1240. (In Russian)
3. Rheinbolt, W.C. A unified Convergence theory for a class of iterative processes. *SIAM J. Numer. Anal.* **1968**, *5*, 42–63. [CrossRef]
4. Argyros, I.K.; Cho, Y.J.; Hilout, S. *Numerical Methods for Equations and its Applications*; CRC Press/Taylor & Francis Group Publ. Comp.: New York, NY, USA, 2012.
5. Argyros, I.K.; Hilout, S. Improved generaliged differentiability conditions for Newton-like methods. *J. Complex.* **2010**, *26*, 316–333. [CrossRef]
6. Argyros, I.K.; Hilout, S. Majorizing sequences for iterative methods. *J. Comput. Appl. Math.* **2012**, *236*, 1947–1960. [CrossRef]
7. Argyros, I.K. An improved error analysis for Newton-like methods under generalized conditions. *J. Comput. Appl. Math.* **2003**, *157*, 169–185. [CrossRef]
8. Argyros, I.K.; Hilout, S. On the convergence of Newton-type methods under mild differentiability conditions. *Number. Algorithms* **2009**, *52*, 701–726. [CrossRef]
9. Sahu, D.R.; Singh, K.K.; Singh, V.K. Some Newton-like methods with sharper error estimates for solving operator equations in Banach spaces. *Fixed Point Theory Appl.* **2012**, *78*, 1–20. [CrossRef]
10. Sahu, D.R.; Singh, K.K.; Singh, V.K. A Newton-like method for generalized operator equations in Banach spaces. *Numer. Algorithms* **2014**, *67*, 289–303. [CrossRef]
11. Sahu, D.R.; Cho, Y.J.; Agarwal, R.P.; Argyros, I.K. Accessibility of solutions of operator equations by Newton-like Methods. *J. Comlex.* **2015**, *31*, 637–657. [CrossRef]
12. Argyros, I.K. A unifying local-semilocal convergence analysis and applications for two-point Newton-like methods in Banach space. *J. Math. Anal. Appl.* **2004**, *298*, 374–397. [CrossRef]
13. Argyros, I.K.; Cho, Y.J.; George, S. On the "Terra incognita" for the Newton-Kantrovich method. *J. Korean Math. Soc.* **2014**, *51*, 251–266. [CrossRef]
14. Argyros, I.K.; Cho, Y.J.; George, S. Local convergence for some third-order iterative methods under weak conditions. *J. Korean Math. Soc.* **2016**, *53*, 781–793. [CrossRef]
15. Ren, H.; Argyros, I.K.; Cho, Y.J. Semi-local convergence of Steffensen-type algorithms for solving nonlinear equations. *Numer. Funct. Anal. Optim.* **2014**, *35*, 1476–1499. [CrossRef]
16. Ezquerro, J.A.; Hernández, M.A.; Salanova, M.A. A discretization scheme for some conservative problems. *J. Comput. Appl. Math.* **2000**, *115*, 181–192. [CrossRef]
17. Ezquerro, J.A.; Hernández, M.A.; Salanova, M.A. A newton like method for solving some boundery value problems. *Numer. Funct. Anal. Optim.* **2002**, *23*, 791–805. [CrossRef]
18. Ezquerro, J.A.; Hernández, M.A. Generalized differentiability conditions for Newton's method. *IMA J. Numer. Anal.* **2002**, *22*, 187–205. [CrossRef]
19. Proinov, P.D. New general convergence theory for iterative process and its applications to Newton-Kantorovich type theores. *J. Complex.* **2010**, *26*, 3–42. [CrossRef]
20. Sahu, D.R.; Yao, J.C.; Singh, V.K.; Kumar, S. Semilocal Convergence Analysis of S-iteration Process of Newton-Kantorovich Like in Banach Spaces. *J. Optim. Theory Appl.* **2017**, *172*, 102–127. [CrossRef]
21. Parhi, S.K.; Gupta, D.K. Convergence of a third order method for fixed points in Banach spaces. *Numer. Algorithms* **2012**, *60*, 419–434. [CrossRef]

22. Agarwal, R.P.; O'Regan, D.; Sahu, D.R. *Fixed Point Theory for Lipschitzian-Type Mappings with Applications*; Topological Fixed Point Theory and its Applications; Springer: New York, NY, USA, 2009; p. 6.
23. Agarwal, R.P.; Meehan, M.; O'Regan, D. *Fixed Point Theory and Applications*; Cambridge University Press: Cambridge, UK, 2004.
24. Picard, E. Memorire sur la theorie des equations aux derivees partielles et la methode aes approximations successive. *J. Math. Pures Appl.* **1980**, *6*, 145–210.
25. Cho, Y.J. Survey on metric fixed point theory and applications. In *Advances on Real and Complex Analysis with Applications*; Trends in Mathematics; Ruzhahsky, M., Cho, Y.J., Agarwal, P., Area, I., Eds.; Birkhäuser, Springer: Basel, Switzerland, 2017.
26. Granas, A.; Dugundji, J. *Fixed Point Theory*; Springer: New York, NY, USA, 2003.
27. Zeidler, E. *Nonlinear Functional Analysis and Its Applications I: Fixed-Point Theorems*; Springer: New York, NY, USA, 1986.
28. Zeidler, E. *Nonlinear Functional Analysis and Its Applications III: Variational Methods and Applications*; Springer: New York, NY, USA, 1985.
29. Argyros, I.K. On Newton's method under mild differentiability conditions and applications. *Appl. Math. Comput.* **1999**, *102*, 177–183. [CrossRef]
30. Ortega, J.M.; Rheinboldt, W.C. *Iterative Solution of Nonlinear Equations in Several Variables*; Academic Press: New York, NY, USA; London, UK, 1970.
31. Bartle, R.G. Newton's method in Banach spaces. *Proc. Am. Math. Soc.* **1955**, *6*, 827–831.
32. Rall, L.B. Convergence of Stirling's method in Banaeh spaces. *Aequa. Math.* **1975**, *12*, 12–20. [CrossRef]
33. Parhi, S.K.; Gupta, D.K. Semilocal convergence of a Stirling-like method in Banach spaces. *Int. J. Comput. Methods* **2010**, *7*, 215–228. [CrossRef]
34. Argyros, I.K.; Muruster, S.; George, S. On the Convergence of Stirling's Method for Fixed Points Under Not Necessarily Contractive Hypotheses. *Int. J. Appl. Comput. Math.* **2017**, *3*, 1071–1081. [CrossRef]
35. Alshomrani, A.S.; Maroju, P.; Behl, R. Convergence of a Stirling-like method for fixed points in Banach spaces. *J. Comput. Appl. Math.* **2018**. [CrossRef]
36. Parhi, S.K.; Gupta, D.K. A third order method for fixed points in Banach spaces. *J. Math. Anal. Appl.* **2009**, *359*, 642–652. [CrossRef]
37. Rall, L.B. *Computational Solution of Nonlinear Operator Equations*; John Wiley and Sons: New York, NY, USA, 1969.
38. Hernández, M.A.; Salanova, M.A. A Newton-like iterative process for the numerical solution of Fredholm nonlinear integral equations. *J. Integr. Equ. Appl.* **2005**, *17*, 1–17. [CrossRef]
39. Kohaupt, L. A Newton-like method for the numerical solution of nonlinear Fredholm-type operator equations. *Appl. Math. Comput.* **2012**, *218*, 10129–10148. [CrossRef]

7

A Fast Derivative-Free Iteration Scheme for Nonlinear Systems and Integral Equations

Mozafar Rostami, Taher Lotfi * and Ali Brahmand

Department of Mathematics, Hamedan Branch, Islamic Azad University, Hamedan 15743-65181, Iran
* Correspondence: lotfi@iauh.ac.ir

Abstract: Derivative-free schemes are a class of competitive methods since they are one remedy in cases at which the computation of the Jacobian or higher order derivatives of multi-dimensional functions is difficult. This article studies a variant of Steffensen's method with memory for tackling a nonlinear system of equations, to not only be independent of the Jacobian calculation but also to improve the computational efficiency. The analytical parts of the work are supported by several tests, including an application in mixed integral equations.

Keywords: integral equation; efficiency index; nonlinear models; iterative methods; higher order

MSC: 65H10; 45D05

---

1. Introductory Notes

1.1. Background

There exist many works handling the approximate solution of linear and nonlinear integral equations. However, tackling nonlinear integral equations would be more challenging due to the presence of nonlinearity which might be expensive for different solvers [1,2].

Some authors discussed the asymptotic error expansion of collocation-type and Nystrom-type methods for Volterra–Fredholm integral equations with nonlinearity, see [3] for a complete discussion on this issue. One class of nonlinear internal equations is the mixed Hammerstein integral equations with several application in engineering problems [2].

Since in the process of finding the solution of such integral equations, most of the time a system of algebraic equation would occur that must be solved quickly and accurately, thus we here bring the attention to develop and study a useful numerical solution scheme for solving nonlinear systems with application in tackling nonlinear integral equations.

Clearly, there are some other nonlinear problems in literature which could yield in tackling nonlinear system of equations, see e.g., [4,5].

1.2. Definition

Consider a nonlinear system of equations of algebraic type as follows [6]:

$$\begin{cases} a_1(x_1, x_2, \ldots, x_m) = 0, \\ a_2(x_1, x_2, \ldots, x_m) = 0, \\ \vdots \\ a_m(x_1, x_2, \ldots, x_m) = 0, \end{cases} \tag{1}$$

which contains m equations with m unknowns and $A(x) = (a_1(x), a_2(x), \ldots, a_m(x))^T$ while $a_1(x)$, $a_2(x)$, $\ldots$, $a_m(x)$ are the functions of coordinate. We can also write (1) using $x = (x_1, x_2, \ldots, x_m)$ in a more compact form as

$$A(x) = 0. \tag{2}$$

The purpose of this work is to study finding the solution of system (1) via iteration process and discuss its application in solving nonlinear integral equations. As such, now let us briefly review some of the existing methods for finding its simple roots in the next subsection.

1.3. Existing Solvers

The Steffensen's scheme for solving nonlinear systems is written as follows [7]:

$$\begin{cases} w^{(n)} = x^{(n)} + A(x^{(n)}), \quad x^{(0)} \in \mathbb{R}^m, \\ x^{(n+1)} = x^{(n)} - [x^{(n)}, w^{(n)}; A]^{-1} A(x^{(n)}), \qquad n = 0, 1, 2, \cdots, \end{cases} \tag{3}$$

which is based upon the divided difference operator (DDO). The 1st order DDO of A for the multidimensional nodes x and y is expressed by a component-to-component procedure as follows [8]:

$$[x, y; A]_{i,j} = \frac{A_i(x_1, \ldots, x_j, y_{j+1}, \ldots, y_m) - A_i(x_1, \ldots, x_{j-1}, y_j, \ldots, y_m)}{x_j - y_j}, \; 1 \le i, j \le m. \tag{4}$$

Recall that first-order divided difference of A on $\mathbb{R}^m$ is a mapping as follows:

$$[\cdot, \cdot; A] : D \subset \mathbb{R}^m \times \mathbb{R}^m \to \mathcal{L}(\mathbb{R}^m), \tag{5}$$

that reads

$$[y, x; A](y - x) = A(y) - A(x), \; \forall x, y \in D. \tag{6}$$

Here $\mathcal{L}(\cdot)$ shows the set of bounded linear functions. By considering $h = y - x$, one we can also express the first-order DDO as follows [8]:

$$[x + h, x; A] = \int_0^1 A'(x + th)dt, \; \forall (x, h) \in \mathbb{R}^m \times \mathbb{R}^m. \tag{7}$$

Traub in [9] investigated another way based on the function $J(x, H)$ for approximating the Jacobian matrix of the Newton's method and to obtain the Steffensen's scheme based on a point-wise definition.

An improvement of (3) was given in [10,11] as follows:

$$\begin{cases} z^{(n)} = x^{(n)} - [x^{(n)}, w^{(n)}; A]^{-1} A(x^{(n)}), \\ x^{(n+1)} = z^{(n)} - [x^{(n)}, w^{(n)}; A]^{-1} A(z^{(n)}), \end{cases} \tag{8}$$

wherein

$$w^{(n)} = x^{(n)} + \beta A(x^{(n)}),\ \beta \in \mathbb{R}. \tag{9}$$

The point in (8) in contrast to (3) is that it applies two steps and of course two m-D functional evaluations to reach a higher rate than quadratic. Here, the idea is to freeze the DDO per cycle and then increase the sub steps so as to gain as much as possible of order improvement, as well as some improvements in the computational efficiency index of the scheme.

Let us also recall some of the iteration schemes having the requirement of Jacobian computation now. The Jarratt's iteration having fourth rate of convergence for solving (1) is given by [12]:

$$\begin{cases} z^{(n)} = x^{(n)} - \frac{2}{3}A'(x^{(n)})^{-1}A(x^{(n)}), \\ x^{(n+1)} = x^{(n)} - \frac{1}{2}(3A'(z^{(n)}) - A'(x^{(n)}))^{-1} \\ \qquad\qquad \times(3A'(z^{(n)}) + A'(x^{(n)}))A'(x^{(n)})^{-1}A(x^{(n)}). \end{cases} \tag{10}$$

This fourth-order iteration expression requires the computation of two matrix inverses (based on the resolution of linear systems) to achieve its rate, which manifest that getting higher rate of convergence in the form of a two-step method is costly.

1.4. Motivation

All methods discussed until now are without memory; some improvements over such schemes can be done by considering additional memory terms.

Our motivation of pursuing this aim is not only limited at tackling nonlinear systems, but a motivation is to apply such schemes for practical engineering problems such as the nonlinear mixed integral equations, see e.g., [13–16].

The goal in our development is to reach a higher computational efficiency using as low as possible number of linear systems of equations and the functional evaluations. This is directly interlinked with the concept of scientific computing and numerical analysis which gives a meaning to the investigation and proposing novel numerical procedures.

1.5. Achievement and Contribution

The objective of this work is to present a two-step higher order scheme to solve system of nonlinear equations. As such, we present an iteration method with memory for finding both real and complex zeros. Our scheme does not require computing the Fréchet derivatives of the function.

1.6. Organization

We unfold this article as follows. In Section 2, the derivation and contribution of an iteration expression is furnished. Section 3 provides an error analysis for its convergence rate. The computational efficiency of different solvers by including not only the number of functional evaluations, but also the number of system of involved linear equations, the number of LU (Lower-Upper) factorizations as well as the other similar operations will be discussed in Section 4 in detail. Section 5 discusses the application of the proposed scheme. Concluding remarks are given in Section 6.

2. A Derivative-Free Scheme

Here our attempt is to increase the computational efficiency index of (8) without imposing several more steps or further DDDs in each cycle. To complete such a task, we rely on the concept of methods with memory which state that the convergence speed and efficiency of iterative methods could be improved by saving and using the already computed values of the functions and nodes.

In fact, the error equation of the uni-parametric family of methods (8) includes a term of the form below:

$$I + \beta A'(\alpha) = 0. \tag{11}$$

The free nonzero parameter β in (11) can clearly affect not only on the domain of convergence (attraction basins of the iterative method) but also to the improvement of the convergence order. When tackling a nonlinear system of equations, and since α is not known, we can use an approximation for $A'(\alpha)$ to make the whole relation (11) approximately zero. Therefore, we may write

$$\beta \simeq -A'(\bar{\alpha})^{-1}, \tag{12}$$

wherein $\bar{\alpha}$ is an approximation of the solution (per cycle).

It is important to discuss how we approximate the matrix $\beta := B^{(n)} (n \geq 1)$ by employing some estimates to $-A'(\alpha)$ computed via the existing data.

To improve the performance of (8) using the notion of methods with memory, we consider the following iteration expression:

$$\begin{cases} w^{(n)} & = x^{(n)} + \beta A(x^{(n)}) \\ z^{(n)} & = x^{(n)} - [x^{(n)}, w^{(n)}; A]^{-1} A(x^{(n)}), \\ x^{(n+1)} & = z^{(n)} - [x^{(n)}, w^{(n)}; A]^{-1} A(z^{(n)}). \end{cases} \tag{13}$$

To ease up the implementation of the scheme with memory, let us first consider

$$\beta := B^{(n)} = -[w^{(n-1)}, x^{(n-1)}; A]^{-1} = -M^{(n-1)^{-1}} \approx -A'(\alpha)^{-1}. \tag{14}$$

and

$$\begin{cases} M^{(n-1)}\delta^{(n)} = A(x^{(n)}), \\ M^{(n-1)}\gamma^{(n)} = A(y^{(n)}). \end{cases} \tag{15}$$

Thus, now we contribute the following scheme:

$$\begin{cases} B^{(n)} = -[w^{(n-1)}, x^{(n-1)}; A]^{-1}, & n \geq 1, \\ w^{(n)} = x^{(n)} + B^{(n)} A(x^{(n)}), n \geq 1, \\ y^{(n)} = x^{(n)} + \delta^{(n)}, \quad n \geq 0, \\ x^{(n+1)} = y^{(n)} + \gamma^{(n)}. \end{cases} \tag{16}$$

Lemma 1. *Let $D \subset \mathbb{R}^m$ be a nonempty convex domain. Suppose that A is thrice Fréchet differentiable on D, and that $[u, v; A] \in \mathcal{L}(D, D)$, for any $u, v \in D (u \neq v)$ and the initial value $x^{(0)}$ and the solution α are close to each other. By considering $B^{(n)} = -[w^{(n-1)}, x^{(n-1)}; A]^{-1}$ and $d^{(n)} := I + B^{(n)} A'(\alpha)$, one obtains the error relation below*

$$d^{(n)} \sim e^{(n-1)}. \tag{17}$$

Proof. See [17] for more details. □

To implement (16), one needs to solve some linear systems of algebraic equations. This means that at each new step, a new LU factorization is needed, and no information can be exploited from the previous steps. However, there exists a body of literature about recycling this kind of information to obtain updated preconditioner for iterative solvers, [18–20]. We leave future discussion about constructing and imposing such a preconditioner for future works in this field.

As long as the coefficient matrices are sparse or large and sparse, a Krylov subspace method can be employed to speed up the process. However, the merit in (16) is that the two linear systems have one same coefficient matrix. Hence, only one LU factorization would be enough and by saving the decomposition, one can act it to two different right-hand-side vectors to get the solution vectors per sub cycles of (16).

A challenging part of the implementation using (16) is the incorporation of $B^{(n)}$. This is a not anymore a constant and it should be defined as a matrix. In this paper, whenever required the initial matrix $B^{(0)}$ is specified by:

$$B^{(0)^{-1}} = \text{diag}\left(-\frac{1}{1000}\right). \tag{18}$$

The choice of the initial matrix $B^{(0)}$ affects directly on the whole process in order to arrive in the convergence phase as quickly as possible. Here, (18) is in agreement with the dynamical studies of Steffensen-type methods with memory at which the basins of attractions are larger as long as the free parameter is close to zero.

Noting also that updating $B^{(n)}$ per cycle is again based on the already computed LU factorization while it should only act on the identity matrix to proceed.

3. Rate of Convergence

It is known that via the Taylor expansion of $A'(x+th)$ in the node x and integrating, one obtains:

$$\int_0^1 A'(x+th)dt = A'(x) + \frac{1}{2}A''(x)h + \frac{1}{6}A'''(x)h^2 + \frac{1}{24}A^{(iv)}(x)h^3 + \mathcal{O}(h^4). \tag{19}$$

It is here assumed that $A'(\alpha)$ is not singular and $e^{(n)} = x^{(n)} - \alpha$ is called the error at the n-th iterate and [6,21]:

$$e^{(n+1)} = He^{(n)^p} + \mathcal{O}(e^{(n)^{p+1}}), \tag{20}$$

(20) is the equation of error, whereas H is a p-linear function. This means that $H \in \mathcal{L}(\mathbb{R}^m, \mathbb{R}^m, \ldots, \mathbb{R}^m)$. Moreover, we consider:

$$e^{(n)^p} = (\underbrace{e^{(n)}, e^{(n)}, \ldots, e^{(n)}}_{p \text{ times}}), \tag{21}$$

which would be a matrix.

Before stating the main theorem, it is pointed out that if A be differentiable in terms of Fréchet concept in D sufficiently. As in [22], the l-th differentiation of A at $u \in \mathbb{R}^m$, $l \geq 1$, is the following l-linear function

$$A^{(l)}(u) : \mathbb{R}^m \times \cdots \times \mathbb{R}^m \longrightarrow \mathbb{R}^m, \tag{22}$$

so that $A^{(l)}(u)(v_1, v_2, \ldots, v_l) \in \mathbb{R}^m$. It is also famous that, for $\alpha + h \in \mathbb{R}^m$ locating in a neighborhood of a root α of (1), the Taylor expansion could be written and we have [22]:

$$A(\alpha + h) = A'(\alpha)\left[h + \sum_{l=2}^{p-1} C_l h^l\right] + \mathcal{O}(h^p), \tag{23}$$

wherein

$$C_l = (1/l!)[A'(\alpha)]^{-1}A^{(l)}(\alpha),\ l \geq 2. \tag{24}$$

One finds $C_l h^l \in \mathbb{R}^m$, because $A^{(l)}(\alpha) \in \mathcal{L}(\mathbb{R}^m \times \cdots \times \mathbb{R}^m, \mathbb{R}^m)$ and $[A'(\alpha)]^{-1} \in \mathcal{L}(\mathbb{R}^m)$. Moreover, for A' we have:

$$A'(\alpha + h) = A'(\alpha)\left[I + \sum_{l=2}^{p-1} lC_l h^{l-1}\right] + \mathcal{O}(h^p), \tag{25}$$

where I is the unit matrix of appropriate size. Here, $lC_l h^{l-1} \in \mathcal{L}(\mathbb{R}^m)$.

Theorem 1. *Assume that in (1), $A : D \subseteq \mathbb{R}^m \longrightarrow \mathbb{R}^m$ is Fréchet differentiable sufficiently at any points of D at $\alpha \in \mathbb{R}^m$. Here we assume $A(\alpha) = 0$ and $det(A'(x)) \neq 0$. Then, (16) with a choice of suitable initial vector has 3.30 R-order of convergence.*

Proof. For the iteration scheme (16) in the case without memory and using (23)–(25), we can obtain:

$$e^{(n+1)} = (\beta A'(\alpha) + I)(\beta A'(\alpha) + 2I)C_2^2 e^{(n)^3} + \mathcal{O}(e^{(n)^4}). \tag{26}$$

Let us now re-write (26) in the asymptotical form as comes next:

$$e^{(n+1)} \sim d_1^{(n)} e^{(n)^3}. \tag{27}$$

Several symbolic calculations by taking into account that the coefficient of the error terms in our m-D case are all matrices, Lemma 1, and their multiplications does not admit commutativity, one obtains that:

$$d_1^{(n)} \sim e^{(n-1)}, \quad \forall n \geq 1. \tag{28}$$

Therefore, one attains:

$$d_1^{(n)^2} \sim e^{(n-1)^2}, \qquad \forall n \geq 0. \tag{29}$$

Combining (28) and (29) into (27), we attain:

$$e^{(n+1)} \sim e^{(n-1)^1} e^{(n)^3}. \tag{30}$$

It shows that

$$\frac{1}{p} + 3 = p, \tag{31}$$

wherein its convergence r-order is given by:

$$p = \frac{1}{2}\left(\sqrt{13} + 3\right) \simeq 3.30278. \tag{32}$$

The proof is ended. □

4. Efficiency

Here we only need to compute one LU factorization per cycle and act it two times for different linear systems with two different right hand sides and one time on an identity matrix for the acceleration matrix $B^{(n)}$ to achieve a higher speed rate.

It is recalled that the classical index of efficiency is defined by [8]:

$$E = p^{\frac{1}{c}}, \tag{33}$$

wherein p is the convergence rate and C is the whole burden per cycle considering the number of functional evaluations.

When dealing with nonlinear system of equations, the cost of functional evaluations per cycle can be expressed as follows:

- To evaluate A, m evaluations of functions are required.
- To evaluate the associated Jacobian matrix A' needs m^2 evaluations of functions.
- To evaluate the first-order DDO, we need $m^2 - m$ evaluations of functions.
- In addition, the LU factorization cost is $\theta(2\frac{m^3}{3})$ plus $\theta(2m^2)$ in tackling the two involved triangular systems.

wherein θ is a weight that connects the cost of 1 evaluation of function with one flops. Here it is assumed that $\theta = 1$. No preconditioning is imposed in each cycle of these methods for solving the linear systems. This is done for all the compared methods.

To be more precise, we consider that the cost for computing each of the scalar functions is unit. The cost for computing other involved calculations are all also a factor of this unity cost. This is the way to give a flops-like efficiency index [23].

Considering only the consumed functional evaluations per cycle might not be a key element for reporting the indices of efficiency when solving nonlinear systems of equations. The number of matrix products, scalar products, decomposition of LU and the solution of the triangular systems of algebraic linear equations are significant in estimating the real cost and superiority of a scheme in comparison to the existing solvers in literature [23].

Hence, the results can be summarized as follows for large m:

$$2^{\frac{1}{\frac{2m^3}{3}+4m^2}} < 2^{\frac{1}{\frac{2m^3}{3}+3m^2+m}} < 3^{\frac{1}{\frac{2m^3}{3}+6m^2+m}} < 3.30^{\frac{1}{\frac{2m^3}{3}+6m^2+m}}. \tag{34}$$

In our comparisons, we applied the Newton's quadratically convergent iteration expression (NM) and also Steffensen's method (SM), the third-order expression of Amat et al. (8) denoted by AM, and the presented approach (16) showed by PM, for tackling our nonlinear systems of algebraic equations. This is also plotted in Figure 1 showing the competitiveness of the scheme with memory (16).

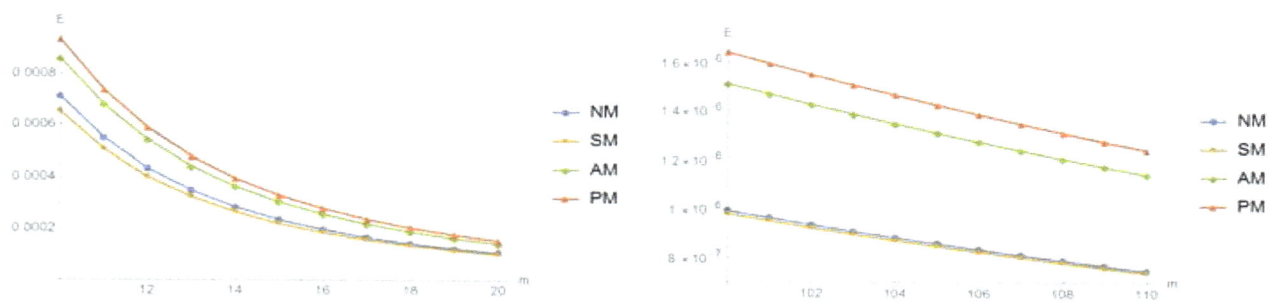

Figure 1. The comparison of flops-like efficiency indices for various schemes with and without memory by changing m.

5. Computational Tests

The aim of this section is to reveal the application of our proposed nonlinear solver for some practical problems. The software Mathematica 11.0, [24,25] was used for doing all calculations regarding the compared methods. We avoided computing any matrix inverse and the linear systems were solved applying the command `LinearSolver[]`. For the implementation of such schemes a possible stopping criterion can be defined based on the residual norm and imposed as follows:

$$||A(x^{(k)})||_2 \leq \varepsilon, \tag{35}$$

wherein ε is the required accuracy. $||\cdot||_2$ is the l^2 norm.

To confirm the theoretical convergence speed in our numerical tests, we obtain the the numerical rate of convergence by employing the following definition:

$$\rho \approx \frac{\ln(||A(x^{(k+1)})||_2/||A(x^{(k)})||_2)}{\ln(||A(x^{(k)})||_2/||A(x^{(k-1)})||_2)}. \tag{36}$$

5.1. An Academical Test

Example 1. *Here a nonlinear system of equations $A(x) = 0$, having complex root is examined as comes next:*

$$A(x) = \begin{cases} 5\exp(x_1-2)x_2 + 2{x_7}^{x_{10}} + 8{x_3}^{x_4} - 5{x_6}^3 - x_9, \\ 5\tan(x_1+2) + \cos({x_9}^{x_{10}}) + {x_2}^3 + 7{x_3}^4 - 2\sin^3(x_6), \\ {x_1}^2 - x_{10}x_5x_6x_7x_8x_9 + \tan(x_2) + 2{x_3}^{x_4} - 5{x_6}^3, \\ 2\tan({x_1}^2) + 2^{x_2} + {x_3}^2 - 5{x_5}^3 - x_6 + {x_8}^{\cos(x_9)}, \\ 10{x_1}^2 - x_{10} + \cos(x_2) + {x_3}^2 - 5{x_6}^3 - 2x_8 - 4^{x_9}, \\ \cos^{-1}({x_1}^2)\sin(x_2) - 2x_{10}{x_5}^4x_6x_9 + {x_3}^2, \\ x_1{x_2}^{x_7} - {x_8}^{x_{10}} + {x_3}^5 - 5{x_5}^3 + x_7, \\ \cos^{-1}(-10x_{10} + x_8 + x_9) + x_4\sin(x_2) + x_3 - 15{x_5}^2 + x_7, \\ 10x_1 + {x_3}^2 - 5{x_5}^2 + 10{x_6}^{x_8} - \sin(x_7) + 2x_9, \\ x_1\sin(x_2) - 2{x_{10}}^{x_8} + x_{10} - 5x_6 - 10x_9, \end{cases} \tag{37}$$

wherein the exact solution just shown up to 10 decimal places as follows:

$$\begin{aligned} \alpha \simeq\ & (1.3273490437 + 0.3502924960i, 1.058599346 - 1.748724664i, \\ & 1.0276186794 - 0.0141308051i, 3.273950008 + 0.127828308i, \\ & 0.8318243937 + 0.0017551949i, -0.4853245912 + 0.6848776400i, \\ & 0.1693667630 + 0.1840917580i, 1.534419958 - 0.321214766i, \\ & 2.086379651 + 0.426342755i, -1.989592331 + 1.478395393i)^*. \end{aligned} \tag{38}$$

The numerical evidences and the computational order of convergence ρ for this experiment are reported forward in Table 1 using 1000 fixed floating point arithmetic and the starting value $x^{(0)} = (1.2 + 0.3I, 1.1 - 1.9I, 1.0 - 0.1I, 2.5 + 0.5I, 0.8 - 0.1I, -0.4 + 1.I, 0.1 + 0.1I, 1.3 - 0.7I, 2.0 + 0.5I, -1.9 + 1.4I)^*$. Here, the residual norm $||\cdot||_2$ is reported.

Table 1. Comparison evidences for Example 1.

Met.	$\|\|A(x^{(3)})\|\|$	$\|\|A(x^{(4)})\|\|$	$\|\|A(x^{(5)})\|\|$	$\|\|A(x^{(6)})\|\|$	$\|\|A(x^{(7)})\|\|$	$\|\|A(x^{(8)})\|\|$	$\|\|A(x^{(9)})\|\|$	ρ
NM	$8.19E-1$	$2.73E-2$	$1.79E-5$	$1.28E-11$	$2.52E-23$	$8.28E-47$	$2.50E-94$	2.02
SM	$7.68E-1$	$1.83E-2$	$7.33E-6$	$5.17E-12$	$1.51E-24$	$2.03E-49$	$3.91E-99$	1.99
AM	$8.50E-2$	$6.50E-6$	$4.98E-17$	$3.68E-50$	$1.26E-149$	$4.83E-448$		3.00
PM	$6.80E-2$	$1.12E-7$	$5.15E-26$	$1.76E-86$	$6.47E-287$			3.31

5.2. An Integral Equation Using a Collocation Approach

Example 2. *The purpose of this test was to examine the performance of the new derivative-free scheme with memory for the following mixed Hammerstein integral equation [6]:*

$$y(s) = 1 + \frac{1}{5}\int_0^1 G(s,t)y(t)^3 dt, \tag{39}$$

wherein $y \in C[0,1]$, $s,t \in [0,1]$ and the kernel G is defined as follows:

$$G(s,t) = \begin{cases} (1-s)t, & t \leq s, \\ s(1-t), & t > s. \end{cases} \tag{40}$$

By employing the well-resulted Gauss-Legendre quadrature formula in discretization of integral equations given in the following form, we will be able to tackle (39):

$$\int_0^1 y(t)dt \approx \sum_{j=1}^{\chi} w_j y(t_j), \tag{41}$$

where the abscissas t_j and the weights w_j were determined via the formula of Gauss–Legendre quadrature. The lower limit of integration in standard Gauss–Legendre quadrature formula is -1. In order to approximate the integral (41) over $[0,1]$, we should map the roots of Legendre polynomials t_j on this segment and scale the weights w_j.

Showing the estimation $y(t_i)$ via x_i $(i = 1,2,\cdots,\chi)$, one would be able to transfigure the process of solving nonlinear mixed integral equations into a set of nonlinear algebraic equations as comes next:

$$5x_i - 5 - \sum_{j=1}^{\chi} c_{ij}x_j^3 = 0, \quad i = 1,2,\cdots,\chi, \tag{42}$$

where

$$c_{ij} = \begin{cases} w_j t_j(1-t_i), & if\ j \leq i, \\ w_j t_i(1-t_j), & if\ i < j. \end{cases} \tag{43}$$

For this example, we employed 200 digits floating point in the computations with the stop termination as the following residual norm (35) with $\varepsilon = 10^{-100}$. The initial vector was selected as $x^{(0)} = (3,3,\ldots,3)^*$, while the results are shown in Figure 2 using $\chi = 40$ as a list log plot of the function values by performing the cycle. It reveals a stable and fast performance of the new scheme with memory in solving integral equations. Recalling that the Figure 2 can be interpreted only as an error of numerical solution of the system (42) but not the error of solution of the source integral Equation (39).

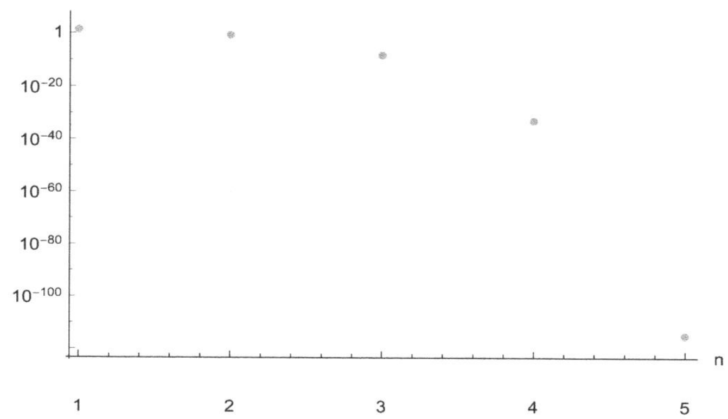

Figure 2. Error history for solving the integral equation in Example 2 using PM (Performed by Mathematica).

6. Summary

For derivative-involved iteration schemes in solving nonlinear systems, we use the $m \times m$ Jacobian matrix, i.e., $F'(x)$, with entries $F'(x)_{jk} = \partial_{x_k} f_j(x)$. Higher order schemes, such as Chebyshev methods, need higher multi dimensional derivatives which make them less practical. To be more precise, the first Fréchet derivative is a matrix with m^2 elements, while the 2nd order Fréchet differentiation has m^3 entries (ignoring the symmetric feature).

In this work, we have developed and introduced a variant of Steffensen's method with memory for tackling nonlinear problems. The scheme consists of two steps and requires the the computation of only one LU factorization which makes its computational efficiency index higher than some of the existing solvers in the literature.

The application of the iteration scheme for nonlinear integral equations via the collocation approach was discussed and its application for other types of nonlinear discretized set of equations obtained from practical problems such as the ones in [26,27] can be investigated similarly.

Author Contributions: All authors contributed equally in preparing and writing this work.

Acknowledgments: We are grateful to three anonymous referees for several comments which improved the readability of this work.

References

1. Qasim, S.; Ali, Z.; Ahmad, F.; Serra-Capizzano, S.; Ullah, M.Z.; Mahmood, A. Solving systems of nonlinear equations when the nonlinearity is expensive. *Comput. Math. Appl.* **2016**, *71*, 1464–1478. [CrossRef]
2. Wazwaz, A.-M. *Linear and Nonlinear Integral Equations*; Higher Education Press: Beijing, China; Springer: Berlin/Heidelberg, Germany, 2011.
3. Mashayekhi, S.; Razzaghi, M.; Tripak, O. Solution of the nonlinear mixed Volterra-Fredholm integral equations by hybrid of block-pulse functions and Bernoulli polynomials. *Sci. World J.* **2014**, *2014*. [CrossRef] [PubMed]
4. Alzahrani, E.O.; Al-Aidarous, E.S.; Younas, A.M.M.; Ahmad, F.; Ahmad, S.; Ahmad, S. A higher order frozen Jacobian iterative method for solving Hamilton-Jacobi equations. *J. Nonlinear Sci. Appl.* **2016**, *9*, 6210–6227. [CrossRef]
5. Soleymani, F. Pricing multi–asset option problems: A Chebyshev pseudo–spectral method. *BIT Numer. Math.* **2019**, *59*, 243–270. [CrossRef]
6. Ortega, J.M.; Rheinboldt, W.C. *Iterative Solution of Nonlinear Equations in Several Variables*; Academic Press: New York, NY, USA, 1970.
7. Noda, T. The Steffensen iteration method for systems of nonlinear equations. *Proc. Jpn. Acad.* **1987**, *63*, 186–189. [CrossRef]
8. Grau-Sánchez, M.; Grau, À.; Noguera, M. On the computational efficiency index and some iterative methods for solving systems of nonlinear equations. *J. Comput. Appl. Math.* **2011**, *236*, 1259–1266. [CrossRef]
9. Traub, J.F. *Iterative Methods for the Solution of Equations*; Prentice Hall: New York, NY, USA, 1964.
10. Amat, S.; Busquier, S. Convergence and numerical analysis of a family of two-step Steffensen's methods. *Comput. Math. Appl.* **2005**, *49*, 13–22. [CrossRef]
11. Soleymani, F.; Sharifi, M.; Shateyi, S.; Haghani, F.K. A class of Steffensen-type iterative methods for nonlinear systems. *J. Appl. Math.* **2014**, *2014*. [CrossRef]
12. Babajee, D.K.R.; Dauhoo, M.Z.; Darvishi, M.T.; Barati, A. A note on the local convergence of iterative methods based on Adomian decomposition method and 3-node quadrature rule. *Appl. Math. Comput.* **2008**, *200*, 452–458. [CrossRef]
13. Alaidarous, E.S.; Ullah, M.Z.; Ahmad, F.; Al-Fhaid, A.S. An efficient higher-order quasilinearization method for solving nonlinear BVPs. *J. Appl. Math.* **2013**, *2013*. [CrossRef]
14. Hanaç, E. The phase plane analysis of nonlinear equation. *J. Math. Anal.* **2018**, *9*, 89–97.
15. Hasan, P.M.A.; Sulaiman, N.A. Numerical treatment of mixed Volterra-Fredholm integral equations using trigonometric functions and Laguerre polynomials. *ZANCO J. Pure Appl. Sci.* **2018**, *30*, 97–106.

16. Qasim, U.; Ali, Z.; Ahmad, F.; Serra-Capizzano, S.; Ullah, M.Z.; Asma, M. Constructing frozen Jacobian iterative methods for solving systems of nonlinear equations, associated with ODEs and PDEs using the homotopy method. *Algorithms* **2016**, *9*, 18. [CrossRef]
17. Ahmad, F.; Soleymani, F.; Khaksar Haghani, F.; Serra-Capizzano, S. Higher order derivative-free iterative methods with and without memory for systems of nonlinear equations. *Appl. Math. Comput.* **2017**, *314*, 199–211. [CrossRef]
18. Bellavia, S.; Bertaccini, D.; Morini, B. Nonsymmetric preconditioner updates in Newton-Krylov methods for nonlinear systems. *SIAM J. Sci. Comput.* **2011**, *33*, 2595–2619. [CrossRef]
19. Bellavia, S.; Morini, B.; Porcelli, M. New updates of incomplete LU factorizations and applications to large nonlinear systems. *Optim. Methods Softw.* **2014**, *29*, 321–340. [CrossRef]
20. Bertaccini, D.; Durastante, F. Interpolating preconditioners for the solution of sequence of linear systems. *Comput. Math. Appl.* **2016**, *72*, 1118–1130. [CrossRef]
21. Sharma, J.R.; Kumar, D.; Argyros, I.K.; Magreñán, Á.A. On a bi-parametric family of fourth order composite Newton-Jarratt methods for nonlinear systems. *Mathematics* **2019**, *7*, 492. [CrossRef]
22. Cordero, A.; Hueso, J.L.; Martínez, E.; Torregrosa, J.R. A modified Newton-Jarratt's composition. *Numer. Algorithms* **2010**, *55*, 87–99. [CrossRef]
23. Montazeri, H.; Soleymani, F.; Shateyi, S.; Motsa, S.S. On a new method for computing the numerical solution of systems of nonlinear equations. *J. Appl. Math.* **2012**, *2012*, 1–15. [CrossRef]
24. Sánchez León, J.G. *Mathematica Beyond Mathematics: The Wolfram Language in the Real World*; Taylor & Francis Group: Boca Raton, FL, USA, 2017.
25. Wagon, S. *Mathematica in Action*, 3rd ed.; Springer: Berlin, Germany, 2010.
26. Soheili, A.R.; Soleymani, F. Iterative methods for nonlinear systems associated with finite difference approach in stochastic differential equations. *Numer. Algorithms* **2016**, *71*, 89–102. [CrossRef]
27. Soleymani, F.; Barfeie, M. Pricing options under stochastic volatility jump model: A stable adaptive scheme. *Appl. Numer. Math.* **2019**, *145*, 69–89. [CrossRef]

8

Calculating the Weighted Moore–Penrose Inverse by a High Order Iteration Scheme

Haifa Bin Jebreen

Mathematics Department, College of Science, King Saud University, Riyadh 11451, Saudi Arabia; hjebreen@ksu.edu.sa

Abstract: The goal of this research is to extend and investigate an improved approach for calculating the weighted Moore–Penrose (WMP) inverses of singular or rectangular matrices. The scheme is constructed based on a hyperpower method of order ten. It is shown that the improved scheme converges with this rate using only six matrix products per cycle. Several tests are conducted to reveal the applicability and efficiency of the discussed method, in contrast with its well-known competitors.

Keywords: iteration scheme; Moore–Penrose; rectangular matrices; rate of convergence; efficiency index

MSC: 15A09; 65F30

---

1. Introduction

1.1. Background

Constructing and discussing different features of iterative schemes for the calculation of outer inverses is an active topic of current research in Applied Mathematics (for more details, refer to [1–3]). Many papers have been published in the field of outer inverses over the past few decades, each having their own domain of validity and usefulness. In fact, in 1920, Moore was a pioneer of this field and published seminal works about the outer inverse [4,5]. However, several deep works were published during the 1950s (as reviewed and observed in [4]). It is also noted that pseudo-inverse operator was first introduced by Fredholm in [6].

The method of partitioning (due to Greville) was a pioneering work in computing generalized inverses, which was re-introduced and re-investigated in [4,7]. This scheme requires a lot of operations and is subject to cancelation and rounding errors. Among the generalized inverses, the weighted Moore–Penrose (WMP) inverse is important, as it can be simplified to a pseudo-inverse, as well as a regular inverse. Several applications of computing the WMP inverse can be observed, with some discussion, in the recent literature [8,9]; including applications to the solution of matrix equations. See [10–13] for further discussions and applications.

Furthermore, for large matrices, or as long as the weight matrices in the process of computing the WMP inverse are ill-conditioned, symbolic computation of the current algorithms may not properly work due to several reasons, such as time consumption, requiring higher memory space, or instability. On the other hand, several numerical methods for the weighted Moore–Penrose (WMP) inverse are not stabie or possess slow convergence rates. Hence, it is necessary to investigate and extend novel and useful iterative matrix methods for such an objective; see, also, the discussions in [14,15].

1.2. Definition

Let us consider that M and N are two square Hermitian positive definite (HPD) matrices of sizes m and n ($m \leq n$) and $A \in \mathbb{C}^{m\times n}$. Then, there is a unique matrix X satisfying the following identities [16]:

1. $AXA = A$,
2. $XAX = X$,
3. $(MAX)^* = MAX$,
4. $(NXA)^* = NXA$.

Then, $X \in \mathbb{C}^{n\times m}$ is called the WMP inverse of A, and is shown by $A^{\dagger}_{MN}$. Noting that, as long as $M = I_{m\times m}$ and $N = I_{n\times n}$, then X is the Moore–Penrose (MP) inverse, or simply the pseudo-inverse of A, and we show it by $A^{\dagger}$ [17]. Furthermore, when the matrix A is non-singular, then the pseudo-inverse will be simplified to the regular inverse.

The weighted singular value decomposition (WSVD), first introduced in [18], is normally applied to define this generalized inverse. Consider that the rank of A is r. Then, we have $U \in \mathbb{C}^{m\times m}$ and $V \in \mathbb{C}^{n\times n}$, satisfying the following relations:

$$U^*MU = I_{m\times m}, \tag{1}$$

and

$$V^*N^{-1}V = I_{n\times n}, \tag{2}$$

such that

$$A = U\begin{pmatrix} D & 0 \\ 0 & 0 \end{pmatrix}V^*. \tag{3}$$

Thus, $A^{\dagger}_{MN}$ is furnished as follows:

$$A^{\dagger}_{MN} = N^{-1}V\begin{pmatrix} D^{-1} & 0 \\ 0 & 0 \end{pmatrix}U^*M, \tag{4}$$

where we have a diagonal matrix $D = \mathrm{diag}(\sigma_1, \sigma_2, \ldots, \sigma_r)$, for $\sigma_1 \geq \sigma_2 \geq \ldots \geq \sigma_r > 0$, while σ_i^2 is the non-zero eigenvalue of $N^{-1}A^*MA$. In addition,

$$\|A\|_{MN} = \sigma_1, \quad \|A^{\dagger}_{MN}\|_{NM} = \frac{1}{\sigma_r}. \tag{5}$$

In this work, $A^{\#} = N^{-1}A^*M$ is used as the weighted matrix of the conjugate transpose of A. See [19] for more details.

1.3. Literature

Schulz-type methods for the calculation of the WMP inverse are sensitive to the choice of the initial value; that is, the initial choice of matrix must be close enough to the generalized inverse so as to guarantee the scheme to converge [20]. More precisely, convergence can only be observed if the starting matrix is chosen carefully. However, this starting value can be chosen simply for the case of the WMP inverse. The pioneering work in [21] gave several suggestions, along with deep discussions, about how to make such a choice quickly.

Let us, now, briefly provide some of the pioneering and most important matrix iterative methods for computing the WMP inverse.

The second–order Schulz scheme for finding the WMP inverse, requiring only two matrix products per computing cycle, is given by [22]:

$$X_{k+1} = X_k(2I - AX_k),\ k = 0, 1, 2, \cdots. \tag{6}$$

Throughout the work, I stands for the identity matrix, unless clearly stated otherwise.

An improvement of (6) with third-order convergence, known as Chebyshev's method, was discussed in [23] for computing $A^{\dagger}_{MN}$ as follows:

$$X_{k+1} = X_k(3I - AX_k(3I - AX_k)),\ k = 0, 1, 2, \cdots. \tag{7}$$

The authors in [23] proposed another third-order iterative formulation, having one more matrix multiplication, as follows:

$$X_{k+1} = X_k \left[I + \frac{1}{2}(I - AX_k)(I + (2I - AX_k)^2) \right],\ k = 0, 1, 2, \cdots. \tag{8}$$

It is necessary to recall that a general class of iteration schemes for computing the WMP inverse and some other kinds of other generalized inverses was discussed and investigated in [24] (Chapter 5) to have p-th order using a total of p matrix products. An example could be the following fourth-order iteration:

$$X_{k+1} = X_k(I + B_k(I + B_k(I + B_k))),\ k = 0, 1, 2, \cdots, \tag{9}$$

where $B_k = I - AX_k$. As another instance, a tenth-order matrix method could be furnished as follows [25]:

$$\begin{aligned} X_{k+1} = \ & X_k(I + B_k(I + B_k(I + B_k(I + B_k(I + B_k(I \\ & + B_k(I + B_k(I + B_k(I + B_k))))))))),\ k = 0, 1, 2, \cdots. \end{aligned} \tag{10}$$

1.4. Motivation and Organization

The main motivation behind proposing and extending new iterative methods for the WMP inverse is to apply them in practical large scale problems [26], as well as to improve the computational efficiency, which is directly linked to the concept of numerical analysis for designing new iterative expressions which are economically useful, in being able to reduce computational complexity and time requirements.

Hence, with this motivation at hand, to increase the computational efficiency index as well as to contribute in this field, the main focus of this work is to investigate a tenth-order method requiring only six matrix multiplications per cycle. We prove that this can provide an improvement of the computational efficiency index in calculating the WMP inverse.

The paper is organized as follows. Section 1 discusses the preliminaries and literature of this topic very briefly, to prepare the reader for the analytical discussions of Section 2, in which we describe an effective iteration formulation for the WMP inverse. It is investigated that the method needs only six matrix multiplications to reach its tenth order of convergence.

Concrete proofs of convergence are furnished in Section 3. Section 4 discusses the application of our formulation to the WMP inverses of many randomly generated matrices of various dimensions. Numerical evidence demonstrates the usefulness of this method for computing the WMP inverse, in terms of the elapsed computation time. Finally, several concluding remarks and comments are given in Section 5.

2. A High Order Scheme for the WMP Inverse

For the use of iterative methods, such as the ones described in Section 1, it is required to employ a starting value when computing the WMP inverse. As in [27], one general procedure to find this starting matrix is of the following form:

$$X_0 = \lambda A^{\#}, \tag{11}$$

where $A^{\#} = N^{-1}A^{*}M$ is the matrix of weighted conjugate transpose (WCT) for the input matrix A and

$$\lambda = \frac{1}{\sigma_1^2}. \tag{12}$$

Recall that, in (12), σ_1 is the the largest eigenvalue of $N^{-1}A^{*}MA$.

2.1. Derivation

Another reason for proposing a higher order method is that methods based on improvements of the Schulz iteration scheme are slow in the initial phase of iteration. This means that the convergence order cannot be observed at the beginning, it can be seen only after performing several iterates. On the other hand, by incorporating a stop condition using matrix norms, we can increase the elapsed time of executing the written programs for finding the WMP inverse.

Accordingly, to contribute and extend a high order matrix iteration scheme in this context, we first take into account a tenth-order scheme having ten matrix multiplications per cycle, as follows:

$$X_{k+1} = X_k(I + B_k + B_k^2 + \cdots + B_k^9). \tag{13}$$

Now, to develop the performance of (13), we factorize (13) to reduce the number of products. So, we can write

$$X_{k+1} = X_k\,(I + B_k)\,[(I - B_k + B_k^2 - B_k^3 + B_k^4)(I + B_k + B_k^2 + B_k^3 + B_k^4)]. \tag{14}$$

This formulation for the matrix iteration requires seven matrix products. However, it is possible to reduce this number of products by considering a more tight formulation for (14). Hence, we write

$$X_{k+1} = X_k\,(I + B_k)\,M_k, \tag{15}$$

where

$$M_k = [(I + \chi B_k^2 + B_k^4)(I + \kappa B_k^2 + B_k^4)]. \tag{16}$$

To find the unknown weighting coefficients in (15) and, more specifically, in (16), we need to solve a symbolic problem. As such, a Mathematica code [28] was employed to do such a task, as follows:

```
ClearAll["Global`*"];
fact1 = (1 + a B^2 + B^4);
fact2 = (1 + b B^2 + B^4);
sol = fact1*fact2 + (c B^2) // Expand
S = Table[
s[i] = Coefficient[sol, B^i], {i, 2, 6, 2}
] // Simplify
Solve[
s[2] == 1 && s[4] == 1 && s[6] == 1, {a, b, c}
] // Simplify
{a, b, c} = {a, b, c} /. %[[1]] // Simplify
Chop@sol // Simplify
```

This was given only to ease understanding of the procedure of obtaining the coefficient. Now, we obtain:

$$\chi = \frac{1}{2}\left(1-\sqrt{5}\right), \qquad \kappa = \frac{1}{2}\left(1+\sqrt{5}\right). \tag{17}$$

This means that (15) requires only six matrix products per cycle to hit a convergence speed of ten.

2.2. Several Lemmas

Before providing the main results concerning the convergence analysis of the proposed scheme, we furnish the following lemmas, inspired by [29], which reveal how the iterates generated by (15) have some specific important relations and, then, show a relation between (4) and (15).

Lemma 1. *For $\{X_k\}_{k=0}^{k=\infty}$ produced by (15) using the starting matrix (11), for any $k \geq 0$, it holds that*

$$\begin{aligned} (MAX_k)^* &= MAX_k, \\ (NX_kA)^* &= NX_kA, \\ X_kAA^{\dagger}_{MN} &= X_k, \\ A^{\dagger}_{MN}AX_k &= X_k. \end{aligned} \tag{18}$$

Proof. The proof can be done by employing mathematical induction. When $k = 0$ and X_0 is the suitable initial matrix, the first two relations in (18) are straightforward. Hence, we discuss the last two relations by applying the following identities:

$$(AA^{\dagger}_{MN})^{\#} = AA^{\dagger}_{MN}, \tag{19}$$

and

$$(A^{\dagger}_{MN}A)^{\#} = A^{\dagger}_{MN}A. \tag{20}$$

Accordingly, we have:

$$\begin{aligned} X_0AA^{\dagger}_{MN} &= \lambda A^{\#}AA^{\dagger}_{MN} \\ &= \lambda A^{\#}(AA^{\dagger}_{MN})^{\#} \\ &= \lambda A^{\#}(A^{\dagger}_{MN})^{\#}A^{\#} \\ &= \lambda(AA^{\dagger}_{MN}A)^{\#} \\ &= \lambda A^{\#} \\ &= X_0, \end{aligned} \tag{21}$$

and also

$$\begin{aligned} A^{\dagger}_{MN}AX_0 &= \lambda A^{\dagger}_{MN}AA^{\#} \\ &= \lambda(A^{\dagger}_{MN}A)^{\#}A^{\#} \\ &= \lambda(A^{\#}(A^{\dagger}_{MN})^{\#}A^{\#}) \\ &= \lambda(A(A^{\dagger}_{MN}A)^{\#} \\ &= \lambda A^{\#} \\ &= X_0. \end{aligned} \tag{22}$$

Subsequently, now the relation is valid for $k > 0$, then we discuss that it will still be true for $k+1$. Taking our matrix iteration (15) into consideration, we have:

$$\begin{aligned} (MAX_{k+1})^* &= (MA(X_k(I+B_k)[(I+\chi B_k^2+B_k^4)(I+\kappa B_k^2+B_k^4)]))^* \\ &= [MAX_k(I+B_k+B_k^2+B_k^3+B_k^4+B_k^5+B_k^6+B_k^7+B_k^8+B_k^9)]^* \\ &= MA[X_k(I+B_k+B_k^2+B_k^3+B_k^4+B_k^5+B_k^6+B_k^7+B_k^8+B_k^9)] \\ &= MAX_{k+1}, \end{aligned}$$

using that

$$(M(AX_k))^* = MAX_k, \tag{23}$$

M is a Hermitian positive definite matrix ($M^* = M$), and similar facts, such as:

$$\begin{aligned}(M(AX_k)^2)^* &= (M(AX_k)(AX_k))^* \\ &= (AX_k)^*(M(AX_k))^* \\ &= (AX_k)^*(M(AX_k)) \\ &= (AX_k)^*M^*(AX_k) \\ &= (M(AX_k))^*(AX_k) \\ &= M(AX_k)(AX_k) \\ &= M(AX_k)^2.\end{aligned} \tag{24}$$

Hence, the first relation in (18) is true for $k+1$, and the 2nd relation could be investigated similarly. For the other relation in (18), by employing the assumption that

$$X_k AA^{\dagger}_{MN} = X_k, \tag{25}$$

and (15), we have:

$$\begin{aligned}X_{k+1}AA^{\dagger}_{MN} &= (X_k\,(I+B_k)\,[(I+\chi B_k^2+B_k^4)(I+\kappa B_k^2+B_k^4)])AA^{\dagger}_{MN} \\ &= (X_k + X_kB_k + X_kB_k^2 + X_kB_k^3 + X_kB_k^4 + X_kB_k^5 + X_kB_k^6 \\ &\quad + X_kB_k^7 + X_kB_k^8 + X_kB_k^9)AA^{\dagger}_{MN} \\ &= X_kAA^{\dagger}_{MN} + X_kB_kAA^{\dagger}_{MN} + X_kB_k^2AA^{\dagger}_{MN} + X_kB_k^3AA^{\dagger}_{MN} + X_kB_k^4AA^{\dagger}_{MN} \\ &\quad + X_kB_k^5AA^{\dagger}_{MN} + X_kB_k^6AA^{\dagger}_{MN} + X_kB_k^7AA^{\dagger}_{MN} + X_kB_k^8AA^{\dagger}_{MN} + X_kB_k^9AA^{\dagger}_{MN} \\ &= (X_k + X_kB_k + X_kB_k^2 + X_kB_k^3 + X_kB_k^4 + X_kB_k^5 \\ &\quad + X_kB_k^6 + X_kB_k^7 + X_kB_k^8 + X_kB_k^9) \\ &= X_{k+1}.\end{aligned}$$

Therefore, the third relation in (18) is valid for $k+1$. The final relation could be investigated in a similar way, and the result follows. The proof is, thus, complete. □

Lemma 2. *Employing the assumptions of Lemma 1 and (3), then for (15) we have:*

$$(V^{-1}N)X_k(M^{-1}(U^*)^{-1}) = diag(T_k, 0), \tag{26}$$

where T_k is a diagonal matrix, $V^*N^{-1}V = I_{n\times n}$, $U^*MU = I_{m\times m}$, $V \in \mathbb{C}^{n\times n}$, $U \in \mathbb{C}^{m\times m}$, and $A = U\Sigma V^*$.

Proof. Assume that $T_0 = \lambda D$ and that σ_i^2 are the non-zero eigenvalues of the matrix $N^{-1}A^*MA$, while $D = \text{diag}(\sigma_1, \sigma_2, \ldots, \sigma_r)$, $\sigma_i > 0$ for any i. Thus, we can write that:

$$\begin{aligned}T_{k+1} := \varphi(T_k) &= T_k\,(I + (I - DT_k))\,[(I + \chi(I-DT_k)^2 + (I-DT_k)^4) \\ &\quad \times (I + \kappa(I-DT_k)^2 + (I-DT_k)^4)].\end{aligned} \tag{27}$$

Applying mathematical induction, one can write that

$$\begin{aligned}(V^{-1}N)X_0(M^{-1}(U^*)^{-1}) &= \lambda(V^{-1}N)A^{\#}(M^{-1}(U^*)^{-1}) \\ &= \lambda(V^{-1}N)N^{-1}A^*(MM^{-1}(U^*)^{-1}) \\ &= \lambda(V^{-1}N)N^{-1}V\text{diag}(D,0)U^*(MM^{-1}(U^*)^{-1}) \\ &= \text{diag}(\lambda D, 0).\end{aligned} \tag{28}$$

In addition, when (31) is satisfied, then using (15), one can get that:

$$\begin{aligned}(V^{-1}N)X_{k+1}(M^{-1}(U^*)^{-1}) &= (V^{-1}N)X_k(M^{-1}(U^*)^{-1}) \\ &\quad \times (2I-(V^{-1}N)AX_k(M^{-1}(U^*)^{-1})) \\ &\quad \times[(I+\chi(V^{-1}N)(I-AX_k)^2(M^{-1}(U^*)^{-1}) \\ &\quad +(V^{-1}N)(I-AX_k)^4(M^{-1}(U^*)^{-1})) \\ &\quad \times(I+\kappa(V^{-1}N)(I-AX_k)^2(M^{-1}(U^*)^{-1}) \\ &\quad +(V^{-1}N)(I-AX_k)^4)(M^{-1}(U^*)^{-1})].\end{aligned} \tag{29}$$

Using the fact that $A = U^*MU\text{diag}(D,0) = V^*NV$, one attains

$$(V^{-1}N)X_{k+1}(M^{-1}(U^*)^{-1}) = \text{diag}(\varphi(T_k),0), \tag{30}$$

which shows that (27) is a diagonal matrix. This completes the proof. □

3. Error Analysis

The objective of this section is to provide a matrix analysis for the convergence of the iteration scheme (15).

Theorem 1. *Let us consider that A is an $m \times n$ matrix whose WSVD is provided by (4). Furthermore, assume that the starting value is given by (11). Thus, the matrix sequence from (15) tends to $A^{\dagger}_{MN}$.*

Proof. In light of (4), to prove our convergence for the WMP inverse, we now just need to prove that

$$\lim_{k\to\infty} (V^{-1}N)X_k(M^{-1}(U^*)^{-1}) = \text{diag}(D^{-1},0). \tag{31}$$

It is obtained, using Lemmas 1 and 2, that

$$T_k = \text{diag}(\tau_1^{(k)}, \tau_2^{(k)}, \ldots, \tau_r^{(k)}), \tag{32}$$

where

$$\tau_i^{(0)} = \lambda\sigma_i \tag{33}$$

and

$$\begin{aligned}\tau_i^{(k+1)} &= \tau_i^{(k)}\left(2I+\sigma_i\tau_i^{(k)}\right)[(I \\ &\quad +\chi(\sigma_i\tau_i^{(k)})^2+(\sigma_i\tau_i^{(k)})^4)(I \\ &\quad +\kappa(\sigma_i\tau_i^{(k)})^2+(\sigma_i\tau_i^{(k)})^4)].\end{aligned} \tag{34}$$

The sequence produced by (34) is the result of employing (15) in calculating the zero σ_i^{-1} of the function

$$\phi(\tau) = \sigma_i - \tau^{-1}, \tag{35}$$

using the starting condition $\tau_i^{(0)}$.

We observe that convergence to σ_i^{-1} can be achieved, as long as

$$0 < \tau_i^{(0)} < \frac{2}{\sigma_i}, \tag{36}$$

which results in a criterion on λ (the selection in formula (12) has now been shown). Hence, $\{T_k\} \to \Sigma^{-1}$, and (31) is satisfied. It is now clear that $\{X_k\}_{k=0}^{k=\infty} \to A^{\dagger}_{MN}$ when $k \to \infty$. This concludes the proof. □

4. Computational Tests

In this section, our aim is to study the efficiency of the proposed approach for calculating the WMP inverse computationally and analytically. To do this, we considered several competitors from the literature in our comparisons, such as those from (6), (7), (10), and (15), denoted by "SM2", "CM3", "KMS10", and "PM10", respectively.

Note that all computations were done in Mathematica 11.0 [30] and the time is reported in seconds. The hardware used was a CPU Intel Core i5 2430-M with 16 GB of RAM.

We know that the efficiency index is expressed by [31]:

$$EI = \rho^{\frac{1}{\kappa}}, \tag{37}$$

where ρ and κ stand for the speed and the whole cost in each cycle, respectively.

As such, the efficiency index of different methods (6–10) and (15) are reported by: $2^{\frac{1}{2}} \simeq 1.414$, $3^{\frac{1}{3}} \simeq 1.442$, $3^{\frac{1}{4}} \simeq 1.316$, $4^{\frac{1}{4}} \simeq 1.414$, $10^{\frac{1}{10}} \simeq 1.258$, and $10^{\frac{1}{6}} \simeq 1.467$, respectively. Clearly, our investigated iterative expression has better a index and can be more useful in finding the WMP inverse.

Example 1. *[29] The purpose of this experiment was to examine the calculation of WMP inverses for 10 uniform randomly provided $m1 \times n1 = 200 \times 210$ matrices, as follows:*

```
SeedRandom[12]; no = 10; m1 = 200; n1 = 210;
ParallelTable[A[k] = RandomReal[{1}, {m1, n1}];, {k, no}];
```

where the ten various HPD matrices M and N were given by:

```
ParallelTable[MM[k] = RandomReal[{2}, {m1, m1}];, {k, no}];
ParallelTable[MM[k] = Transpose[MM[k]].MM[k];, {k, no}];
ParallelTable[NN[k] = RandomReal[{3}, {n1, n1}];, {k, no}];
ParallelTable[NN[k] = Transpose[NN[k]].NN[k];, {k, no}];
```

The results by applying the stop termination

$$||X_{k+1} - X_k||_2 \leq 10^{-10}, \tag{38}$$

are reported in Tables 1 and 2, based on the number of iterations, elapsed CPU time (in seconds), and $X_0 = \frac{1}{\sigma_1^2} A^{\#}$. As can be observed from the results, the best scheme in terms of number of iterations and time was (15).

Table 1. Comparison based on the number of iterations and the required mean in Experiment 1.

Methods	SM2	CM3	KMS10	PM10
A_1	68	43	22	22
A_2	69	44	22	22
A_3	67	43	21	21
A_4	71	46	23	23
A_5	72	46	23	23
A_6	72	46	23	23
A_7	66	42	21	21
A_8	78	50	25	25
A_9	63	41	20	20
A_{10}	69	44	22	22
Mean	69.5	44.5	22.2	22.2

Table 2. Comparison based on the elapsed CPU time and its mean in Experiment 1.

Methods	SM2	CM3	KMS10	PM10
A_1	1.4954	1.04826	0.996155	0.755317
A_2	1.4563	1.08006	0.984057	0.767785
A_3	1.37301	1.03847	0.967427	0.720294
A_4	1.53201	1.10927	1.0365	0.789994
A_5	1.50908	1.10164	0.998853	0.794098
A_6	1.51215	1.11421	1.03361	0.823177
A_7	1.39481	1.00116	0.915244	0.743779
A_8	1.62742	1.24438	1.12434	0.87007
A_9	1.32916	0.999683	0.903072	0.709523
A_{10}	1.49738	1.05736	0.985084	0.764156
Mean	1.47267	1.07945	0.994434	0.773819

Example 2. *The iterative methods were compared for five randomly generated dense* $m1 \times n1 = 500 \times 500$ *matrices produced in Mathematica environment by the following piece of code:*

```
m1 = 500; n1 = 500; no = 5; SeedRandom[12];
ParallelTable[A[k] = RandomReal[{0, 1}, {m1, n1}];, {k, no}];
ParallelTable[MM[k] = RandomReal[{0, 1}, {m1, m1}];, {k, no}];
ParallelTable[MM[k] = Transpose[MM[k]].MM[k];, {k, no}];
ParallelTable[NN[k] = RandomReal[{0, 1}, {n1, n1}];, {k, no}];
ParallelTable[NN[k] = Transpose[NN[k]].NN[k];, {k, no}];
```

Here, we applied the stopping condition

$$||X_{k+1} - X_k||_\infty \leq 10^{-10}, \tag{39}$$

with a change in the initial approximation as $X_0 = \frac{1.5}{\sigma_1^2} A^{\#}$. Noting that the weights M and N were very ill-conditioned, as we had produced them to be. We report the results in Tables 3 and 4, which reveal that the novel approach was superior to the existing solvers.

Table 3. Comparison based on the number of iterations and the required mean in Experiment 2.

Methods	SM2	CM3	KMS10	PM10
A_1	98	61	30	30
A_2	86	55	27	27
A_3	83	53	26	26
A_4	85	54	27	27
A_5	81	52	26	26
Mean	86.6	55.	27.2	27.2

Table 4. Comparison based on the elapsed time and its mean in Experiment 2.

Methods	SM2	CM3	KMS10	PM10
A_1	7.89745	7.20885	12.1801	7.11963
A_2	6.90346	6.6397	10.933	6.50042
A_3	2.34013	2.23622	3.75341	2.20977
A_4	2.23133	2.15679	3.78848	2.23819
A_5	2.44316	2.26733	3.79153	2.2391
Mean	4.36311	4.10178	6.88929	4.06142

One other application of (15), aside from computing the WMP inverse, is in finding good approximate inverse pre-conditioners for Krylov methods when tackling large sparse linear system of equations (see, e.g., [29]). In fact, to apply our scheme in such environments, we can employ several commands, such as `SparseArray[]` for handling sparse matrices.

The main advantage of the proposed method is the improvement of convergence order obtained by improving the computational efficiency index. Although this computational efficiency index improvement was not observed to be drastic, in solving practical problems in higher dimensions it leads to a clear reduction of computation time.

5. Ending Notes

We have investigated a tenth order iterative method for computing the WMP inverse requiring only six matrix products. The WMP inverse has many applications, from the numerical solution of non-linear equations (those involving singular linear systems [32]) to direct engineering applications. Clearly, the efficiency index will reach $10^{1/6} \simeq 1.46$, which is better than the Newton–Schulz and Chebyshev methods for calculating the WMP inverse. The convergence order of the scheme was supported and upheld analytically. The extension of this improved version of the hyperpower family for computing other types generalized inverses, such as outer and inner inverses, under special criteria and initial matrices provides a direction for future works in this active topic of research.

References

1. Bin Jebreen, H.; Chalco-Cano, Y. An improved computationally efficient method for finding the Drazin inverse. *Disc. Dyn. Nat. Soc.* **2018**, *2018*, 6758302. [CrossRef]
2. Niazi Moghani, Z.; Khanehgir, M.; Mohammadzadeh Karizaki, M. Explicit solution to the operator equation $AXD + FX^*B = C$ over Hilbert C*-modules. *J. Math. Anal.* **2019**, *10*, 52–64.
3. Stanimirović, P.S.; Katsikis, V.N.; Srivastava, S.; Pappas, D. A class of quadratically convergent iterative methods. *RACSAM* **2019**, 1–22. [CrossRef]
4. Ben-Israel, A.; Greville, T.N.E. *Generalized Inverses: Theory and Applications*, 2nd ed.; Springer: New York, NY, USA, 2003.
5. Godunov, S.K.; Antonov, A.G.; Kiriljuk, O.P.; Kostin, V.I. *Guaranteed Accuracy in Numerical Linear Algebra*; Springer: Dordrecht, The Netherlands, 1993.
6. Fredholm, I. Sur une classe d'équations fonctionnelles. *Acta Math.* **1903**, *27*, 365–390. [CrossRef]
7. Wang, G.R. A new proof of Grevile's method for computing the weighted M-P inverse. *J. Shangai Norm. Univ.* **1985**, *3*, 32–38.
8. Bakhtiari, Z.; Mansour Vaezpour, S. Positive solutions to the system of operator equations $T_iX = U_i$ and $T_iXV_i = U_i$. *J. Math. Anal.* **2016**, *7*, 102–117.
9. Xia, Y.; Chen, T.; Shan, J. A novel iterative method for computing generalized inverse. *Neural Comput.* **2014**, *26*, 449–465. [CrossRef]
10. Courriee, P. Fast computation of Moore-Penrose inverse matrices. *arXiv preprint* **2008**, arXiv:0804.4809.
11. Lu, S.; Wang, X.; Zhang, G.; Zhou, X. Effective algorithms of the Moore-Penrose inverse matrices for extreme learning machine. *Intell. Data Anal.* **2015**, *19.4*, 743–760. [CrossRef]
12. Sheng, X.; Chen, G. The generalized weighted Moore-Penrose inverse. *J. Appl. Math. Comput.* **2007**, *25*, 407–413. [CrossRef]
13. Soleymani, F.; Soheili, A.R. A revisit of stochastic theta method with some improvements. *Filomat* **2017**, *31*, 585–596. [CrossRef]
14. Söderström, T.; Stewart, G.W. On the numerical properties of an iterative method for computing the Moore-Penrose generalized inverse. *SIAM J. Numer. Anal.* **1974**, *11*, 61–74. [CrossRef]
15. Stanimirović, P.S.; Ciric, M.; Stojanović, I.; Gerontitis, D. Conditions for existence, representations, and computation of matrix generalized inverses. *Complexity* **2017**, *2017*, 6429725. [CrossRef]
16. Gulliksson, M.E.; Wedin, P.A.; Wei, Y. Perturbation identities for regularized Tikhonov inverse and weighted pseudo inverse. *BIT* **2000**, *40*, 513–523. [CrossRef]
17. Roy, F.; Gupta, D.K.; Stanimirović, P.S. An interval extension of SMS method for computing weighted Moore-Penrose inverse. *Calcolo* **2018**, *55*, 15. [CrossRef]

18. Van Loan, C.F. Generalizing the singular value decomposition. *SIAM J. Numer. Anal.* **1976**, *13*, 76–83. [CrossRef]
19. Zhang, N.; Wei, Y. A note on the perturbation of an outer inverse. *Calcolo* **2008**, *45*, 263–273. [CrossRef]
20. Ghorbanzadeh, M.; Mahdiani, K.; Soleymani, F.; Lotfi, T. A class of Kung-Traub-type iterative algorithms for matrix inversion. *Int. J. Appl. Comput. Math.* **2016**, *2*, 641–648. [CrossRef]
21. Pan, V.Y. *Structured Matrices and Polynomials: Unified Superfast Algorithms*; BirkhWauser: Boston, MA, USA; Springer: New York, NY, USA, 2001.
22. Schulz, G. Iterative Berechnung der Reziproken matrix. *Z. Angew. Math. Mech.* **1933**, *13*, 57–59. [CrossRef]
23. Li, H.-B.; Huang, T.-Z.; Zhang, Y.; Liu, X.-P.; Gu, T.-X. Chebyshev-type methods and preconditioning techniques. *Appl. Math. Comput.* **2011**, *218*, 260–270. [CrossRef]
24. Krishnamurthy, E.V.; Sen, S.K. *Numerical Algorithms—Computations in Science and Engineering*; Affiliated East-West Press: New Delhi, India, 1986.
25. Sen, S.K.; Prabhu, S.S. Optimal iterative schemes for computing Moore-Penrose matrix inverse. *Int. J. Syst. Sci.* **1976**, *8*, 748–753. [CrossRef]
26. Grevile, T.N.E. Some applications of the pseudo-inverse of matrix. *SIAM Rev.* **1960**, *3*, 15–22. [CrossRef]
27. Huang, F.; Zhang, X. An improved Newton iteration for the weighted Moore-Penrose inverse. *Appl. Math. Comput.* **2006**, *174*, 1460–1486. [CrossRef]
28. Sánchez León, J.G. *Mathematica Beyond Mathematics: The Wolfram Language in the Real World*; Taylor & Francis Group: Boca Raton, FL, USA, 2017.
29. Zaka Ullah, M.; Soleymani, F.; Al-Fhaid, A.S. An efficient matrix iteration for computing weighted Moore-Penrose inverse. *Appl. Math. Comput.* **2014**, *226*, 441–454. [CrossRef]
30. Trott, M. *The Mathematica Guide-Book for Numerics*; Springer: New York, NY, USA, 2006.
31. Ostrowski, A.M. Sur quelques transformations de la serie de LiouvilleNewman. *CR Acad. Sci. Paris* **1938**, *206*, 1345–1347.
32. Soheili, A.R.; Soleymani, F. Iterative methods for nonlinear systems associated with finite difference approach in stochastic differential equations. *Numer. Algor.* **2016**, *71*, 89–102. [CrossRef]

9

Design and Complex Dynamics of Potra–Pták-Type Optimal Methods for Solving Nonlinear Equations and its Applications

Prem B. Chand [1,†], Francisco I. Chicharro [2,†], Neus Garrido [3,*,†] and Pankaj Jain [1,†]

1 Department of Mathematics, South Asian University, Akbar Bhawan, Chanakya Puri, New Delhi 110021, India; premchand13@gmail.com (P.B.C.); pankaj.jain@sau.ac.in (P.J.)
2 Escuela Superior de Ingeniería y Tecnología, Universidad Internacional de La Rioja, Av. La Paz 137, 26006 Logroño, Spain; francisco.chicharro@unir.net
3 Institute for Multidisciplinary Mathematics, Universitat Politècnica de València, Cno. de Vera s/n, 46022 València, Spain
* Correspondence: neugarsa@upvnet.upv.es
† These authors contributed equally to this work.

Abstract: In this paper, using the idea of weight functions on the Potra–Pták method, an optimal fourth order method, a non optimal sixth order method, and a family of optimal eighth order methods are proposed. These methods are tested on some numerical examples, and the results are compared with some known methods of the corresponding order. It is proved that the results obtained from the proposed methods are compatible with other methods. The proposed methods are tested on some problems related to engineering and science. Furthermore, applying these methods on quadratic and cubic polynomials, their stability is analyzed by means of their basins of attraction.

Keywords: nonlinear equations; Potra–Pták method; optimal methods; weight function; basin of attraction; engineering applications

PACS: 65H05

1. Introduction

For solving nonlinear equations iteratively, the Newton's method given by

$$x_{n+1} = x_n - \frac{f(x_n)}{f'(x_n)}$$

is one of the most commonly used methods. The efficiency index as defined by Ostroswki in [1], which relates the order of convergence of a method p with the number of function evaluations per iteration d, is given by the expression $p^{1/d}$. Newton's method is quadratically convergent and requires two function evaluations per iteration and, thereby, has the efficiency index value of $2^{1/2} \approx 1.414$. Numerous methods have appeared giving higher order of convergence or better efficiency. One of the recent strategies to increase the order of the methods is the use of weight functions [2–5]. In this regard, Sharma and Behl [6] presented the fourth order method:

$$\begin{aligned} y_n &= x_n - \frac{2}{3}\frac{f(x_n)}{f'(x_n)}, \\ x_{n+1} &= x_n - \left(-\frac{1}{2} + \frac{3}{8}\frac{f'(y_n)}{f'(x_n)} + \frac{9}{8}\frac{f'(x_n)}{f'(y_n)}\right)\frac{f(x_n)}{f'(x_n)}. \end{aligned} \tag{1}$$

Similarly, Sharifi et al. [7] used weight functions on the third order Heun's method and proposed the fourth order method

$$\begin{aligned} y_n &= x_n - \frac{2}{3}\frac{f(x_n)}{f'(x_n)}, \\ x_{n+1} &= x_n - \frac{f(x_n)}{4}\left(\frac{1}{f'(x_n)} + \frac{3}{f'(y_n)}\right)\left(1 + \frac{3}{8}\left(\frac{f'(y_n)}{f'(x_n)} - 1\right)^2 - \frac{69}{64}\left(\frac{f'(y_n)}{f'(x_n)} - 1\right)^3 + \frac{f(x_n)}{f'(y_n)}\right). \end{aligned} \tag{2}$$

According to Kung and Traub [8], an iterative method is said to be optimal if its order is 2^{d-1}, where d is the number of function evaluations per iteration. Notice that Newton's method as well as (1) and (2) are all optimal.

Potra and Pták [9], as an attempt to improve Newton's method, gave the method

$$\begin{aligned} y_n &= x_n - \frac{f(x_n)}{f'(x_n)}, \\ x_{n+1} &= x_n - \frac{f(x_n)+f(y_n)}{f'(x_n)}. \end{aligned} \tag{3}$$

This method is cubically convergent but is not optimal, as it requires three function evaluations per iteration.

The aim, in the present paper, is to further investigate the method (3). Precisely, we use weight functions and improve the order of convergence of (3). We do it in three ways which correspond to the methods of orders 4, 6 and 8. Out of these, the methods with orders 4 and 8 are optimal.

Dynamics of a rational operator give important information about the convergence, efficiency and stability of the iterative methods. During the last few decades, many researchers, e.g., [10–16] and references therein, study the dynamical behavior of rational operators associated with iterative methods. Furthermore, there is an extensive literature [17–21] to understand and implement further results on the dynamics of rational functions. In this paper, we also analyze the dynamical behavior of the methods that we have developed in this paper. Furthermore, at the end of this work, the basins of attraction are also presented and compared among the proposed and other methods of the corresponding order.

The remaining part of the paper is organized as follows. In Section 2, the development of the methods and their convergence analysis are given. In Section 3, the proposed methods are tested on some functions, and the results are compared with other methods in the head of Numerical Examples. In Section 4, the proposed methods are tested on some engineering and science related designs. Section 5 is devoted to analyze the stability of the introduced methods by means of complex dynamics. In this sense, the study of the rational function resulting from the application of the methods to several nonlinear functions is developed, and their basins of attraction are represented. Finally, Section 6 covers the conclusions of the research.

2. Development of Methods and Their Convergence Analysis

In this section, the methods of order four, six and eight are introduced, and its convergence is analyzed.

2.1. Optimal Fourth Order Method

Based on the Potra–Pták method (3), we propose the following two-step method using a weight function, whose iterative expression is

$$\begin{aligned} y_n &= x_n - \frac{f(x_n)}{f'(x_n)}, \\ x_{n+1} &= x_n - w(t_n)\frac{f(x_n)+f(y_n)}{f'(x_n)}, \end{aligned} \tag{4}$$

where $w(t_n) = a_1 + a_2 t_n + a_3 t_n^2$ and $t_n = \frac{f(y_n)}{f(x_n)}$. The convergence of (4) is proved in the following theorem.

Theorem 1. *Let f be a real or complex valued function defined in the interval I having a sufficient number of smooth derivatives. Let α be a simple root of the equation $f(x) = 0$ and the initial point x_0 is close enough to α. Then, the method (4) is fourth order of convergence if $a_1 = 1$, $a_2 = 0$ and $a_3 = 2$.*

Proof. We denote $c_j = \frac{f^{(j)}(\alpha)}{j!\,f'(\alpha)}$. Let $e_n = x_n - \alpha$ be the error in x_n. Then, Taylor's series expansion of $f(x_n)$ and $f'(x_n)$ about α gives

$$f(x_n) = f'(\alpha)\left(e_n + c_2e_n^2 + c_3e_n^3 + c_4e_n^4 + c_5e_n^5 + c_6e_n^6 + c_7e_n^7 + c_8e_n^8 + O(e_n^9)\right) \tag{5}$$

and

$$f'(x_n) = f'(\alpha)\left(1 + 2c_2e_n + 3c_3e_n^2 + 4c_4e_n^3 + 5c_5e_n^4 + 6c_6e_n^5 + 7c_7e_n^6 + 8c_8e_n^7 + O(e_n^8)\right). \tag{6}$$

Let $d_n = y_n - \alpha$, then, from the first equation of (4), we get

$$\begin{aligned} d_n &= c_2e_n^2 - 2\left(c_2^2 - c_3\right)e_n^3 + \left(4c_2^3 - 7c_2c_3 + 3c_4\right)e_n^4 + \left(-8c_2^4 + 20c_3c_2^2 - 10c_4c_2 - 6c_3^2 + 4c_5\right)e_n^5 \\ &+ \left(16c_2^5 - 52c_3c_2^3 + 28c_4c_2^2 + \left(33c_3^2 - 13c_5\right)c_2 - 17c_3c_4 + 5c_6\right)e_n^6 \\ &-2\left(16c_2^6 - 64c_3c_2^4 + 36c_4c_2^3 + 9\left(7c_3^2 - 2c_5\right)c_2^2 + \left(8c_6 - 46c_3c_4\right)c_2 - 9c_3^3 + 6c_4^2 + 11c_3c_5 - 3c_7\right)e_n^7 \\ &+\Big(64c_2^7 - 304c_3c_2^5 + 176c_4c_2^4 + \left(408c_3^2 - 92c_5\right)c_2^3 + \left(44c_6 - 348c_3c_4\right)c_2^2 \\ &+75c_3^2c_4 - 31c_4c_5 - 27c_3c_6 + c_2\left(-135c_3^3 + 118c_5c_3 + 64c_4^2 - 19c_7\right) + 7c_8\Big)e_n^8 + O(e_n^9) \end{aligned}$$

so that, using Taylor's series expansion of $f(y_n)$ about α, we get

$$\begin{aligned} f(y_n) &= f(d_n + \alpha) \\ &= f'(\alpha)\Big[c_2e_n^2 - 2(c_2^2 - c_3)e_n^3 + \left(5c_2^3 - 7c_2c_3 + 3c_4\right)e_n^4 - 2\left(6c_2^4 - 12c_3c_2^2 + 5c_4c_2 + 3c_3^2 - 2c_5\right)e_n^5 \\ &+\left(28c_2^5 - 73c_3c_2^3 + 34c_4c_2^2 + \left(37c_3^2 - 13c_5\right)c_2 - 17c_3c_4 + 5c_6\right)e_n^6 \\ &-2\Big(32c_2^6 - 103c_3c_2^4 + 52c_4c_2^3 + \left(80c_3^2 - 22c_5\right)c_2^2 + \left(8c_6 - 52c_3c_4\right)c_2 \\ &-9c_3^3 + 6c_4^2 + 11c_3c_5 - 3c_7\Big)e_n^7 + \Big(144c_2^7 - 552c_3c_2^5 + 297c_4c_2^4 + 2\left(291c_3^2 - 67c_5\right)c_2^3 \\ &+\left(54c_6 - 455c_3c_4\right)c_2^2 + 75c_3^2c_4 - 31c_4c_5 - 27c_3c_6 + c_2\left(-147c_3^3 + 134c_5c_3 + 73c_4^2 - 19c_7\right) \\ &+7c_8\Big)e_n^8 + O(e_n^9)\Big]. \end{aligned} \tag{7}$$

Now, from (5) and (7), we get

$$\begin{aligned} t_n &= \frac{f(y_n)}{f(x_n)} \\ &= c_2e_n + \left(-3c_2^2 + 2c_3\right)e_n^2 + \left(8c_2^3 - 10c_2c_3 + 3c_4\right)e_n^3 + \left(-20c_2^4 + 37c_2^2c_3 - 14c_2c_4 - 8c_3^2 + 4c_5\right)e_n^4 \\ &+\left(48c_2^5 - 118c_2^3c_3 + 51c_2^2c_4 + 55c_2c_3^2 - 18c_2c_5 - 22c_3c_4 + 5c_6\right)e_n^5 \\ &+\Big(-112c_2^6 + 344c_3c_2^4 - 163c_4c_2^3 + \left(65c_5 - 252c_3^2\right)c_2^2 + 2\left(75c_3c_4 - 11c_6\right)c_2 \\ &+26c_3^3 - 15c_4^2 - 28c_3c_5 + 6c_7\Big)e_n^6 + O(e_n^7). \end{aligned} \tag{8}$$

Therefore, using the results obtained above in the second equation of (4), we get

$$\begin{aligned} e_{n+1} &= \left(1 - a_1\right)e_n - a_2c_2e_n^2 + \left(2a_1c_2^2 + 3a_2c_2^2 - 2a_2c_3 - a_3c_2^2\right)e_n^3 \\ &+ \left(-9a_1c_2^3 + 7a_1c_2c_3 - 6a_2c_2^3 + 10a_2c_2c_3 - 3a_2c_4 + 6a_3c_2^3 - 4a_3c_2c_3\right)e_n^4 + O(e_n^5). \end{aligned} \tag{9}$$

In order to obtain fourth order of convergence, in view of (9), we must have

$$\begin{aligned} 1-a_1 &= 0, \\ a_2 &= 0, \\ 2a_1c_2^2+3a_2c_2^2-2a_2c_3-a_3c_2^2 &= 0, \end{aligned}$$

which gives $a_1 = 1$, $a_2 = 0$ and $a_3 = 2$. Therefore, from (9), the error equation of the method (4) becomes

$$e_{n+1} = (3c_2^3 - c_2c_3)e_n^4 + \mathcal{O}\left(e_n^5\right),$$

and the assertion follows. □

In view of Theorem 1, the proposed fourth order method is

$$\begin{aligned} y_n &= x_n - \frac{f(x_n)}{f'(x_n)}, \\ x_{n+1} &= x_n - \left(1+2\left(\frac{f(y_n)}{f(x_n)}\right)^2\right)\frac{f(x_n)+f(y_n)}{f'(x_n)}, \end{aligned} \tag{10}$$

which requires three function evaluations per iteration and consequently is optimal. In addition, the efficiency index of (10) is 1.5874, which is higher than that of (3) having an efficiency index of 1.442.

2.2. Sixth Order Method

Using the results obtained in (10), we propose a new method defined by

$$\begin{aligned} y_n &= x_n - \frac{f(x_n)}{f'(x_n)}, \\ z_n &= x_n - \left(1+2\left(\frac{f(y_n)}{f(x_n)}\right)^2\right)\frac{f(x_n)+f(y_n)}{f'(x_n)}, \\ x_{n+1} &= z_n - w_1(t_n)\frac{f(z_n)}{f'(x_n)}, \end{aligned} \tag{11}$$

where $w_1(t_n) = b_1 + b_2t_n$ is a new weight function and t_n is as in (4). The order of convergence is shown in the following result.

Theorem 2. *Let f be a real or complex valued function defined in an interval I having a sufficient number of smooth derivatives. Let α be a simple root of the equation $f(x) = 0$ and the initial point x_0 is close enough to α. Then, (11) has a sixth order of convergence if $b_1 = 1$ and $b_2 = 2$.*

Proof. Let $\theta_n = z_n - \alpha$. Then, from second equation of (11), we obtain

$$\begin{aligned} \theta_n &= (3c_2^3 - c_2c_3)e_n^4 - 2\left(8c_2^4 - 10c_2^2c_3 + c_3^2 + c_2c_4\right)e_n^5 \\ &\quad + (46c_2^5 - 114c_2^3c_3 + 30c_2^2c_4 + 42c_2c_3^2 - 3c_2c_5 - 7c_3c_4)e_n^6 + O(e_n^7). \end{aligned} \tag{12}$$

Now, by expanding $f(z_n)$ about α using Equation (12), we obtain

$$\begin{aligned} f(z_n) &= f(\theta_n + \alpha) \\ &= f'(\alpha)[(3c_2^3 - c_2c_3)\,e_n^4 - 2\left(8c_2^4 - 10c_2^2c_3 + c_3^2 + c_2c_4\right)e_n^5 \\ &\quad + \left(46c_2^5 - 114c_2^3c_3 + 30c_2^2c_4 + 42c_2c_3^2 - 3c_2c_5 - 7c_3c_4\right)e_n^6 + O(e_n^7)]. \end{aligned} \tag{13}$$

Therefore, using (6), (8) and (13) in the third equation of (11), we obtain

$$\begin{aligned} e_{n+1} &= (1-b_1)c_2(3c_2^2 - c_3)e_n^4 \\ &\quad + \left(c_2^4(22b_1 - 3b_2 - 16) + c_2^2c_3(-22b_1 + b_2 + 20) + 2(b_1-1)c_2c_4 + 2(b_1-1)c_3^2\right)e_n^5 \\ &\quad + (c_2^5(-90b_1 + 31b_2 + 46) + c_2^3c_3(167b_1 - 31b_2 - 114) + 2c_2^2c_4(-17b_1 + b_2 + 15) \\ &\quad + c_2(c_3^2(-49b_1 + 4b_2 + 42) + 3(b_1-1)c_5) + 7(b_1-1)c_3c_4)e_n^6 + O(e_n^7). \end{aligned} \tag{14}$$

In order to obtain sixth order of convergence, the coefficients of e_n^4 and e_n^5 must vanish in (14), i.e., $b_1 = 1$ and $b_2 = 2$. Therefore, the error equation of the method (11) becomes

$$e_{n+1} = c_2\left(18c_2^4 - 9c_2^2c_3 + c_3^2\right)e_n^6 + \mathcal{O}\left(e_n^7\right),$$

and the assertion follows. □

In view of Theorem 2, the following is the sixth order method

$$\begin{aligned} y_n &= x_n - \tfrac{f(x_n)}{f'(x_n)}, \\ z_n &= x_n - \left(1 + 2\left(\tfrac{f(y_n)}{f(x_n)}\right)^2\right)\tfrac{f(x_n)+f(y_n)}{f'(x_n)}, \\ x_{n+1} &= z_n - \left(1 + 2\tfrac{f(y_n)}{f(x_n)}\right)\tfrac{f(z_n)}{f'(x_n)}. \end{aligned} \tag{15}$$

2.3. Optimal Eighth Order Method

Notice that the method (15) is not optimal as it requires four function evaluation per iteration to achieve sixth order of convergence. Its efficiency index is 1.5651, which is less than that of the fourth order method (10). However, an eighth order method is obtained by (10) using an additional Newton step. The resulting iterative scheme is

$$\begin{aligned} y_n &= x_n - \tfrac{f(x_n)}{f'(x_n)}, \\ z_n &= x_n - \left(1 + 2\left(\tfrac{f(y_n)}{f(x_n)}\right)^2\right)\tfrac{f(x_n)+f(y_n)}{f'(x_n)}, \\ x_{n+1} &= z_n - \tfrac{f(z_n)}{f'(z_n)}. \end{aligned} \tag{16}$$

Nevertheless, this method requires five function evaluation per iteration, so that its efficiency index reduces to 1.5157, and, moreover, it is not optimal. Towards making the method (16) more efficient and optimal, we approximate $f'(z)$ as

$$f'(z_n) \approx \frac{f'(x_n)}{J(t_n, u_n)\cdot G(s_n)}, \tag{17}$$

where

$$t_n = \frac{f(y_n)}{f(x_n)}, \quad u_n = \frac{f(z_n)}{f(x_n)}, \quad s_n = \frac{f(z_n)}{f(y_n)}.$$

Here, J and G are some appropriate weight functions of two variables and one variable, respectively. This type of approximations was done by Matthies et al. in [22]. Accordingly, we propose the following method:

$$\begin{aligned} y_n &= x_n - \tfrac{f(x_n)}{f'(x_n)}, \\ z_n &= x_n - \left(1 + 2\left(\tfrac{f(y_n)}{f(x_n)}\right)^2\right)\tfrac{f(x_n)+f(y_n)}{f'(x_n)}, \\ x_{n+1} &= z_n - \tfrac{f(z_n)}{f'(x_n)}\cdot J(t_n, u_n)\cdot G(s_n), \end{aligned} \tag{18}$$

where t_n, u_n, and s_n, are as in (17). For the method (18), we take the functions J and G as

$$J(t_n, u_n) = \frac{1 + 2t_n + (\beta + 2)u_n + 3t_n^2}{1 + \beta u_n} \tag{19}$$

and

$$G(s_n) = \frac{1 + \lambda s_n}{1 + (\lambda - 1)s_n}, \tag{20}$$

where β and λ belong to $\mathbb{C}$. We prove the following result.

Theorem 3. *Let f be a real or complex valued function defined on some interval I having a sufficient number of smooth derivatives. Let α be a simple root of the equation $f(x) = 0$ and the initial point x_0 is close enough to α. Then, (18) is an eighth order of convergence for the functions J and G given by (19) and (20), respectively.*

Proof. In view of (5) and (13), we obtain

$$\begin{aligned} u_n &= \frac{f(z_n)}{f(x_n)} \\ &= \left(3c_2^3 - c_2c_3\right)e_n^3 + \left(-19c_2^4 + 21c_2^2c_3 - 2c_2c_4 - 2c_3^2\right)e_n^4 \\ &\quad + \left(65c_2^5 - 138c_2^3c_3 + 32c_2^2c_4 + 45c_2c_3^2 - 3c_2c_5 - 7c_3c_4\right)e_n^5 + O(e_n^6). \end{aligned}$$

Similarly, (7) and (13) yield

$$\begin{aligned} s_n &= \frac{f(z_n)}{f(y_n)} \\ &= \left(3c_2^2 - c_3\right)e_n^2 - 2\left(5c_2^3 - 6c_2c_3 + c_4\right)e_n^3 + \left(11c_2^4 - 44c_2^2c_3 + 17c_2c_4 + 11c_3^2 - 3c_5\right)e_n^4 \\ &\quad + \left(56c_2^5 + 28c_2^3c_3 - 56c_2^2c_4 - 60c_2c_3^2 + 22c_2c_5 + 30c_3c_4 - 4c_6\right)e_n^5 + O(e_n^6). \end{aligned}$$

Consequently, (19) gives

$$\begin{aligned} J(t_n, u_n) &= 1 + 2c_2e_n + \left(4c_3 - 3c_2^2\right)e_n^2 + \left(4c_2^3 - 10c_2c_3 + 6c_4\right)e_n^3 \\ &\quad + \left(-3(2\beta+1)c_2^4 + 2(\beta+10)c_2^2c_3 - 14c_2c_4 - 8c_3^2 + 8c_5\right)e_n^4 \\ &\quad + \Big((47\beta - 38)c_2^5 - (57\beta + 14)c_2^3c_3 + 4(\beta+7)c_2^2c_4 \\ &\quad + 2c_2\left(4(\beta+4)c_3^2 - 9c_5\right) - 22c_3c_4 + 10c_6\Big)e_n^5 + O(e_n^6), \end{aligned} \tag{21}$$

and (20) gives

$$\begin{aligned} G(s_n) &= 1 + \left(3c_2^2 - c_3\right)e_n^2 - 2\left(5c_2^3 - 6c_2c_3 + c_4\right)e_n^3 \\ &\quad + \left((20 - 9\lambda)c_2^4 + 2(3\lambda - 25)c_2^2c_3 - (\lambda - 12)c_3^2 + 17c_2c_4 - 3c_5\right)e_n^4 \\ &\quad + 2\Big((30\lambda - 2)c_2^5 + (60 - 46\lambda)c_2^3c_3 + 2(3\lambda - 17)c_2^2c_4 + c_2\left(6(2\lambda - 7)c_3^2 + 11c_5\right) \\ &\quad + (17 - 2\lambda)c_3c_4 - 2c_6\Big)e_n^5 + O(e_n^6). \end{aligned} \tag{22}$$

Now, using the values from (6), (12), (13), (21), and (22) in (18), the error equation of the method is

$$e_{n+1} = c_2\left(3c_2^2 - c_3\right)\left(c_2^4(6\beta + 9\lambda + 9) - 2c_2^2c_3(\beta + 3\lambda + 4) + c_2c_4 + c_3^2\lambda\right)e_n^8 + O\left(e_n^9\right),$$

which gives the eighth order of convergence. □

3. Numerical Examples

In this section, we test the performance of the methods proposed in Section 2 with the help of some numerical examples. We compare the results obtained with the known methods of the corresponding order. We consider the following nonlinear equations and initial guesses:

- $f_1(x) = \sin^2 x - x^2 + 1, x_0 = 2,$
- $f_2(x) = \ln(1 + x^2) + \exp(x^2 - 3x)\sin x, x_0 = 2,$
- $f_3(x) = x^2 - (1 - x)^5, x_0 = 1,$
- $f_4(x) = x^2 - \exp(x) - 3x + 2, x_0 = 1,$
- $f_5(x) = \sqrt{x^2 + 2x + 5} - 2\sin x - x^2 + 3, x_0 = 2.$

In the previous section, we have proved the theoretical order of convergence of various methods. For practical purposes, we can test numerically the order of convergence of these methods by using

Approximated Computational Order of Convergence (or ACOC), defined by Cordero and Torregrosa [23]. They defined the ACOC of a sequence $\{x_k\}, k \geq 0$ as

$$ACOC = \frac{\log\left(|x_{k+1} - x_k| \,/\, |x_k - x_{k-1}|\right)}{\log\left(|x_k - x_{k-1}| \,/\, |x_{k-1} - x_{k-2}|\right)}. \tag{23}$$

The use of $ACOC$, given by (23), serves as a practical check on the theoretical error calculations.

We apply our proposed methods and other existing methods as discussed in the following subsections on each of the test functions. Various results of up to four iterations are observed, and we compare the results obtained at the 4th iteration among different methods of the corresponding order and shown in Tables 1–3. For a particular test function, we take the same initial guess x_0 for each of the methods under consideration. We compare the approximate error $\Delta x_n \equiv |x_n - x_{n-1}|$, the approximate solution x_n, the absolute value of corresponding functional value $|f(x_n)|$, and approximated computational order of convergence $(ACOC)$ at $n = 4$. In the tables, "NC" stands for no convergence of the method. We use Mathematica 9.0 for the calculations.

3.1. Comparison of the Fourth Order Method

Let us denote our method (10) by M_{41}. We shall compare this method with

- Sharma and Behl method (1), denoted by M_{42},
- Sharifi et al. method (2), denoted by M_{43},
- Jarratt's method [24], denoted by M_{44} and given by

$$\begin{aligned} y_n &= x_n - \frac{2}{3}\frac{f(x_n)}{f'(x_n)}, \\ x_{n+1} &= x_n - \left(\frac{3f'(y_n)+f'(x_n)}{6f'(y_n)-2f'(x_n)}\right)\frac{f(x_n)}{f'(x_n)}, \end{aligned}$$

- Kung–Traub [8] method, denoted by M_{45}, and given by

$$\begin{aligned} y_n &= x_n - \frac{f(x_n)}{f'(x_n)}, \\ x_{n+1} &= y_n - \left(\frac{f(x_n)\cdot f(y_n)}{(f(x_n)-f(y_n))^2}\right)\frac{f(x_n)}{f'(x_n)}. \end{aligned}$$

All the methods M_{4i}, $i = 1,2,3,4,5$ are optimal. Table 1 records the performance of all these methods.

Table 1. Comparison of numerical results of fourth order methods at the 4th iteration.

		f_1	f_2	f_3	f_4	f_5
Δx_n	M_{41}	8.7309×10^{-26}	2.7730×10^{-55}	9.9454×10^{-30}	1.2399×10^{-65}	9.2139×10^{-82}
	M_{42}	1.1188×10^{-27}	2.9815×10^{-28}	1.0915×10^{-24}	7.7434×10^{-72}	3.5851×10^{-101}
	M_{43}	1.1523×10^{-23}	NC	6.1887×10^{-13}	1.3049×10^{-15}	3.6376×10^{-49}
	M_{44}	2.0493×10^{-32}	2.0594×10^{-31}	1.1971×10^{-20}	1.5448×10^{-71}	1.1488×10^{-97}
	M_{45}	4.0043×10^{-28}	2.8464×10^{-57}	2.4018×10^{-30}	4.7295×10^{-65}	2.8215×10^{-81}
x_n	M_{41}	1.4045	-7.8835×10^{-218}	0.3460	0.2575	2.3320
	M_{42}	1.4045	-6.9805×10^{-110}	0.3460	0.2575	2.3320
	M_{43}	1.4045	NC	0.3460	0.2575	2.3320
	M_{44}	1.4045	3.2977×10^{-123}	0.3460	0.2575	2.3320
	M_{45}	1.4045	-3.5010×10^{-226}	0.3460	0.2575	2.3320
$\|f(x_n)\|$	M_{41}	1.9828×10^{-100}	7.8835×10^{-218}	1.9230×10^{-116}	2.5756×10^{-262}	1.1861×10^{-326}
	M_{42}	4.0436×10^{-108}	6.9805×10^{-110}	1.1758×10^{-96}	6.8107×10^{-287}	1.9034×10^{-404}
	M_{43}	3.6237×10^{-93}	NC	6.4877×10^{-49}	7.5782×10^{-62}	2.9990×10^{-196}
	M_{44}	1.7439×10^{-127}	3.2977×10^{-123}	4.4608×10^{-80}	1.3131×10^{-285}	2.5652×10^{-390}
	M_{45}	5.7027×10^{-110}	3.5010×10^{-226}	9.4841×10^{-120}	6.9959×10^{-260}	1.1952×10^{-324}
$ACOC$	M_{41}	3.9919	4.0000	4.0184	4.0000	4.0000
	M_{42}	3.9935	3.9953	4.0646	4.0000	4.0000
	M_{43}	4.1336	NC	3.5972	4.6265	4.0214
	M_{44}	3.9978	4.0069	3.9838	4.0000	4.0000
	M_{45}	3.9946	4.0001	3.9878	4.0000	4.0000

3.2. Comparison of Sixth Order Methods

We denote our sixth order method (15) by M_{61}. We shall compare this method with

- M_{62} : Method of Neta [25] with $a = 1$, given by

$$\begin{aligned} y_n &= x_n - \frac{f(x_n)}{f'(x_n)}, \\ z_n &= y_n - \frac{f(x_n)+af(y_n)}{f(x_n)+(a-2)f(y_n)} \frac{f(y_n)}{f'(x_n)}, \\ x_{n+1} &= z_n - \frac{f(x_n)-f(y_n)}{f(x_n)-3f(y_n)} \frac{f(z_n)}{f'(x_n)}, \end{aligned}$$

- M_{63} : Method of Grau et al. [26] given by

$$\begin{aligned} y_n &= x_n - \frac{f(x_n)}{f'(x_n)}, \\ z_n &= y_n - \frac{f(y_n)}{f'(x_n)} \frac{f(x_n)}{f(x_n)-2f(y_n)}, \\ x_{n+1} &= z_n - \frac{f(z_n)}{f'(x_n)} \frac{f(x_n)}{f(x_n)-2f(y_n)}. \end{aligned}$$

- M_{64} : Method of Sharma and Guha [27] with $a = 2$, given by

$$\begin{aligned} y_n &= x_n - \frac{f(x_n)}{f'(x_n)}, \\ z_n &= y_n - \frac{f(y_n)}{f'(x_n)} \frac{f(x_n)}{f(x_n)-2f(y_n)}, \\ x_{n+1} &= z_n - \frac{f(z_n)}{f'(x_n)} \frac{f(x_n)+af(y_n)}{f(x_n)+(a-2)f(y_n)}, \end{aligned}$$

- M_{65} : Method of Chun and Neta [28] given by

$$\begin{aligned} y_n &= x_n - \frac{f(x_n)}{f'(x_n)}, \\ z_n &= y_n - \frac{f(y_n)}{f'(x_n)} \frac{1}{\left(1-\frac{f(y_n)}{f(x_n)}\right)^2}, \\ x_{n+1} &= z_n - \frac{f(y_n)}{f'(x_n)} \frac{1}{\left(1-\frac{f(y_n)}{f(x_n)}-\frac{f(z_n)}{f(x_n)}\right)^2}. \end{aligned}$$

The comparison of the methods M_{6i}, $i = 1, 2, 3, 4, 5$ is tabulated in Table 2. From the table, we observe that the proposed method M_{61} is compatible with the other existing methods. We can see that method M_{63} gives different results for the test functions f_2 and f_5 with given initial guesses.

Table 2. Comparison of numerical results of sixth order methods at the 4th iteration.

		f_1	f_2	f_3	f_4	f_5
Δx_n	M_{61}	1.8933×10^{-73}	1.8896×10^{-148}	5.1627×10^{-90}	1.3377×10^{-199}	9.5891×10^{-261}
	M_{62}	1.6801×10^{-106}	2.9382×10^{-152}	2.4137×10^{-64}	1.7893×10^{-191}	3.75383×10^{-255}
	M_{63}	2.9803×10^{-95}	2.9803×10^{-95}	2.9815×10^{-82}	2.9815×10^{-82}	2.9803×10^{-95}
	M_{64}	5.0012×10^{-85}	2.4246×10^{-153}	4.9788×10^{-69}	4.6397×10^{-198}	4.0268×10^{-259}
	M_{65}	9.9516×10^{-88}	2.1737×10^{-154}	3.3993×10^{-86}	2.7764×10^{-193}	3.4903×10^{-256}
x_n	M_{61}	1.4045	-1.1331×10^{-884}	0.3460	0.2575	2.3320
	M_{62}	1.4045	4.5753×10^{-908}	0.3460	0.2575	2.3320
	M_{63}	1.4045	1.4045	0.3460	0.2575	1.4045
	M_{64}	1.4045	1.0114×10^{-914}	0.3460	0.2575	2.3320
	M_{65}	1.4045	-3.7511×10^{-921}	0.3460	0.2575	2.3320

Table 2. *Cont.*

		f_1	f_2	f_3	f_4	f_5
	M_{61}	5.6523×10^{-436}	1.1331×10^{-884}	1.8046×10^{-535}	0.0	0.0
	M_{62}	6.7308×10^{-636}	4.5753×10^{-908}	1.0347×10^{-381}	0.0	0.0
$\lvert f(x_n)\rvert$	M_{63}	8.1802×10^{-568}	8.1802×10^{-568}	8.2004×10^{-490}	8.2004×10^{-490}	8.1802×10^{-568}
	M_{64}	5.7605×10^{-506}	1.0114×10^{-914}	1.8726×10^{-409}	0.0	0.0
	M_{65}	3.7794×10^{-522}	3.7511×10^{-921}	4.8072×10^{-514}	0.0	0.0
	M_{61}	5.9980	6.0000	5.9980	6.0000	6.0000
	M_{62}	5.9992	6.0000	5.9854	6.0000	6.000
ACOC	M_{63}	5.9997	5.9997	5.9992	5.9992	5.9997
	M_{64}	5.9991	6.0000	5.9984	6.0000	6.0000
	M_{65}	5.9993	6.0000	6.0088	6.0000	6.0000

3.3. Comparison of Eighth Order Methods

Consider the eighth order method (18), which involves the parameter pair (β, λ). We denote

- M_{81} the case where $(\beta, \lambda) = (0, 0)$, whose iterative expression results in

$$\begin{aligned} y_n &= x_n - \frac{f(x_n)}{f'(x_n)}, \\ z_n &= x_n - \frac{f(x_n)+f(y_n)}{f'(x_n)} \left(1 + 2\left(\frac{f(y_n)}{f(x_n)}\right)^2\right), \\ x_{n+1} &= z_n - \frac{f(z_n)}{f'(x_n)} \left(\frac{1+2t_n+2u_n+3t_n^2}{1-s_n}\right), \end{aligned}$$

- M_{82} for $(\beta, \lambda) = (1, 1)$, resulting in the iterative scheme given by M_{81} :

$$\begin{aligned} y_n &= x_n - \frac{f(x_n)}{f'(x_n)}, \\ z_n &= x_n - \frac{f(x_n)+f(y_n)}{f'(x_n)} \left(1 + 2\left(\frac{f(y_n)}{f(x_n)}\right)^2\right), \\ x_{n+1} &= z_n - \frac{f(z_n)}{f'(x_n)} \left(\frac{1+2t_n+3u_n+3t_n^2}{1+u_n}(1+s_n)\right), \end{aligned}$$

- M_{83} for $(\beta, \lambda) = (0, 1)$, whose iterative method is

$$\begin{aligned} y_n &= x_n - \frac{f(x_n)}{f'(x_n)}, \\ z_n &= x_n - \frac{f(x_n)+f(y_n)}{f'(x_n)} \left(1 + 2\left(\frac{f(y_n)}{f(x_n)}\right)^2\right), \\ x_{n+1} &= z_n - \frac{f(z_n)}{f'(x_n)} \left((1 + 2t_n + 2u_n + 3t_n^2)(1 + s_n)\right). \end{aligned}$$

Along with these, we take the following methods for the comparison of numerical results:

- Matthies et al. in [22] presented an optimal class of $8th$ order method from the Kung–Traub method [8]. For some particular values of the parameters, one of the methods denoted by M_{84} is given by

$$\begin{aligned} y_n &= x_n - \frac{f(x_n)}{f'(x_n)}, \\ z_n &= y_n - \left(\frac{f(x_n)f(y_n)}{(f(x_n)-f(y_n))^2}\right) \frac{f(x_n)}{f'(x_n)}, \\ x_{n+1} &= z_n - \frac{f(z_n)}{f'(x_n)} \left(\frac{2+t_n+5u_n+4t_n^2+4t_n^3}{2-3t_n+u_n+2t_n^2} \cdot \frac{2+s_n}{2-s_n}\right). \end{aligned}$$

- Babajee et al. in [11] presented a family of eighth order methods. For some fixed values of parameters, the method denoted by M_{85} is given by

$$\begin{aligned} y_n &= x_n - \frac{f(x_n)}{f'(x_n)}\left(1+\left(\frac{f(x_n)}{f'(x_n)}\right)^5\right), \\ z_n &= y_n - \frac{f(y_n)}{f'(x_n)}\left(1-\frac{f(y_n)}{f(x_n)}\right)^{-2}, \\ x_{n+1} &= z_n - \frac{f(z_n)}{f'(x_n)}\left(\frac{(1+t_n^2+5t_n^4+s_n}{(1-t_n-u_n)^2}\right). \end{aligned}$$

- Chun and Lee in [29] presented a family of optimal eighth order methods. For some particular values of parameters, the method denoted by M_{86} is given by

$$\begin{aligned} y_n &= x_n - \frac{f(x_n)}{f'(x_n)}, \\ z_n &= y_n - \frac{f(y_n)}{f'(x_n)}\frac{1}{\left(1-\frac{f(y_n)}{f(x_n)}\right)^2}, \\ x_{n+1} &= z_n - \frac{f(z_n)}{f'(x_n)}\frac{1}{\left(1-t_n-\frac{t_n^2}{2}+\frac{t_n^3}{2}-\frac{u_n}{2}-\frac{s_n}{2}\right)^2}. \end{aligned}$$

In all the above methods, t_n, u_n and s_n are as given in (17). The performance of the methods M_{8i}, $i = 1,2,3,4,5,6$ are recorded in Table 3.

Table 3. Comparison of numerical results of eighth order methods at the 4th iteration.

		f_1	f_2	f_3	f_4	f_5
Δx_n	M_{81}	5.8768×10^{-187}	1.5404×10^{-393}	2.5345×10^{-165}	6.1099×10^{-495}	4.4344×10^{-658}
	M_{82}	2.0563×10^{-165}	9.0158×10^{-321}	1.1101×10^{-167}	5.4494×10^{-421}	4.0437×10^{-598}
	M_{83}	4.5429×10^{-170}	1.5139×10^{-324}	2.9710×10^{-168}	2.8838×10^{-421}	2.9107×10^{-604}
	M_{84}	2.4469×10^{-187}	4.9438×10^{-351}	4.3825×10^{-171}	1.8592×10^{-438}	4.3404×10^{-614}
	M_{85}	2.6744×10^{-204}	NC	1.7766×10^{-177}	6.5231×10^{-192}	9.8976×10^{-553}
	M_{86}	4.1482×10^{-235}	1.3271×10^{-380}	5.6991×10^{-175}	2.5934×10^{-455}	7.1011×10^{-617}
x_n	M_{81}	1.4045	0.0	0.3460	0.2575	2.3320
	M_{82}	1.4045	0.0	0.3460	0.2575	2.3320
	M_{83}	1.4045	0.0	0.3460	0.2575	2.3320
	M_{84}	1.4045	0.0	0.3460	0.2575	2.3320
	M_{85}	1.4045	NC	0.3460	0.2575	2.3320
	M_{86}	1.4045	0.0	0.3460	0.2575	2.3320
$\|f(x_n)\|$	M_{81}	0.0	0.0	0.0	0.0	0.0
	M_{82}	0.0	0.0	0.0	0.0	0.0
	M_{83}	0.0	0.0	0.0	0.0	0.0
	M_{84}	0.0	0.0	0.0	0.0	0.0
	M_{85}	0.0	0.0	0.0	0.0	0.0
	M_{86}	0.0	0.0	0.0	0.0	0.0
ACOC	M_{81}	7.9999	8.0000	7.9993	8.0000	8.0000
	M_{82}	7.9996	8.0000	8.0000	8.0000	8.0000
	M_{83}	7.9997	8.0000	7.9996	8.0000	8.0000
	M_{84}	7.9998	8.0000	8.0047	8.0000	8.0000
	M_{85}	7.9995	NC	8.0020	8.0004	8.0000
	M_{86}	8.0000	8.0000	8.0023	8.0000	8.0000

From Tables 1–3, we observe that the proposed methods are compatible with other existing methods (and sometimes perform better than other methods) of the corresponding order. Not any particular method is superior to others for all examples. Among the family of eighth order methods (18), from Table 3, we observe that the method M_{81} performs better than other two. For more understanding about the iterative methods, we study the dynamics of these methods in the next section.

4. Applications

The applications discussed in Sections 4.1–4.3 are based on standard engineering examples, and we refer to [30]. We use the proposed methods M_{41}, M_{61}, and M_{8i}, $i = 1, 2, 3$ to obtain the various results from the first three iterations of these examples. In particular, we compute the value of the unknowns x_{n-1} and x_n, absolute value of the function $f(x_n)$ and absolute value of the difference d of unknown in two consecutive iterations, i.e., $d = |x_n - x_{n-1}|$, $n = 1, 2, 3$.

4.1. Pipe Friction Problem

Determining fluid flow through pipes and tubes has great relevance in many areas of engineering and science. In engineering, typical applications include the flow of liquids and gases through pipelines and cooling systems. Scientists are interested in topics ranging from flow in blood vessels to nutrient transmission through a plant's vascular system. The resistance to flow in such conduits is parameterized by a dimensionless number called the friction factor f. For a flow with turbulence, the Colebrook equation [31] provides a means to calculate the friction factor:

$$0 = \frac{1}{\sqrt{f}} + 2.0 \log\left(\frac{\epsilon}{3.7D} + \frac{2.51}{Re\sqrt{f}}\right), \tag{24}$$

where ϵ is the roughness (m), D is the diameter (m) and Re is the Reynolds number

$$Re = \frac{\rho v D}{\mu}.$$

Here, ρ denotes the fluid density (kg/m^3), v the velocity of the fluid (m/s) and μ the dynamical viscosity (N·s/m^2). A flow is said to be turbulent if $Re > 4000$.

To determine f for air flow through a smooth and thin tube, the parameters are taken to be $\rho = 1.23$ kg/m^3, $\mu = 1.79 \times 10^{-5}$ N·s/m^2, $D = 0.005$ m, $V = 40$ m/s and $\epsilon = 0.0000015$ m. Since the friction factors range from about 0.008 to 0.08, we choose initial guess $f_0 = 0.023$. To determine the approximate value of f, we use the function

$$g(f) = \frac{1}{\sqrt{f}} + 2.0 \log\left(\frac{\epsilon}{3.7D} + \frac{2.51}{Re\sqrt{f}}\right). \tag{25}$$

The results obtained by the various methods are presented in Table 4.

Table 4. Results of pipe friction problem.

# Iter	Value	M_{41}	M_{61}	M_{81}	M_{82}	M_{83}
	f	0.0169	0.0170	0.0170	0.0170	0.0170
1	$g(f)$	0.0240	0.0104	0.0009	0.0005	0.0008
	d	0.0061	0.0060	0.0060	0.0060	0.0060
	f	0.0170	0.0170	0.0170	0.0170	0.0170
2	$g(f)$	3.0954×10^{-9}	2.6645×10^{-15}	8.8818×10^{-16}	8.8818×10^{-16}	8.8818×10^{-16}
	d	0.0001	4.1700×10^{-5}	3.7223×10^{-6}	2.0962×10^{-6}	3.3172×10^{-6}
	f	0.0170	0.0170	0.0170	0.0170	0.0170
3	$g(f)$	8.8818×10^{-16}	8.8818×10^{-16}	8.8818×10^{-16}	8.8818×10^{-16}	8.8818×10^{-16}
	d	1.2442×10^{-11}	1.0408×10^{-17}	6.9389×10^{-18}	0.0	0.0

4.2. Open-Channel Flow

An open problem in civil engineering is to relate the flow of water with other factors affecting the flow in open channels such as rivers or canals. The flow rate is determined as the volume of water passing a particular point in a channel per unit time. A further concern is related to what happens when the channel is slopping.

Under uniform flow conditions, the flow of water in an open channel is given by Manning's equation

$$Q = \frac{\sqrt{S}}{n} A R^{2/3}, \tag{26}$$

where S is the slope of the channel, A is the cross-sectional area of the channel, R is the hydraulic radius of the channel and n is the Manning's roughness coefficient. For a rectangular channel having the width B and the defth of water in the channel y, it is known that

$$A = By$$

and

$$R = \frac{By}{B + 2y}.$$

With these values, (26) becomes

$$Q = \frac{\sqrt{S}}{n} By \left(\frac{By}{B + 2y} \right)^{2/3}. \tag{27}$$

Now, if it is required to determine the depth of water in the channel for a given quantity of water, (27) can be rearranged as

$$f(y) = \frac{\sqrt{S}}{n} By \left(\frac{By}{B + 2y} \right)^{2/3} - Q. \tag{28}$$

In our work, we estimate y when the remaining parameters are assumed to be given as $Q = 14.15$ m^3/s, $B = 4.572$ m, $n = 0.017$ and $S = 0.0015$. We choose as an initial guess $y_0 = 4.5$ m. The results obtained by the various methods are presented in Table 5.

Table 5. Results of an open channel problem.

# Iter	Value	M_{41}	M_{61}	M_{81}	M_{82}	M_{83}
1	y	1.4804	1.4666	1.4652	1.4653	1.4653
	$f(y)$	0.2088	0.0204	0.0016	0.0029	0.0028
	d	3.0200	3.0334	3.0348	3.0347	3.0347
2	y	1.4651	1.4651	1.4651	1.4651	1.4651
	$f(y)$	4.5027×10^{-9}	1.7764×10^{-15}	$\times 10^{-15}$	3.5527×10^{-15}	3.5527×10^{-15}
	d	0.0154	0.0015	0.0001	0.0002	0.0002
3	y	1.4651	1.4651	1.4651	1.4651	1.4651
	$f(y)$	3.5527×10^{-15}	7.1054×10^{-14}	6.5725×10^{-14}	5.3291×10^{-15}	1.7764×10^{-15}
	d	3.3152×10^{-10}	5.1070×10^{-15}	5.1070×10^{-15}	6.6613×10^{-16}	2.2204×10^{-16}

4.3. *Ideal and Non-Ideal Gas Laws*

The ideal gas law is

$$PV = nRT,$$

where P is the absolute pressure, V is the volume, n is the number of moles, R is the universal gas constant and T is the absolute temperature. Due to its limited use in engineering, an alternative equation of state for gases is the given van der Waals equation [32–35]

$$\left(P + \frac{a}{v^2}\right)(v - b) = RT,$$

where $v = \frac{V}{n}$ is the molal volume and a, b are empirical constants that depend on the particular gas. The computation of the molal volume is done by solving

$$f(v) = \left(P + \frac{a}{v^2}\right)(v - b) - RT. \tag{29}$$

We take the remaining parameters as $R = 0.082054$ L atm/(mol K), for carbon dioxide $a = 3.592$, $b = 0.04267$, $T = 300$ K, $p = 1$ atm, and the initial guess for the molal volume is taken as $v_0 = 3$. The results obtained by the various methods are presented in Table 6. In this table, IND stands for indeterminate form.

Table 6. Numerical results of ideal and non-ideal gas law.

# Iter	Value	M_{41}	M_{61}	M_{81}	M_{82}	M_{83}
	v	26.4881	27.0049	23.9583	24.1631	24.0274
1	$f(v)$	1.9647	2.4788	0.5509	0.3474	0.4823
	d	23.4881	24.0049	20.9583	21.1631	21.0274
	v	24.5126	24.5126	24.5126	24.5126	24.5126
2	$f(v)$	2.7340×10^{-8}	3.3573×10^{-12}	0.0	0.0	0.0
	d	1.9756	2.4923	0.5543	0.3495	0.4852
	v	24.5126	24.5126	IND	IND	IND
3	$f(v)$	0.0	0.0	IND	IND	IND
	d	2.7503×10^{-8}	3.3786×10^{-12}	IND	IND	IND

5. Dynamical Analysis

The stability analysis of the methods M_{41}, M_{61} and M_{8i}, $i = 1, 2, 3$, is performed in this section. The dynamics of the proposed methods on a generic quadratic polynomial will be studied, analyzing the associated rational operator for each method. This analysis shows their performance depending on the initial estimations. In addition, method M_{41} is analyzed for cubic polynomials. First, we recall some basics on complex dynamics.

5.1. Basics on Complex Dynamics

Let $R : \hat{\mathbb{C}} \longrightarrow \hat{\mathbb{C}}$ be a rational function defined on the Riemann sphere. Let us recall that every holomorphic function from the Riemann sphere to itself is in fact a rational function $R(z) = \frac{P(z)}{Q(z)}$, where P and Q are complex polynomials (see [36]). For older work on dynamics on the Riemann sphere, see, e.g., [37].

The orbit of a point $z_0 \in \hat{\mathbb{C}}$ is composed by the set of its images by R, i.e.,

$$\{z_0, R(z_0), R^2(z_0), \ldots, R^n(z_0), \ldots\}.$$

A point $z^F \in \hat{\mathbb{C}}$ is a fixed point if $R(z^F) = z^F$. Note that the roots z^* of an equation $f(z) = 0$ are fixed points of the associated operator of the iterative method. Fixed points that do not agree with a root of $f(x) = 0$ are strange fixed points.

The asymptotical behavior of a fixed point z^F is determined by the value of its multiplier $\mu = |R'(z^F)|$. Then, z^F is attracting, repelling or neutral if μ is lower, greater or equal to 1, respectively. In addition, it is superattracting when $\mu = 0$.

For an attracting fixed point z^F, its basin of attraction is defined as the set of its pre-images of any order:

$$\mathcal{A}(z^F) = \{z_0 \in \hat{\mathbb{C}} : R^n(z_0) \longrightarrow z^F, n \to \infty\}.$$

The dynamical plane represents the basins of attraction of a method. By iterating a set of initial guesses, their convergence is analyzed and represented. The points $z^C \in \hat{\mathbb{C}}$ that satisfy $R'(z^C) = 0$ are called critical points of R. When a critical point does not agree with a solution of $f(x) = 0$, it is a free critical point. A classical result [21] establishes that there is at least one critical point associated with each immediate invariant Fatou component.

5.2. Rational Operators

Let $p(z)$ be a polynomial defined on $\hat{\mathbb{C}}$. Corresponding to the methods developed in this paper, i.e., methods (10), (15) and family (18), we define the operators $R_4(z)$, $R_6(z)$ and $R_8(z)$, respectively, in $\hat{\mathbb{C}}$ as follows:

$$\begin{aligned} R_4(z) &= z - \left(1 + 2\left(\frac{p(y(z))}{p(z)}\right)^2\right)\frac{p(z)+p(y(z))}{p'(z)}, \\ R_6(z) &= R_4(z) - \left(1 + 2\frac{p(y(z))}{p(z)}\right)\frac{p(R_4(z))}{p'(z)}, \\ R_8(z) &= R_4(z) - \frac{p(R_4(z))}{p'(z)} J(z)G(z), \end{aligned} \tag{30}$$

where $y(z) = z - \frac{p(z)}{p'(z)}$ and

$$\begin{aligned} J(z) &= \frac{1+2\frac{p(y(z))}{p(z)}+(\beta+2)\frac{p(R_4(z))}{p(z)}+3\left(\frac{p(y(z))}{p(z)}\right)^2}{1+\beta\frac{p(R_4(z))}{p(z)}}, \\ G(z) &= \frac{1+\lambda\frac{p(R_4(z))}{p(y(z))}}{1+(\lambda-1)\frac{p(R_4(z))}{p(y(z))}}. \end{aligned}$$

First, we recall the following result for the generalization of the dynamics of M_{41}.

Theorem 4 (Scaling Theorem for method M_{41}). *Let $f(z)$ be an analytic function in the Riemann sphere and let $A(z) = \eta z + \sigma$, with $\eta \neq 0$, be an affine map. Let $h(z) = \mu(f \circ A)(z)$ with $\mu \neq 0$. Then, the fixed point operator R_4^f is affine conjugated to R_4^h by A, i.e.,*

$$(A \circ R_4^h \circ A^{-1})(z) = R_4^f(z).$$

Proof. From (30), let the fixed point operators associated with f and h be, respectively,

$$\begin{aligned} R_4^f(z) &= z - \left(1 + 2\left(\frac{f(y(z))}{f(z)}\right)^2\right)\frac{f(z)+f(y(z))}{f'(z)}, \\ R_4^h(z) &= z - \left(1 + 2\left(\frac{h(y(z))}{h(z)}\right)^2\right)\frac{h(z)+h(y(z))}{h'(z)}. \end{aligned}$$

Thus, we have

$$(R_4^f \circ A)(z) = A(z) - \left(1 + 2\frac{f^2(A(y))}{f^2(A(z))}\right)\frac{f(A(z))+f(A(y))}{f'(A(z))}. \tag{31}$$

Being $h'(z) = \eta\mu f'(A(z))$, we obtain

$$\begin{aligned} R_4^h(z) &= z - \left(1 + 2\frac{\mu^2 f^2(A(y))}{\mu^2 f^2(A(z))}\right)\frac{\mu f(A(z))+\mu f(A(y))}{\eta\mu f'(A(z))} \\ &= z - \left(1 + 2\frac{f^2(A(y))}{f^2(A(z))}\right)\frac{f(A(z))+f(A(y))}{\eta f'(A(z))}. \end{aligned}$$

The affine map A satisfies $A(z_1 - z_2) = A(z_1) - A(z_2) + \sigma, \forall z_1, z_2$. Then, from (32), we have

$$\begin{aligned}(A \circ R_4^h)(z) &= A(z) - A\left(\left(1 + 2\frac{f^2(A(y))}{f^2(A(z))}\right)\frac{f(A(z))+f(A(y))}{\eta f'(A(z))}\right) + \sigma \\ &= A(z) - \left(\eta\left(1 + 2\frac{f^2(A(y))}{f^2(A(z))}\right)\frac{f(A(z))+f(A(y))}{\eta f'(A(z))} + \sigma\right) + \sigma \\ &= A(z) - \left(1 + 2\frac{f^2(A(y))}{f^2(A(z))}\right)\frac{f(A(z))+f(A(y))}{f'(A(z))}.\end{aligned}$$

Thus, it proves that $(R_4^f \circ A)(z) = (A \circ R_4^h)(z)$ and then method M_{41} satisfies the Scaling Theorem. □

Theorem 4 allows for generalizing the dynamical study of a specific polynomial to a generic family of polynomials by using an affine map. Analogous to the way we proved the Scaling Theorem for the operator R_4, it also follows that the fixed point operators R_6 and R_8 obey the Scaling Theorem.

5.3. Dynamics on Quadratic Polynomials

The application of the rational functions on a generic quadratic polynomial $p(z) = (z-a)(z-b)$, $a, b \in \hat{\mathbb{C}}$ is studied below. Let $R_{4,a,b}$ be the rational operator associated with method M_{41} on $p(z)$. When the Möbius transformation $h(u) = \frac{a-u}{b-u}$ is applied to $R_{4,a,b}$, we obtain

$$S_4(z) = (h \circ R_{4,a,b} \circ h^{-1})(z) = \frac{z^4\left(z^4 + 6z^3 + 14z^2 + 14z + 3\right)}{3z^4 + 14z^3 + 14z^2 + 6z + 1}. \quad (32)$$

The rational operator associated with M_{41} on $p(z)$ does not depend on a and b. Then, the dynamical analysis of the method on all quadratic polynomials can be studied through the analysis of (32). In addition, the Möbius transformation h maps its roots a and b to $z_1^* = 0$ and $z_2^* = \infty$, respectively.

The fixed point operator $S_4(z)$ has nine fixed points: $z_1^F = 0$ and $z_2^F = \infty$, which are superattracting, and $z_3^F = 1$, $z_{4,5}^F = \frac{1}{2}(-3 \pm \sqrt{5})$, $z_{6-7}^F = \frac{-2+\sqrt{2}}{2} \pm i\sqrt{\frac{3}{2} - \sqrt{2}}$, $z_{8-9}^F = \frac{-2-\sqrt{2}}{2} \pm i\sqrt{\frac{3}{2} + \sqrt{2}}$, all of them being repelling. Computing $S_4'(z) = 0$, 5 critical points can be found. $z_{1,2}^C = z_{1,2}^*$ and the free critical points $z_3^C = -1$ and $z_{4,5}^C = \frac{1}{6}(-13 \pm \sqrt{133})$.

Following the same procedure, when Möbius transformation is applied to methods M_6 and M_{8i}, $i = 1, 2, 3$, on polynomial $p(z)$, the respective fixed point operators turn into

$$S_6(z) = \frac{z^6\left(z^{12}+16z^{11}+119z^{10}+544z^9+1700z^8+3808z^7+6206z^6+7288z^5+5973z^4+3248z^3+1111z^2+216z+18\right)}{18z^{12}+216z^{11}+1111z^{10}+3248z^9+5973z^8+7288z^7+6206z^6+3808z^5+1700z^4+544z^3+119z^2+16z+1},$$

$$S_{81}(z) = \frac{P_{30}(z)}{P_{22}(z)}, \qquad S_{82}(z) = \frac{P_{42}(z)}{P_{34}(z)}, \qquad S_{83}(z) = \frac{Q_{42}(z)}{Q_{34}(z)},$$

where P_k and Q_k denote polynomials of degree k.

The fixed point operator S_6 has 19 fixed points: the two superattracting fixed points $z_{1,2}^F = z_{1,2}^*$, the repelling fixed point $z_3^F = 1$ and the repelling fixed points $z_4^F, \ldots, z_{19}^F$, which are the roots of a sixteenth-degree polynomial.

Regarding the critical points of S_6, the roots of $p(z)$ are critical points, and S_6 has the free critical points $z_3^C = -1$ and the roots of a tenth-degree polynomial, $z_4^C, \ldots, z_{11}^C$.

The dynamical planes are a useful tool in order to analyze the stability of an iterative method. Taking each point of the plane as initial estimation to start the iterative process, they represent the convergence of the method depending on the initial guess. In this sense, the dynamical planes show the basins of attraction of the attracting points.

Figure 1 represents the dynamical planes of the methods S_4 and S_6. The generation of the dynamical planes follows the guidelines established in [38]. A mesh of 500×500 complex values has been set as initial guesses in the intervals $-5 < \Re\{z\} < 5$, $-5 < \Im\{z\} < 5$. The roots $z_1^* = 0$ and $z_2^* = \infty$ are mapped with orange and blue colors, respectively. The regions where the colors are darker represent that more iterations are necessary to converge than with the lighter colors, with a maximum of 40 iterations of the methods and a stopping criteria of a difference between two consecutive iterations lower than 10^{-6}.

As Figure 1 illustrates, there is convergence to the roots for every initial guess. Let us remark that, when the order of the method increases, the basin of attraction of $z_1^* = 0$ becomes more intricate.

Finally, for the fixed point operators associated with family M_8, the solutions of $S_{8i}(z) = z$ for $i = 1,2,3$ give the superattracting fixed points $z_{1,2}^F = z_{1,2}^*$ and the repelling point $z_3^F = 1$. In addition, S_{81} has 28 repelling points. S_{82} and S_{83} have 38 repelling points, corresponding to the roots of polynomials of 28 and 38 degree, respectively, and the strange fixed points $z_{4,5}^F = \frac{1}{2}(-1 \pm \sqrt{5})$.

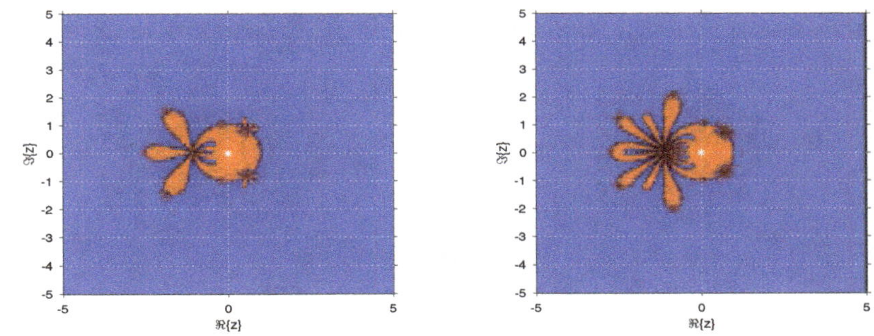

(a) S_4 (b) S_6

Figure 1. Dynamical planes of methods S_4 and S_6.

The number of critical points of the fixed point operators S_{8i} are collected in Table 7. In addition, the number of strange fixed points and free critical points are also included in the table for all of the methods.

Table 7. Number of strange fixed points (SFP) and free critical points (FCP) for the methods on quadratic polynomials.

	S_4	S_6	S_{81}	S_{82}	S_{83}
Strange fixed points	7	17	29	41	41
Free critical points	3	29	29	43	29

Figure 2 represents the dynamical planes of the methods S_{81}, S_{82} and S_{83}. Since the original methods satisfy the Scaling Theorem, the generation of one dynamical plane involves the study of every quadratic polynomial.

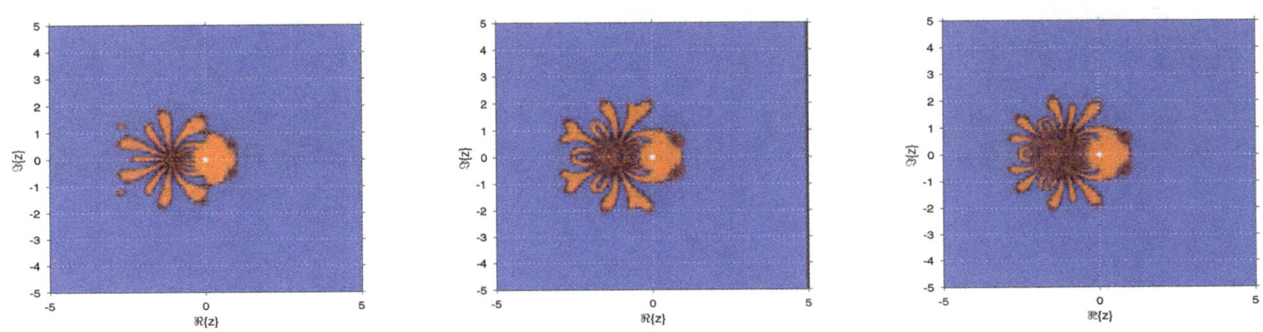

(a) S_{81} (b) S_{82} (c) S_{83}

Figure 2. Dynamical planes of methods $S_{8i}, i = 1,2,3$.

There is an intricate region around $z = -1$ in Figure 2a, becoming wider in Figure 2b,c around $z = -1.5$. However, for every initial guess in the three dynamical planes of Figure 2, there is convergence to the roots.

5.4. Dynamics on Cubic Polynomials

The stability of method M_{41} on cubic polynomials is analyzed below. As stated by the authors in [39], the Scaling Theorem reduces the dynamical analysis on cubic polynomials to the study of dynamics on the cubic polynomials $p_0(z) = z^3$, $p_+(z) = z^3 + z$, $p_-(z) = z^3 - z$ and the family of polynomials $p_\gamma(z) = z^3 + \gamma z + 1$. Let us recall that the first one only has the root $z_1^* = 0$, while $p_+(z)$ and $p_-(z)$ have three simple roots: $z_1^* = 0$ and $z_{2,3}^* = \mp i$ or $z_{2,3}^* = \mp 1$, respectively. For each $\gamma \in \mathbb{C}$, the polynomial $p_\gamma(z)$ also has three simple roots that depend on the value of γ. They will be denoted by $z_{1,2,3}^*(\gamma)$.

By applying method M_{41} to polynomials $p_0(z)$, $p_+(z)$ and $p_-(z)$, the fixed point operators obtained are, respectively,

$$S_{4,0}(z) = \tfrac{46z}{81}, \quad S_{4,+}(z) = \tfrac{6z^5+36z^7+46z^9}{(1+3z^2)^4}, \quad S_{4,-}(z) = \tfrac{6z^5-36z^7+46z^9}{(1-3z^2)^4}.$$

The only fixed point of $S_{4,0}(z)$ agrees with the root of the polynomial, so it is superattracting, and the operator does not have critical points.

The rest of the fixed point operators have six repelling fixed points, in addition to the roots of the corresponding polynomials: $z_{4,5}^F = \pm\frac{i\sqrt{5}}{5}$ and $z_{6-9}^F = \pm i\sqrt{\frac{1}{7}(3 \pm \sqrt{2})}$ for $S_{4,+}(z)$, and $z_{4,5}^F = \pm\frac{\sqrt{5}}{5}$ and $z_{6-9}^F = \pm\sqrt{\frac{1}{7}(3 \pm \sqrt{2})}$ for $S_{4,-}(z)$.

Regarding the critical points of $S_{4,+}(z)$ and $S_{4,-}(z)$, they match with the roots of the polynomials. Moreover, there is the presence of free critical points with values $z_{4,5}^C = \pm i\sqrt{\frac{5}{23}}$ for $S_{4,+}(z)$ and $z_{4,5}^C = \pm\sqrt{\frac{5}{23}}$ for $S_{4,-}(z)$.

As for quadratic polynomials, the dynamical planes of method M_{41} when it is applied to the cubic polynomials have been represented in Figure 3. Depending on the roots of each polynomial, the convergence to $z_1^* = 0$ is represented in orange, while the convergence to z_2^* and z_3^* is represented in blue and green, respectively. It can be see in Figure 3 that there is full convergence to a root in the three cases. However, there are regions with darker colors that indicate a higher number of iterations until the convergence is achieved.

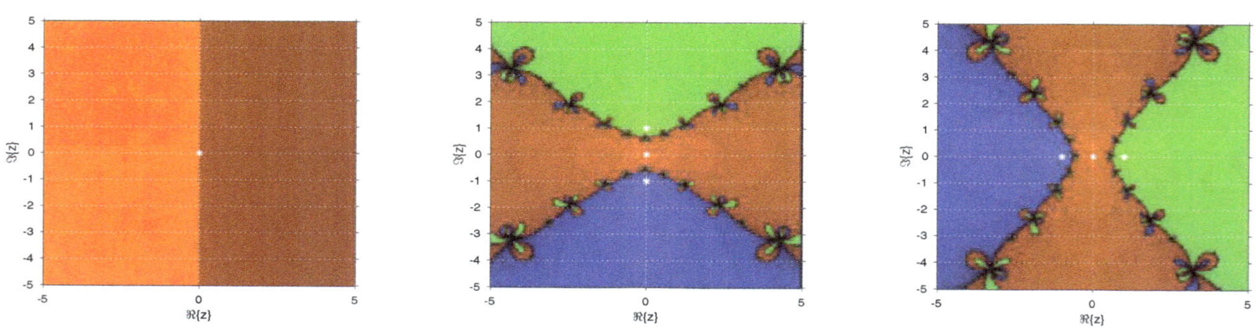

(a) $p_0(z)$ **(b)** $p_+(z)$ **(c)** $p_-(z)$

Figure 3. Dynamical planes of method M_{41} on polynomials $p_0(z)$, $p_+(z)$ and $p_-(z)$.

When method M_{41} is applied on $p_\gamma(z)$, the fixed point function turns into

$$S_{4,\gamma}(z) = -\frac{\gamma^3 - 46z^9 - 36\gamma z^7 + 42z^6 - 6\gamma^2 z^5 + 45\gamma z^4 + 6z^3 + 12\gamma^2 z^2 - 1}{(\gamma + 3z^2)^4}.$$

The fixed points of $S_{4,\gamma}(z)$ are the roots of the polynomial $z^*_{1,2,3}(\gamma)$, being superattracting, and the strange fixed points $z^F_{4-9}(\gamma)$ that are the roots of the sixth-degree polynomial $q(z,\gamma) = 35z^6 + 37\gamma z^4 + 7z^3 + 11\gamma^2 z^2 + \gamma z + \gamma^3 - 1$.

As the asymptotical behavior of $z^F_4(\gamma), \ldots, z^F_9(\gamma)$ depends on the value of γ, the stability planes corresponding to these points are represented in Figure 4. For each strange fixed point, a mesh of 100×100 points covers the values of $\Re(\gamma) \in [-5,5]$ and $\Im(\gamma) \in [-5,5]$. The stability plane shows the values for the parameter where $|S'_{4,\gamma}(z^F)|$ is lower or greater than 1, represented in red or green, respectively.

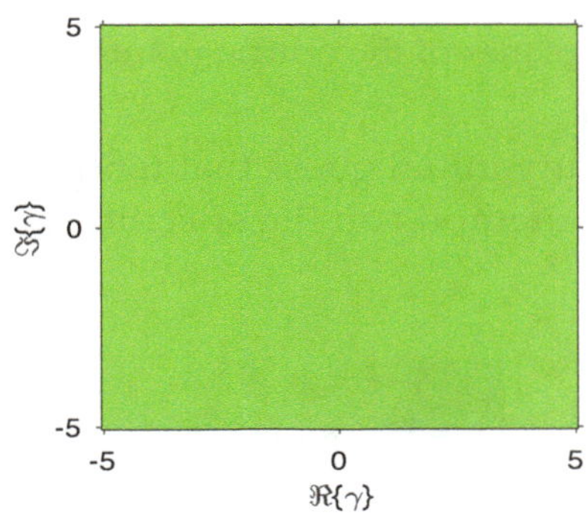

Figure 4. Stability planes of $z^F_{4-9}(\gamma)$.

From Figure 4, the strange fixed points are always repelling for $(\Re(\gamma), \Im(\gamma)) \in [-5,5] \times [-5,5]$. Then, the only attracting fixed points are the roots of the polynomial. This fact guarantees a better stability of the method.

The solutions of $S'_{4,\gamma}(z) = 0$ are the critical points $z^C_{1,2,3}(\gamma) = z^*_{1,2,3}(\gamma)$ and the free critical points $z^C_4 = 0$ and

$$\begin{aligned} z^C_5(\gamma) &= \frac{\left(\sqrt{69}\sqrt{125\gamma^3+2484}+414\right)^{2/3} - 5\sqrt[3]{69}\gamma}{69^{2/3}\sqrt[3]{\sqrt{69}\sqrt{125\gamma^3+2484}+414}}, \\ z^C_{6,7}(\gamma) &= \frac{(-1\pm i\sqrt{3})\left(\sqrt{69}\sqrt{125\gamma^3+2484}+414\right)^{2/3} + 5\sqrt[3]{69}(1\pm i\sqrt{3})\gamma}{2\ 69^{2/3}\sqrt[3]{\sqrt{69}\sqrt{125\gamma^3+2484}+414}}. \end{aligned}$$

When the fixed point function has dependence on a parameter, another useful representation is the parameters' plane. This plot is generated in a similar way to the dynamical planes, but, in this case, by iterating the method taking as an initial guess a free critical point and varying the value of γ in a complex mesh of values, so each point in the plane represents a method of the family. The parameters' plane helps to select the values for the parameter that give rise to the methods of the family with more stability.

The parameters' planes of the four free critical points are shown in Figure 5. Parameter γ takes the values of 500×500 points in a complex mesh in the square $[-5,5] \times [-5,5]$. Each point is represented in orange, green or blue when the corresponding method converges to an attracting fixed point. The iterative process ends when the maximum number of 40 iterations is reached, in which case the point is represented in black, or when the method converges as soon as, by the stopping criteria, a difference between two consecutive iterations lower than 10^{-6} is reached.

For the parameters' planes in Figure 5, there is not any black region. This guarantees that the corresponding iterative schemes converge to a root of $p_\gamma(z)$ for all the values of γ.

In order to visualize the basins of attraction of the fixed points, several values of γ have been chosen to perform the dynamical planes of method M_{41}. These values have been selected from the different regions of convergence observed in the parameters planes. Figure 6, following the same code of colours and stopping criteria as in the other representations, shows the dynamical planes obtained when these values of γ are fixed.

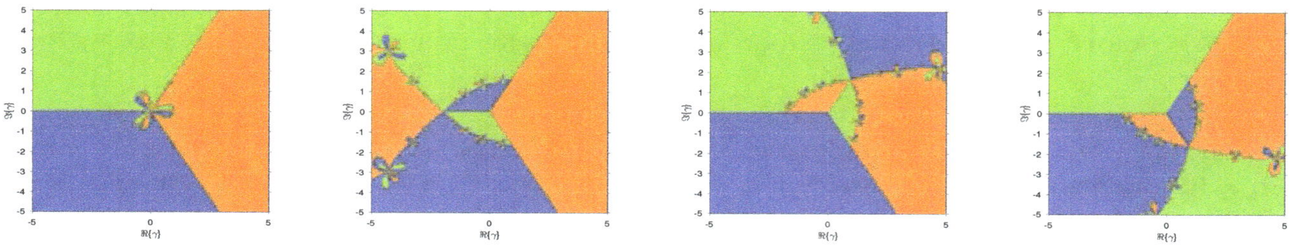

(a) z_4^C (b) $z_5^C(\gamma)$ (c) $z_6^C(\gamma)$ (d) $z_7^C(\gamma)$

Figure 5. Parameter planes of the critical points of method M_{41} on $p_\gamma(z)$.

As Figure 6 shows, there is not any initial guess that tends to a point different than the roots. This fact guarantees the stability of these methods on the specific case of any cubic polynomial.

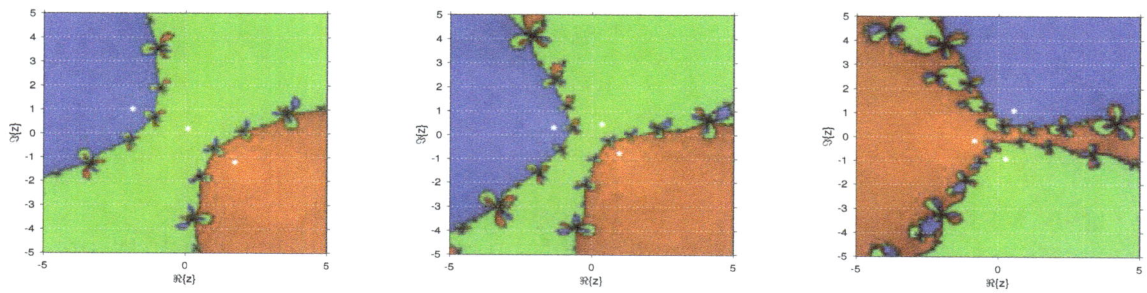

(a) $\gamma = -2 + 4i$ (b) $\gamma = -1 + i$ (c) $\gamma = 0.5 - 0.5i$

Figure 6. Dynamical planes for method M_{41} on $p_\gamma(z)$ for different values of γ.

6. Conclusions

Two iterative schemes of orders of convergence four and six, and a family of methods of order eight have been introduced. The method of order four and the family of order eight are optimal in the sense of Kung–Traub's conjecture. The development of the order of convergence of every method has been performed. For every method, we have made a numerical experiment, over both test functions and real engineering problems. In order to analyze the stability of the introduced methods, the dynamical behavior of them has been studied. The results confirm that the methods have wide basins of attraction, guaranteeing the stability over some nonlinear problems.

Author Contributions: The individual contributions of the authors are as follows: conceptualization, P.J.; validation, P.B.C. and P.J.; formal analysis, F.I.C.; writing, original draft preparation, N.G. and F.I.C.; numerical experiments, N.G. and P.B.C.

Acknowledgments: The second and third authors have been partially supported by PGC2018-095896-B-C22 (MCIU/AEI/FEDER/UE) and Generalitat Valenciana PROMETEO/2016/089.

References

1. Ostrowski, A.M. *Solution of Equations and Systems of Equations*; Prentice-Hall: Englewood Cliffs, NJ, USA, 1964.
2. Cordero, A.; Fardi, M.; Ghasemi, M.; Torregrosa, J.R. Accelerated iterative methods for finding solutions of nonlinear equations and their dynamical behavior. *Calcolo* **2012**, *51*, 17–30. [CrossRef]
3. Lotfi, T.; Sharifi, S.; Salimi, M.; Siegmund, S. A new class of three-point methods with optimal convergence order eight and its dynamics. *Numer. Algor.* **2015**, *68*, 261–288. [CrossRef]
4. Chicharro, F.I.; Cordero, A.; Garrido, N.; Torregrosa, J.R. Wide stability in a new family of optimal fourth-order iterative methods. *Comput. Math. Meth.* **2019**, *1*, e1023. [CrossRef]

5. Chicharro, F.I.; Cordero, A.; Torregrosa, J.R. Dynamics of iterative families with memory based on weight functions procedure. *J. Comput. Appl. Math.* **2019**, *354*, 286–298. [CrossRef]
6. Sharma, R.; Bahl, A. An optimal fourth order iterative method for solving nonlinear equations and its dynamics. *J. Complex Anal.* **2015**, *2015*, 259167. [CrossRef]
7. Sharifi, M.; Babajee, D.K.R.; Soleymani, F. Finding solutions of nonlinear equations by a class of optimal methods. *Comput. Math. Appl.* **2012**, *63*, 764–774. [CrossRef]
8. Kung, H.T.; Traub, J.F. Optimal order of one-point and multipoint iterations. *J. Assoc. Comput. Mach.* **1974**, *21*, 643–651. [CrossRef]
9. Potra, F.A.; Pták, V. Nondiscrete induction and iterative processes. *Res. Notes Math.* **1984**, *103*, 112–119.
10. Amat, S.; Busquier, S.; Plaza, S. Review of some iterative root finding methods from a dynamical point of view. *Scientia* **2004**, *10*, 3–35.
11. Babajee, D.K.R.; Cordero, A.; Soleymani, F.; Torregrosa, J.R. On improved three-step schemes with high efficiency index and their dynamics. *Numer. Algor.* **2014**, *65*, 153–169. [CrossRef]
12. Chicharro, F.I.; Cordero, A.; Torregrosa, J.R.; Vassileva, M.P. King-type derivative-free iterative families: Real and memory dynamics. *Complexity* **2017**, *2017*, 2713145. [CrossRef]
13. Cordero, A.; Feng, L.; Magreñán, Á.A.; Torregrosa, J.R. A new fourth order family for solving nonlinear problems and its dynamics. *J. Math. Chem.* **2015**, *53*, 893–910. [CrossRef]
14. Scott, M.; Neta, B.; Chun, C. Basin attractors for various methods. *Appl. Math. Comput.* **2011**, *218*, 2584–2599. [CrossRef]
15. Varona, J.L. Graphic and numerical comparison between iterative methods. *Math. Intell.* **2002**, *24*, 37–46. [CrossRef]
16. Vrscay, E.R.; Gilbert, W.J. Extraneous fixed points, basin boundaries and chaotic dynamics for Schroder and Konig rational iteration functions. *Numer. Math.* **1988**, *52*, 1–16. [CrossRef]
17. Alexander, D.S. *A History of Complex Dynamics: From Schröder to Fatou and Julia*; Vieweg & Teubner Verlag: Wiesbaden, Germany, 1994.
18. Beardon, A.F. *Iteration of Rational Functions*; Springer: New York, NY, USA, 2000.
19. Blanchard, P. The dynamics of Newton's method. *Proc. Sympos. Appl. Math.* **1994**, *49*, 139–154.
20. Carleson, L.; Gamelin, T.W. *Complex Dynamics*; Springer-Verlag: New York, NY, USA, 1992.
21. Devaney, R.L. *An Introduction to Chaotic Dynamical Systems*; Addison-Wesley: Redwood City, CA, USA, 1989.
22. Matthies, G.; Salimi, M.; Sharifi, S.; Varona, J.L. An optimal class of eighth-order iterative methods based on Kung and Traub's method with its dynamics. *arXiv* **2015**, arXiv:1508.01748v1.
23. Cordero, A.; Torregrosa, J.R. Variants of Newton's method using fifth order quadrature formulas. *Appl. Math. Comput.* **2007**, *190*, 686–698. [CrossRef]
24. Jarratt, P. Some efficient fourth order multipoint methods for solving equations. *BIT* **1969**, *9*, 119–124. [CrossRef]
25. Neta, B. A sixth-order family of methods for nonlinear equations. *Int. J. Comput. Math.* **1979**, *7*, 157–161. [CrossRef]
26. Grau, M.; Díaz-Barrero, J.L. An improvement to Ostrowski root-finding method. *Appl. Math. Comput.* **2006**, *173*, 450–456. [CrossRef]
27. Sharma, J.R.; Guha, R.K. A family of modified Ostrowski's methods with accelerated sixth order convergence. *Appl. Math. Comput.* **2007**, *190*, 111–115.
28. Chun, C.; Neta, B. A new sixth-order scheme for nonlinear equations. *Appl. Math. Lett.* **2012**, *25*, 185–189. [CrossRef]
29. Chun, C.; Lee, M.Y. A new optimal eighth-order family of iterative methods for the solution of nonlinear equations. *Appl. Math. Comput.* **2013**, *223*, 506–519. [CrossRef]
30. Chapra, S.C.; Canale, R.C. *Numerical Methods for Engineers*; McGraw Hills Education: New York, NY, USA, 2015.
31. Brkić, D. A note on explicit approximations to Colebrook's friction factor in rough pipes under highly turbulent cases. *Int. J. Heat Mass tramsf.* **2016**, *93*, 513–515. [CrossRef]
32. Wang, J.; Pang, Y.; Zhang, Y. Limits of solutions to the isentropic Euler equations for van der Waals gas. *Int. J. Nonlinear Sci. Numer. Simul.* **2019**, *20*, 461–473. [CrossRef]
33. Gates, D.J.; Penrose, O. The van der Waals limit for classical systems. I. A variational principle. *Comm. Math. Phys.* **1969**, *15*, 255–276. [CrossRef]

34. Gates, D.J.; Penrose, O. The van der Waals limit for classical systems. II. Existence and continuity of the canonical pressure. *Comm. Math. Phys.* **1970**, *16*, 231–237. [CrossRef]
35. Gates, D.J.; Penrose, O. The van der Waals limit for classical systems. III. Deviation from the van der Waals-Maxwell theory. *Comm. Math. Phys.* **1970**, *17*, 194–209. [CrossRef]
36. Blanchard, P. Complex analytic dynamics on the Riemann sphere. *Bull. Am. Math. Soc.* **1984**, *11*, 85–141. [CrossRef]
37. Schlag, W. *A Course in Complex Analysis and Riemann Surfaces*; American Mathematical Society: Providence, RI, USA, 2014.
38. Chicharro, F.I.; Cordero, A.; Torregrosa, J.R. Drawing dynamical and parameters planes of iterative families and methods. *Sci. World J.* **2013**, *2013*, 780153. [CrossRef] [PubMed]
39. Amat, S.; Busquier, S.; Plaza, S. Chaotic dynamics of a third-order Newton-type method. *J. Math. Anal. Appl.* **2010**, *366*, 24–32. [CrossRef]

10

An Efficient Conjugate Gradient Method for Convex Constrained Monotone Nonlinear Equations with Applications †

Auwal Bala Abubakar [1,2], Poom Kumam [1,3,4,*], Hassan Mohammad [2] and Aliyu Muhammed Awwal [1,5]

1 KMUTTFixed Point Research Laboratory, SCL 802 Fixed Point Laboratory, Science Laboratory Building, Department of Mathematics, Faculty of Science, King Mongkut's University of Technology Thonburi (KMUTT), 126 Pracha-Uthit Road, Bang Mod, Thrung Khru, Bangkok 10140, Thailand

2 Department of Mathematical Sciences, Faculty of Physical Sciences, Bayero University, Kano 700241, Nigeria

3 Center of Excellence in Theoretical and Computational Science (TaCS-CoE), Science Laboratory Building, King Mongkut's University of Technology Thonburi (KMUTT), 126 Pracha-Uthit Road, Bang Mod, Thrung Khru, Bangkok 10140, Thailand

4 Department of Medical Research, China Medical University Hospital, China Medical University, Taichung 40402, Taiwan

5 Department of Mathematics, Faculty of Science, Gombe State University, Gombe 760214, Nigeria

* Correspondence: poom.kum@kmutt.ac.th

† This project was supported by Petchra Pra Jom Klao Doctoral Academic Scholarship for Ph.D. Program at KMUTT. Moreover, this project was partially supported by the Thailand Research Fund (TRF) and the King Mongkut's University of Technology Thonburi (KMUTT) under the TRF Research Scholar Award (Grant No. RSA6080047).

Abstract: This research paper proposes a derivative-free method for solving systems of nonlinear equations with closed and convex constraints, where the functions under consideration are continuous and monotone. Given an initial iterate, the process first generates a specific direction and then employs a line search strategy along the direction to calculate a new iterate. If the new iterate solves the problem, the process will stop. Otherwise, the projection of the new iterate onto the closed convex set (constraint set) determines the next iterate. In addition, the direction satisfies the sufficient descent condition and the global convergence of the method is established under suitable assumptions. Finally, some numerical experiments were presented to show the performance of the proposed method in solving nonlinear equations and its application in image recovery problems.

Keywords: nonlinear monotone equations; conjugate gradient method; projection method; signal processing

MSC: 65K05; 90C52; 90C56; 92C55

---

1. Introduction

In this paper, we consider the following constrained nonlinear equation

$$F(x) = 0, \quad \text{subject to} \quad x \in \Psi, \tag{1}$$

where $F : \mathbb{R}^n \to \mathbb{R}^n$ is continuous and monotone. The constraint set $\Psi \subset \mathbb{R}^n$ is nonempty, closed and convex.

Monotone equations appear in many applications [1–3], for example, the subproblems in the generalized proximal algorithms with Bregman distance [4], reformulation of some ℓ_1-norm regularized problems arising in compressive sensing [5] and variational inequality problems are also converted into nonlinear monotone equations via fixed point maps or normal maps [6], (see References [7–9] for more examples). Among earliest methods for the case $\Psi = \mathbb{R}^n$ is the hyperplane projection Newton method proposed by Solodov and Svaiter in Reference [10]. Subsequently, many methods were proposed by different authors. Among the popular methods are spectral gradient methods [11,12], quasi-Newton methods [13–15] and conjugate gradient methods (CG) [16,17].

To solve the constrained case (1), the work of Solodov and Svaiter was extended by Wang et al. [18] which also involves solving a linear system in each iteration but it was shown later by some authors that the computation of the linear system is not necessary. For examples, Xiao and Zhu [19] presented a CG method, which is a combination the well known CG-DESCENT method in Reference [20] with the projection strategy by Solodov and Svaiter. Liu et al. [21] presented two CG method with sufficiently descent directions. In Reference [22], a modified version of the method in Reference [19] was presented by Liu and Li. The modification improves the numerical performance of the method in Reference [19]. Another extension of the Dai and Kou (DK) CG method combined with the projection method to solve (1) was proposed by Ding et al. in Reference [23]. Just recently, to popularize the Dai-Yuan (DY) CG method, Liu and Feng [24] modified the DY such that the direction will be sufficiently descent. A new hybrid spectral gradient projection method for solving convex constraints nonlinear monotone equations was proposed by Awwal et al. in Reference [25]. The method is a convex combination of two different positive spectral parameters together with the projection strategy. In addition, Abubakar et al. extended the method in Reference [17] to solve (1) and also solve some sparse signal recovery problems.

Inspired by some the above methods, we propose a descent conjugate gradient method to solve problem (1). Under appropriate assumptions, the global convergence is established. Preliminary numerical experiments were given to compare the proposed method with existing methods to solve nonlinear monotone equations and some signal and image reconstruction problems arising from compressive sensing.

The remaining part of this paper is organized as follows. In Section 2, we state the proposed algorithm as well as its convergence analysis. Finally, Section 3 reports some numerical results to show the performance of the proposed method in solving Equation (1), signal recovery problems and image restoration problems.

2. Algorithm: Motivation and Convergence Result

This section starts by defining the projection map together with some of its properties.

Definition 1. *Let $\Psi \subset \mathbb{R}^n$ be a nonempty closed convex set. Then for any $x \in \mathbb{R}^n$, its projection onto Ψ, denoted by $P_\Psi(x)$, is defined by*

$$P_\Psi(x) = \arg\min\{\|x - y\| \ : y \in \Psi\}.$$

Moreover, P_Ψ is nonexpansive, That is,

$$\|P_\Psi(x) - P_\Psi(y)\| \leq \|x - y\|, \quad \forall x, y \in \mathbb{R}^n. \tag{2}$$

All through this article, we assume the followings

(G_1) The mapping F is monotone, that is,

$$(F(x) - F(y))^T(x - y) \geq 0, \quad \forall x, y \in \mathbb{R}^n.$$

(G_2) The mapping F is Lipschitz continuous, that is there exists a positive constant L such that

$$\|F(x) - F(y)\| \leq L\|x - y\|, \ \forall x, y \in \mathbb{R}^n.$$

(G_3) The solution set of (1), denoted by Ψ', is nonempty.

An important property that methods for solving Equation (1) must possess is that the direction d_k satisfy

$$F(x_k)^T d_k \leq -c\|F(x_k)\|^2, \tag{3}$$

where $c > 0$ is a constant. The inequality (3) is called sufficient descent property if $F(x)$ is the gradient vector of a real valued function $f : \mathbb{R}^n \to \mathbb{R}$.

In this paper, we propose the following search direction

$$d_k = \begin{cases} -F(x_k), & \text{if } k = 0, \\ -F(x_k) + \beta_k d_{k-1} - \theta_k F(x_k), & \text{if } k \geq 1, \end{cases} \tag{4}$$

where

$$\beta_k = \frac{\|F(x_k)\|}{\|d_{k-1}\|} \tag{5}$$

and θ_k is determined such that Equation (3) is satisfied. It is easy to see that for $k = 0$, the equation holds with $c = 1$. Now for $k \geq 1$,

$$\begin{aligned} F(x_k)^T d_k &= -F(x_k)^T F(x_k) + F(x_k)^T \frac{\|F(x_k)\|}{\|d_{k-1}\|} d_{k-1} - \theta_k F(x_k)^T F(x_k) \\ &= -\|F(x_k)\|^2 + \frac{\|F(x_k)\|}{\|d_{k-1}\|} F(x_k)^T d_{k-1} - \theta_k \|F(x_k)\|^2 \\ &= \frac{-\|F(x_k)\|^2 \|d_{k-1}\|^2 + \|F(x_k)\| \|d_{k-1}\| F(x_k)^T d_{k-1} - \theta_k \|F(x_k)\|^2 \|d_{k-1}\|^2}{\|d_{k-1}\|^2}. \end{aligned} \tag{6}$$

Taking $\theta_k = 1$ we have

$$F(x_k)^T d_k \leq -\|F(x_k)\|^2. \tag{7}$$

Thus, the direction defined by (4) satisfy condition (3) $\forall k$ where $c = 1$.

To prove the global convergence of Algorithm 1, the following lemmas are needed.

Algorithm 1: (DCG)

Step 0. Given an arbitrary initial point $x_0 \in \mathbb{R}^n$, parameters $\sigma > 0$, $0 < \beta < 1$, $Tol > 0$ and set $k := 0$.

Step 1. If $\|F(x_k)\| \leq Tol$, stop, otherwise go to **Step 2**.

Step 2. Compute d_k using Equation (4).

Step 3. Compute the step size $\alpha_k = \max\{\beta^i \ : i = 0, 1, 2, \cdots\}$ such that

$$-F(x_k + \alpha_k d_k)^T d_k \geq \sigma \alpha_k \|F(x_k + \alpha_k d_k)\| \|d_k\|^2. \tag{8}$$

Step 4. Set $z_k = x_k + \alpha_k d_k$. If $z_k \in \Psi$ and $\|F(z_k)\| \leq Tol$, stop. Else compute

$$x_{k+1} = P_\Psi[x_k - \zeta_k F(z_k)]$$

where

$$\zeta_k = \frac{F(z_k)^T (x_k - z_k)}{\|F(z_k)\|^2}.$$

Step 5. Let $k = k + 1$ and go to **Step 1**.

Lemma 1. *The direction defined by Equation (4) satisfies the sufficient descent property, that is, there exist constants $c > 0$ such that (3) holds.*

Lemma 2. *Suppose that assumptions (G_1)–(G_3) holds, then the sequences $\{x_k\}$ and $\{z_k\}$ generated by Algorithm 1 (CGD) are bounded. Moreover, we have*

$$\lim_{k\to\infty} \|x_k - z_k\| = 0 \tag{9}$$

and

$$\lim_{k\to\infty} \|x_{k+1} - x_k\| = 0. \tag{10}$$

Proof. We will start by showing that the sequences $\{x_k\}$ and $\{z_k\}$ are bounded. Suppose $\bar{x} \in \Psi'$, then by monotonicity of F, we get

$$F(z_k)^T(x_k - \bar{x}) \geq F(z_k)^T(x_k - z_k). \tag{11}$$

Also by definition of z_k and the line search (8), we have

$$F(z_k)^T(x_k - z_k) \geq \sigma\alpha_k^2\|F(z_k)\|\|d_k\|^2 \geq 0. \tag{12}$$

So, we have

$$\begin{aligned}
\|x_{k+1} - \bar{x}\|^2 &= \|P_\Psi[x_k - \zeta_k F(z_k)] - \bar{x}\|^2 \leq \|x_k - \zeta_k F(z_k) - \bar{x}\|^2 \\
&= \|x_k - \bar{x}\|^2 - 2\zeta_k F(z_k)^T(x_k - \bar{x}) + \|\zeta F(z_k)\|^2 \\
&\leq \|x_k - \bar{x}\|^2 - 2\zeta_k F(z_k)^T(x_k - z_k) + \|\zeta F(z_k)\|^2 \\
&= \|x_k - \bar{x}\|^2 - \left(\frac{F(z_k)^T(x_k - z_k)}{\|F(z_k)\|}\right)^2 \\
&\leq \|x_k - \bar{x}\|^2
\end{aligned} \tag{13}$$

Thus the sequence $\{\|x_k - \bar{x}\|\}$ is non increasing and convergent and hence $\{x_k\}$ is bounded. Furthermore, from Equation (13), we have

$$\|x_{k+1} - \bar{x}\|^2 \leq \|x_k - \bar{x}\|^2, \tag{14}$$

and we can deduce recursively that

$$\|x_k - \bar{x}\|^2 \leq \|x_0 - \bar{x}\|^2, \quad \forall k \geq 0.$$

Then from Assumption (G_2), we obtain

$$\|F(x_k)\| = \|F(x_k) - F(\bar{x})\| \leq L\|x_k - \bar{x}\| \leq L\|x_0 - \bar{x}\|.$$

If we let $L\|x_0 - \bar{x}\| = \kappa$, then the sequence $\{F(x_k)\}$ is bounded, that is,

$$\|F(x_k)\| \leq \kappa, \quad \forall k \geq 0. \tag{15}$$

By the definition of z_k, Equation (12), monotonicity of F and the Cauchy-Schwatz inequality, we get

$$\sigma\|x_k - z_k\| = \frac{\sigma\|\alpha_k d_k\|^2}{\|x_k - z_k\|} \leq \frac{F(z_k)^T(x_k - z_k)}{\|x_k - z_k\|} \leq \frac{F(z_k)^T(x_k - z_k)}{\|x_k - z_k\|} \leq \|F(x_k)\|. \tag{16}$$

The boundedness of the sequence $\{x_k\}$ together with Equations (15) and (16), implies the sequence $\{z_k\}$ is bounded.

Since $\{z_k\}$ is bounded, then for any $\bar{x} \in \Psi$, the sequence $\{z_k - \bar{x}\}$ is also bounded, that is, there exists a positive constant $\nu > 0$ such that

$$\|z_k - \bar{x}\| \leq \nu.$$

This together with Assumption (G_2) yields

$$\|F(z_k)\| = \|F(z_k) - F(\bar{x})\| \leq L\|z_k - \bar{x}\| \leq L\nu.$$

Therefore, using Equation (13), we have

$$\frac{\sigma^2}{(L\nu)^2}\|x_k - z_k\|^4 \leq \|x_k - \bar{x}\|^2 - \|x_{k+1} - \bar{x}\|^2,$$

which implies

$$\frac{\sigma^2}{(L\nu)^2}\sum_{k=0}^{\infty}\|x_k - z_k\|^4 \leq \sum_{k=0}^{\infty}(\|x_k - \bar{x}\|^2 - \|x_{k+1} - \bar{x}\|^2) \leq \|x_0 - \bar{x}\| < \infty. \tag{17}$$

Equation (17) implies

$$\lim_{k\to\infty}\|x_k - z_k\| = 0.$$

However, using Equation (2), the definition of ζ_k and the Cauchy-Schwartz inequality, we have

$$\begin{aligned}\|x_{k+1} - x_k\| &= \|P_\Psi[x_k - \zeta_k F(z_k)] - x_k\| \\ &\leq \|x_k - \zeta_k F(z_k) - x_k\| \\ &= \|\zeta_k F(z_k)\| \\ &= \|x_k - z_k\|,\end{aligned} \tag{18}$$

which yields

$$\lim_{k\to\infty}\|x_{k+1} - x_k\| = 0.$$

□

Equation (9) and definition of z_k implies that

$$\lim_{k\to\infty}\alpha_k\|d_k\| = 0. \tag{19}$$

Lemma 3. *Suppose d_k is generated by Algorithm 1 (CGD), then there exist $M > 0$ such the $\|d_k\| \leq M$*

Proof. By definition of d_k and Equation (15)

$$\begin{aligned} \|d_k\| &= \| -2F(x_k) + \frac{\|F(x_k)\|}{\|d_{k-1}\|} d_{k-1}\| \\ &\leq 2\|F(x_k)\| + \frac{\|F(x_k)\|}{\|d_{k-1}\|} \|d_{k-1}\| \\ &\leq 3\|F(x_k)\| \\ &\leq 3\kappa. \end{aligned} \tag{20}$$

Letting $M = 3\kappa$, we have the desired result. □

Theorem 1. *Suppose that assumptions (G_1)–(G_3) hold and let the sequence $\{x_k\}$ be generated by Algorithm 1, then*

$$\liminf_{k\to\infty} \|F(x_k)\| = 0, \tag{21}$$

Proof. To prove the Theorem, we consider two cases;
Case 1
Suppose $\liminf_{k\to\infty} \|d_k\| = 0$, we have $\liminf_{k\to\infty} \|F(x_k)\| = 0$. Then by continuity of F, the sequence $\{x_k\}$ has some accumulation point $\bar{x}$ such that $F(\bar{x}) = 0$. Because $\{\|x_k - \bar{x}\|\}$ converges and $\bar{x}$ is an accumulation point of $\{x_k\}$, therefore $\{x_k\}$ converges to $\bar{x}$.
Case 2
Suppose $\liminf_{k\to\infty} \|d_k\| > 0$, we have $\liminf_{k\to\infty} \|F(x_k)\| > 0$. Then by (19), it holds that $\lim_{k\to\infty} \alpha_k = 0$. Also from Equation (8),

$$-F(x_k + \beta^{i-1} d_k)^T d_k < \sigma \beta^{i-1} \|F(x_k + \beta^{i-1} d_k)\| \|d_k\|^2$$

and the boundedness of $\{x_k\}, \{d_k\}$, we can choose a sub-sequence such that allowing k to go to infinity in the above inequality results

$$F(\bar{x})^T \bar{d} > 0. \tag{22}$$

On the other hand, allowing k to approach ∞ in (7), implies

$$F(\bar{x})^T \bar{d} \leq 0. \tag{23}$$

(22) and (23) imply contradiction. Hence, $\liminf_{k\to\infty} \|F(x_k)\| > 0$ is not true and the proof is complete. □

3. Numerical Examples

This section gives the performance of the proposed method with existing methods such as PCG and PDY proposed in References [22,24], respectively, to solve monotone nonlinear equations using 9 benchmark test problems. Furthermore Algorithm 1 is applied to restore a blurred image. All codes were written in MATLAB R2018b and run on a PC with intel COREi5 processor with 4 GB of RAM and CPU 2.3 GHZ. All runs were stopped whenever $\|F(x_k)\| < 10^{-5}$.
The parameters chosen for the existing algorithm are as follows:

PCG method: All parameters are chosen as in Reference [22].

PDY method: All parameters are chosen as in Reference [24].

Algorithm 1: We have tested several values of $\beta \in (0, 1)$ and found that $\beta = 0.7$ gives the best result. In addition, to implement most of the optimization algorithms, the parameter σ is chosen as a very small number. Therefore, we chose $\beta = 0.7$ and $\sigma = 0.0001$ for the implementation of the proposed algorithm.

We test 9 different problems with dimensions ranging from $n = 1000, 5000, 10,000, 50,000, 100,000$ and 6 initial points: $x_1 = (0.1, 0.1, \cdots, 1)^T$, $x_2 = (0.2, 0.2, \cdots, 0.2)^T$, $x_3 = (0.5, 0.5, \cdots, 0.5)^T$, $x_4 = (1.2, 1.2, \cdots, 1.2)^T$, $x_5 = (1.5, 1.5, \cdots, 1.5)^T$, $x_6 = (2, 2, \cdots, 2)^T$. In Tables 1–9, the number of iterations (ITER), number of function evaluations (FVAL), CPU time in seconds (TIME) and the norm at the approximate solution (NORM) were reported. The symbol '−' is used when the number of iterations exceeds 1000 and/or the number of function evaluations exceeds 2000.

The test problems are listed below, where the function F is taken as $F(x) = (f_1(x), f_2(x), \ldots, f_n(x))^T$.

Problem 1 ([26]). *Exponential Function.*

$$\begin{aligned}
f_1(x) &= e^{x_1} - 1, \\
f_i(x) &= e^{x_i} + x_i - 1, \text{ for } i = 2, 3, ..., n, \\
\text{and } \Psi &= \mathbb{R}^n_+.
\end{aligned}$$

Problem 2 ([26]). *Modified Logarithmic Function.*

$$\begin{aligned}
f_i(x) &= \ln(x_i + 1) - \frac{x_i}{n}, \text{ for } i = 2, 3, ..., n, \\
\text{and } \Psi &= \{x \in \mathbb{R}^n : \sum_{i=1}^{n} x_i \le n, x_i > -1, i = 1, 2, \ldots, n\}.
\end{aligned}$$

Problem 3 ([13]). *Nonsmooth Function.*

$$\begin{aligned}
f_i(x) &= 2x_i - \sin|x_i|, \ i = 1, 2, 3, ..., n, \\
\text{and } \Psi &= \{x \in \mathbb{R}^n : \sum_{i=1}^{n} x_i \le n, x_i \ge 0, i = 1, 2, \ldots, n\}.
\end{aligned}$$

It is clear that Problem 3 is nonsmooth at $x = 0$.

Problem 4 ([26]). *Strictly Convex Function I.*

$$\begin{aligned}
f_i(x) &= e^{x_i} - 1, \text{ for } i = 1, 2, ..., n, \\
\text{and } \Psi &= \mathbb{R}^n_+.
\end{aligned}$$

Problem 5 ([26]). *Strictly Convex Function II.*

$$\begin{aligned}
f_i(x) &= \frac{i}{n} e^{x_i} - 1, \text{ for } i = 1, 2, ..., n, \\
\text{and } \Psi &= \mathbb{R}^n_+.
\end{aligned}$$

Problem 6 ([27]). *Tridiagonal Exponential Function*

$$\begin{aligned}
f_1(x) &= x_1 - e^{\cos(h(x_1 + x_2))}, \\
f_i(x) &= x_i - e^{\cos(h(x_{i-1} + x_i + x_{i+1}))}, \text{ for } i = 2, ..., n-1, \\
f_n(x) &= x_n - e^{\cos(h(x_{n-1} + x_n))}, \\
h &= \frac{1}{n+1} \text{ and } \Psi = \mathbb{R}^n_+.
\end{aligned}$$

Problem 7 ([28]). *Nonsmooth Function*

$$f_i(x) = x_i - \sin|x_i - 1|, \ i = 1, 2, 3, ..., n.$$
$$and \ \Psi = \{x \in \mathbb{R}^n : \sum_{i=1}^{n} x_i \leq n, x_i \geq -1, i = 1, 2, \ldots, n\}.$$

Problem 8 ([23]). *Penalty 1*

$$t_i = \sum_{i=1}^{n} x_i^2, \quad c = 10^{-5}$$
$$f_i(x) = 2c(x_i - 1) + 4(t_i - 0.25)x_i, \ i = 1, 2, 3, ..., n.$$
$$and \ \Psi = \mathbb{R}^n_+.$$

Problem 9 ([29]). *Semismooth Function*

$$f_1(x) = x_1 + x_1^3 - 10,$$
$$f_2(x) = x_2 - x_3 + x_2^3 + 1,$$
$$f_3(x) = x_2 + x_3 + 2x_3^3 - 3,$$
$$f_4(x) = 2x_4^3,$$
$$and \ \Psi = \{x \in \mathbb{R}^4 : \sum_{i=1}^{4} x_i \leq 3, x_i \geq 0, i = 1, 2, 3, 4\}.$$

In addition, we employ the performance profile developed in Reference [30] to obtain Figures 1–3, which is a helpful process of standardizing the comparison of methods. The measure of the performance profile considered are; number of iterations, CPU time (in seconds) and number of function evaluations. Figure 1 reveals that Algorithm 1 most performs better in terms of number of iterations, as it solves and wins 90 percent of the problems with less number of iterations, while PCG and PDY solves and wins less than 10 percent. In Figure 2, Algorithm 1 performed a little less by solving and winning over 80 percent of the problems with less CPU time as against PCG and PDY with similar performance of less than 10 percent of the problems considered. The translation of Figure 3 is identical to Figure 1. Figure 4 is the plot of the decrease in residual norm against number of iterations on problem 9 with x_4 as initial point. It shows the speed of the convergence of each algorithm using the convergence tolerance 10^{-5}, it can be observed that Algorithm 1 converges faster than PCG and PDY.

Table 1. Numerical Results for **Algorithm 1 (DCG)**, **PCG** and **PDY** for Problem 1 with given initial points and dimensions.

		Algorithm 1				**PCG**				**PDY**			
DIMENSION	INITIAL POINT	ITER	FVAL	TIME	NORM	ITER	FVAL	TIME	NORM	ITER	FVAL	TIME	NORM
1000	x_1	11	49	0.025557	8.88×10^{-6}	18	73	0.019295	5.72×10^{-6}	12	49	0.16248	9.18×10^{-6}
	x_2	12	53	0.014164	4.78×10^{-6}	18	73	0.011648	9.82×10^{-6}	13	53	0.03780	6.35×10^{-6}
	x_3	12	53	0.008524	8.75×10^{-6}	19	77	0.011197	7.1×10^{-6}	14	57	0.01550	5.59×10^{-6}
	x_4	13	57	0.011333	6.68×10^{-6}	18	73	0.022197	8.27×10^{-6}	15	61	0.01746	4.07×10^{-6}
	x_5	13	57	0.014202	6.09×10^{-6}	63	254	0.046072	9.58×10^{-6}	14	57	0.02193	9.91×10^{-6}
	x_6	13	57	0.011045	8.14×10^{-6}	61	246	0.031608	9.15×10^{-6}	40	162	0.03472	9.70×10^{-6}
5000	x_1	12	53	0.024311	5.82×10^{-6}	18	73	0.11431	7.42×10^{-6}	13	53	0.03158	6.87×10^{-6}
	x_2	13	57	0.027361	3.13×10^{-6}	19	77	0.03997	6.53×10^{-6}	14	57	0.04270	4.62×10^{-6}
	x_3	13	57	0.02541	5.73×10^{-6}	20	81	0.056159	5.2×10^{-6}	15	61	0.05433	4.18×10^{-6}
	x_4	14	61	0.032038	4.38×10^{-6}	19	77	0.038381	8.1×10^{-6}	15	61	0.04357	9.08×10^{-6}
	x_5	14	61	0.039044	3.98×10^{-6}	62	250	0.15836	9.53×10^{-6}	15	61	0.08960	7.30×10^{-6}
	x_6	14	61	0.027231	5.33×10^{-6}	60	242	0.13276	9.1×10^{-6}	39	158	0.11284	9.86×10^{-6}
10,000	x_1	12	53	0.05434	8.23×10^{-6}	18	73	0.073207	9.5×10^{-6}	13	53	0.06371	9.70×10^{-6}
	x_2	13	57	0.045664	4.43×10^{-6}	19	77	0.090771	8.15×10^{-6}	14	57	0.06336	6.53×10^{-6}
	x_3	13	57	0.041922	8.09×10^{-6}	20	81	0.070859	6.74×10^{-6}	15	61	0.06414	5.90×10^{-6}
	x_4	14	61	0.047641	6.2×10^{-6}	20	81	0.087357	5.11×10^{-6}	16	65	0.07920	4.28×10^{-6}
	x_5	14	61	0.045734	5.62×10^{-6}	62	250	0.24646	8.87×10^{-6}	39	158	0.22101	7.97×10^{-6}
	x_6	14	61	0.057104	7.54×10^{-6}	59	238	0.19949	9.96×10^{-6}	87	351	0.36237	9.93×10^{-6}
50,000	x_1	13	57	0.16384	5.41×10^{-6}	19	77	0.25487	8.8×10^{-6}	14	57	0.27607	7.12×10^{-6}
	x_2	13	57	0.18633	9.9×10^{-6}	20	81	0.32689	7.39×10^{-6}	15	61	0.26220	4.91×10^{-6}
	x_3	14	61	0.20801	5.32×10^{-6}	21	85	0.33649	6.31×10^{-6}	16	65	0.28260	4.37×10^{-6}
	x_4	15	65	0.1946	4.08×10^{-6}	21	85	0.32779	5.1×10^{-6}	38	154	0.60650	7.54×10^{-6}
	x_5	15	65	0.19799	3.69×10^{-6}	61	246	0.82615	8.85×10^{-6}	177	712	2.52330	9.44×10^{-6}
	x_6	15	65	0.22418	4.95×10^{-6}	59	238	0.79992	8.5×10^{-6}	361	1449	5.97950	9.74×10^{-6}
100,000	x_1	13	57	0.32291	7.65×10^{-6}	20	81	0.53846	5.52×10^{-6}	15	61	0.39342	3.39×10^{-6}
	x_2	14	61	0.33329	4.12×10^{-6}	21	85	0.61533	4.62×10^{-6}	15	61	0.42154	6.94×10^{-6}
	x_3	14	61	0.37048	7.52×10^{-6}	21	85	0.53638	8.78×10^{-6}	16	65	0.45851	6.18×10^{-6}
	x_4	15	65	0.36058	5.76×10^{-6}	21	85	0.62002	7.21×10^{-6}	175	704	4.36100	9.47×10^{-6}
	x_5	15	65	0.34975	5.22×10^{-6}	60	242	1.4564	9.73×10^{-6}	176	708	4.29180	9.91×10^{-6}
	x_6	15	65	0.3621	7.01×10^{-6}	58	234	1.4155	9.42×10^{-6}	360	1445	9.71190	9.99×10^{-6}

Table 2. Numerical Results for **Algorithm 1 (DCG)**, **PCG** and **PDY** for Problem 2 with given initial points and dimensions.

		Algorithm 1				PCG				PDY			
DIMENSION	INITIAL POINT	ITER	FVAL	TIME	NORM	ITER	FVAL	TIME	NORM	ITER	FVAL	TIME	NORM
1000	x_1	9	38	3.1744	5.84×10^{-6}	15	59	0.049899	8.59×10^{-6}	10	39	0.01053	6.96×10^{-6}
	x_2	10	42	0.014633	6.25×10^{-6}	11	42	0.015089	9.07×10^{-6}	11	43	0.00937	9.23×10^{-6}
	x_3	9	38	0.017067	7.4×10^{-6}	17	66	0.016935	6.44×10^{-6}	13	51	0.01111	6.26×10^{-6}
	x_4	7	30	0.006392	6.53×10^{-6}	18	69	0.01436	6×10^{-6}	14	55	0.02154	9.46×10^{-6}
	x_5	11	46	0.011954	3.47×10^{-6}	13	48	0.00907	7.58×10^{-6}	15	59	0.01850	4.60×10^{-6}
	x_6	12	50	0.68666	6.74×10^{-6}	18	68	0.01352	5.4×10^{-6}	15	59	0.01938	7.71×10^{-6}
5000	x_1	10	42	0.11241	3.53×10^{-6}	16	63	0.041151	9.35×10^{-6}	11	43	0.03528	4.86×10^{-6}
	x_2	11	46	0.028723	3.81×10^{-6}	12	46	0.028706	8.8×10^{-6}	12	47	0.04032	6.89×10^{-6}
	x_3	10	42	0.029367	4.3×10^{-6}	18	70	0.047532	6.98×10^{-6}	14	55	0.04889	4.61×10^{-6}
	x_4	13	54	0.036231	3.67×10^{-6}	19	73	0.052164	6.45×10^{-6}	15	59	0.04826	6.96×10^{-6}
	x_5	11	46	0.04963	7.21×10^{-6}	14	52	0.040529	6.71×10^{-6}	16	63	0.05969	3.37×10^{-6}
	x_6	13	54	0.054971	4.05×10^{-6}	19	72	0.12303	5.71×10^{-6}	16	63	0.06253	5.64×10^{-6}
10,000	x_1	10	42	0.049614	4.98×10^{-6}	17	67	0.074779	6.6×10^{-6}	11	43	0.06732	6.85×10^{-6}
	x_2	11	46	0.061595	5.36×10^{-6}	13	50	0.08308	6.11×10^{-6}	12	47	0.12232	9.72×10^{-6}
	x_3	10	42	0.054587	6.02×10^{-6}	18	70	0.085554	9.83×10^{-6}	14	55	0.08288	6.51×10^{-6}
	x_4	13	54	0.073333	5.16×10^{-6}	19	73	0.10579	9.07×10^{-6}	15	59	0.08413	9.82×10^{-6}
	x_5	12	50	0.06306	2.83×10^{-6}	14	52	0.074982	9.18×10^{-6}	16	63	0.09589	4.75×10^{-6}
	x_6	13	54	0.062259	5.69×10^{-6}	19	72	0.099167	8.02×10^{-6}	16	64	0.11499	8.55×10^{-6}
50,000	x_1	11	46	0.20703	3.1×10^{-6}	18	71	0.39473	7.37×10^{-6}	12	47	0.27826	5.23×10^{-6}
	x_2	12	50	0.23251	3.35×10^{-6}	14	54	0.27346	6.74×10^{-6}	13	51	0.29642	7.11×10^{-6}
	x_3	11	46	0.21338	3.73×10^{-6}	20	78	0.37249	5.5×10^{-6}	15	59	0.35602	4.82×10^{-6}
	x_4	14	58	0.3232	3.22×10^{-6}	21	81	0.37591	5.07×10^{-6}	35	141	0.69470	6.69×10^{-6}
	x_5	12	50	0.22703	6.27×10^{-6}	16	60	0.26339	5.02×10^{-6}	35	141	0.68488	9.12×10^{-6}
	x_6	14	58	0.25979	3.54×10^{-6}	20	76	0.33814	8.93×10^{-6}	35	141	0.70973	9.91×10^{-6}
100,000	x_1	11	46	0.55511	4.38×10^{-6}	19	75	0.65494	5.22×10^{-6}	12	47	0.44541	7.39×10^{-6}
	x_2	12	50	0.54694	4.73×10^{-6}	14	54	0.4944	9.52×10^{-6}	14	55	0.53299	3.39×10^{-6}
	x_3	11	46	0.40922	5.27×10^{-6}	20	78	0.78319	7.78×10^{-6}	15	60	0.58603	8.71×10^{-6}
	x_4	14	58	0.62049	4.55×10^{-6}	21	81	0.76051	7.17×10^{-6}	72	290	2.70630	8.31×10^{-6}
	x_5	12	50	0.47039	8.86×10^{-6}	16	60	0.58545	7.07×10^{-6}	72	290	2.72220	8.68×10^{-6}
	x_6	14	58	0.71174	5.01×10^{-6}	21	80	0.77051	6.32×10^{-6}	72	290	2.75850	8.96×10^{-6}

Table 3. Numerical Results for **Algorithm 1 (DCG)**, **PCG** and **PDY** for Problem 3 with given initial points and dimensions.

		Algorithm 1 (DCG)				**PCG**				**PDY**			
DIMENSION	INITIAL POINT	ITER	FVAL	TIME	NORM	ITER	FVAL	TIME	NORM	ITER	FVAL	TIME	NORM
1000	x_1	10	43	0.75322	9.9×10^{-6}	19	76	0.55752	5.62×10^{-6}	12	48	0.01255	4.45×10^{-6}
	x_2	11	47	0.006933	5.46×10^{-6}	20	80	0.010936	5.58×10^{-6}	12	48	0.01311	9.02×10^{-6}
	x_3	12	51	0.00676	3.48×10^{-6}	21	84	0.011048	6.58×10^{-6}	13	52	0.01486	8.34×10^{-6}
	x_4	12	51	0.009664	4.41×10^{-6}	22	88	0.011058	5.67×10^{-6}	14	56	0.01698	8.04×10^{-6}
	x_5	11	47	0.010487	9.06×10^{-6}	22	88	0.012198	5.64×10^{-6}	14	56	0.01551	9.72×10^{-6}
	x_6	13	55	0.012702	3.15×10^{-6}	21	84	0.018231	8.36×10^{-6}	14	56	0.01534	9.42×10^{-6}
5000	x_1	11	47	0.019458	6.19×10^{-6}	20	80	0.040808	6.29×10^{-6}	12	48	0.03660	9.94×10^{-6}
	x_2	12	51	0.021562	3.42×10^{-6}	21	84	0.06688	6.25×10^{-6}	13	52	0.03616	6.85×10^{-6}
	x_3	12	51	0.024274	7.79×10^{-6}	22	88	0.04144	7.37×10^{-6}	14	56	0.04594	6.14×10^{-6}
	x_4	12	51	0.026771	9.86×10^{-6}	23	92	0.052214	6.35×10^{-6}	15	60	0.04342	6.01×10^{-6}
	x_5	12	51	0.026814	5.67×10^{-6}	23	92	0.041444	6.31×10^{-6}	15	60	0.04296	7.25×10^{-6}
	x_6	13	55	0.023903	7.03×10^{-6}	22	88	0.040135	9.37×10^{-6}	32	129	0.10081	8.85×10^{-6}
10,000	x_1	11	47	0.044134	8.75×10^{-6}	20	80	0.064312	8.9×10^{-6}	13	52	0.06192	4.77×10^{-6}
	x_2	12	51	0.051947	4.83×10^{-6}	21	84	0.088102	8.84×10^{-6}	13	52	0.06442	9.68×10^{-6}
	x_3	13	55	0.057291	3.08×10^{-6}	23	92	0.07296	5.22×10^{-6}	14	56	0.09499	8.69×10^{-6}
	x_4	13	55	0.055134	3.9×10^{-6}	23	92	0.075265	8.99×10^{-6}	15	60	0.07696	8.5×10^{-6}
	x_5	12	51	0.047551	8.02×10^{-6}	23	92	0.073937	8.93×10^{-6}	33	133	0.18625	6.45×10^{-6}
	x_6	13	55	0.055069	9.95×10^{-6}	23	92	0.099888	6.64×10^{-6}	33	133	0.15548	7.51×10^{-6}
50,000	x_1	12	51	0.19938	5.47×10^{-6}	21	84	0.27031	9.97×10^{-6}	14	56	0.23642	3.51×10^{-6}
	x_2	13	55	0.22499	3.02×10^{-6}	22	88	0.2657	9.9×10^{-6}	14	56	0.24813	7.12×10^{-6}
	x_3	13	55	0.19396	6.89×10^{-6}	24	96	0.3246	5.85×10^{-6}	15	60	0.27049	6.53×10^{-6}
	x_4	13	55	0.20259	8.72×10^{-6}	25	100	0.32373	5.04×10^{-6}	34	137	0.54545	7.13×10^{-6}
	x_5	13	55	0.19452	5.01×10^{-6}	25	100	0.33764	5.01×10^{-6}	68	274	1.02330	9.99×10^{-6}
	x_6	14	59	0.22015	6.22×10^{-6}	24	96	0.33687	7.44×10^{-6}	69	278	1.03810	8.05×10^{-6}
100,000	x_1	12	51	0.39983	7.74×10^{-6}	22	88	0.63809	7.06×10^{-6}	14	56	0.45475	4.96×10^{-6}
	x_2	13	55	0.32765	4.28×10^{-6}	23	92	0.63458	7.02×10^{-6}	15	60	0.49018	3.39×10^{-6}
	x_3	13	55	0.30133	9.75×10^{-6}	24	96	0.71422	8.27×10^{-6}	15	60	0.49016	9.24×10^{-6}
	x_4	14	59	0.42865	3.45×10^{-6}	25	100	0.73524	7.13×10^{-6}	139	559	4.03110	9.01×10^{-6}
	x_5	13	55	0.34512	7.09×10^{-6}	25	100	0.70625	7.09×10^{-6}	70	282	2.07100	8.54×10^{-6}
	x_6	14	59	0.40387	8.8×10^{-6}	25	100	0.76777	5.27×10^{-6}	139	559	4.02440	9.38×10^{-6}

Table 4. Numerical Results for **Algorithm 1 (DCG)**, **PCG** and **PDY** for Problem 4 with given initial points and dimensions.

		Algorithm 1				PCG				PDY			
DIMENSION	INITIAL POINT	ITER	FVAL	TIME	NORM	ITER	FVAL	TIME	NORM	ITER	FVAL	TIME	NORM
1000	x_1	10	43	0.15461	8.33×10^{-6}	18	72	0.11853	9.93×10^{-6}	12	48	0.00989	4.60×10^{-6}
	x_2	11	47	0.006276	3.84×10^{-6}	19	76	0.014318	8.75×10^{-6}	12	48	0.00966	9.57×10^{-6}
	x_3	11	47	0.009859	3.91×10^{-6}	20	80	0.0093776	7.15×10^{-6}	13	52	0.00887	8.49×10^{-6}
	x_4	11	47	0.007976	5.21×10^{-6}	47	189	0.023321	7.83×10^{-6}	12	48	0.01207	5.83×10^{-6}
	x_5	12	51	0.008382	4.09×10^{-6}	46	185	0.047105	9.76×10^{-6}	29	117	0.05371	9.43×10^{-6}
	x_6	12	51	0.008645	3.32×10^{-6}	41	165	0.027719	8.77×10^{-6}	29	117	0.02396	6.65×10^{-6}
5000	x_1	11	47	0.022024	5.21×10^{-6}	20	80	0.029445	5.57×10^{-6}	13	52	0.02503	3.49×10^{-6}
	x_2	11	47	0.020587	8.59×10^{-6}	20	80	0.033115	9.8×10^{-6}	13	52	0.02626	7.24×10^{-6}
	x_3	11	47	0.023714	8.75×10^{-6}	21	84	0.033318	8.01×10^{-6}	14	56	0.03349	6.29×10^{-6}
	x_4	12	51	0.024728	3.26×10^{-6}	49	197	0.071715	9.46×10^{-6}	13	52	0.02258	4.25×10^{-6}
	x_5	12	51	0.031015	9.14×10^{-6}	49	197	0.068565	8.68×10^{-6}	31	125	0.05471	7.59×10^{-6}
	x_6	12	51	0.030012	7.43×10^{-6}	44	177	0.070862	7.79×10^{-6}	63	254	0.10064	8.54×10^{-6}
10,000	x_1	11	47	0.041476	7.37×10^{-6}	20	80	0.043013	7.88×10^{-6}	13	52	0.03761	4.93×10^{-6}
	x_2	12	51	0.047866	3.4×10^{-6}	21	84	0.051685	6.94×10^{-6}	14	56	0.04100	3.37×10^{-6}
	x_3	12	51	0.042607	3.46×10^{-6}	22	88	0.050422	5.67×10^{-6}	14	56	0.03919	8.90×10^{-6}
	x_4	12	51	0.036406	4.61×10^{-6}	50	201	0.17563	9.84×10^{-6}	32	129	0.09613	6.02×10^{-6}
	x_5	13	55	0.041374	3.61×10^{-6}	50	201	0.20035	9.03×10^{-6}	32	129	0.09177	6.44×10^{-6}
	x_6	13	55	0.039847	2.94×10^{-6}	45	181	0.12214	8.11×10^{-6}	64	258	0.20791	9.39×10^{-6}
50,000	x_1	12	51	0.13928	4.61×10^{-6}	21	84	0.27145	8.83×10^{-6}	14	56	0.17193	3.63×10^{-6}
	x_2	12	51	0.18031	7.6×10^{-6}	22	88	0.23149	7.78×10^{-6}	14	56	0.15237	7.54×10^{-6}
	x_3	12	51	0.12526	7.74×10^{-6}	23	92	0.28789	6.36×10^{-6}	15	60	0.16549	6.66×10^{-6}
	x_4	13	55	0.14322	2.88×10^{-6}	53	213	0.61624	8.75×10^{-6}	67	270	0.76283	7.81×10^{-6}
	x_5	13	55	0.17904	8.08×10^{-6}	53	213	0.7119	8.02×10^{-6}	67	270	0.76157	8.80×10^{-6}
	x_6	13	55	0.13635	6.57×10^{-6}	47	189	0.48192	9.8×10^{-6}	269	1080	2.92510	9.41×10^{-6}
100,000	x_1	12	51	0.24293	6.52×10^{-6}	22	88	0.60822	6.25×10^{-6}	14	56	0.30229	5.13×10^{-6}
	x_2	13	55	0.27433	3.01×10^{-6}	23	92	0.52965	5.51×10^{-6}	15	60	0.31648	3.59×10^{-6}
	x_3	13	55	0.2714	3.06×10^{-6}	23	92	0.57064	8.99×10^{-6}	32	129	0.72838	9.99×10^{-6}
	x_4	13	55	0.26819	4.08×10^{-6}	54	217	1.1805	9.1×10^{-6}	135	543	2.86780	9.73×10^{-6}
	x_5	14	59	0.31696	3.2×10^{-6}	54	217	1.107	8.34×10^{-6}	272	1092	5.74140	9.91×10^{-6}
	x_6	13	55	0.2698	9.29×10^{-6}	49	197	1.0617	7.49×10^{-6}	548	2197	11.44130	9.87×10^{-6}

Table 5. Numerical Results for **Algorithm 1 (DCG)**, **PCG** and **PDY** for Problem 5 with given initial points and dimensions.

		Algorithm 1				PCG				PDY			
DIMENSION	INITIAL POINT	ITER	FVAL	TIME	NORM	ITER	FVAL	TIME	NORM	ITER	FVAL	TIME	NORM
1000	x_1	19	78	0.71709	8.63×10^{-6}	22	83	0.099338	7.48×10^{-6}	16	63	0.07575	6.03×10^{-6}
	x_2	21	86	0.017127	7.65×10^{-6}	23	88	0.016014	7.31×10^{-6}	16	63	0.01470	5.42×10^{-6}
	x_3	23	95	0.013909	7.23×10^{-6}	23	90	0.016328	9.31×10^{-6}	33	132	0.02208	6.75×10^{-6}
	x_4	22	92	0.0165	8.64×10^{-6}	49	197	0.030124	8.45×10^{-6}	30	121	0.01835	8.39×10^{-6}
	x_5	35	145	0.024702	8.26×10^{-6}	53	213	0.039321	8.38×10^{-6}	32	129	0.02700	8.47×10^{-6}
	x_6	43	182	0.027471	8.7×10^{-6}	46	185	0.033627	8.8×10^{-6}	30	121	0.01712	6.95×10^{-6}
5000	x_1	146	592	0.23803	9.45×10^{-6}	24	91	0.060158	6.36×10^{-6}	17	67	0.04394	5.64×10^{-6}
	x_2	21	86	0.04337	9.46×10^{-6}	25	95	0.060385	6.24×10^{-6}	17	67	0.04635	5.07×10^{-6}
	x_3	24	99	0.054619	8.27×10^{-6}	25	98	0.040015	5.86×10^{-6}	35	140	0.08311	9.74×10^{-6}
	x_4	24	100	0.066424	6.66×10^{-6}	53	213	0.098097	9.11×10^{-6}	33	133	0.08075	6.02×10^{-6}
	x_5	38	157	0.071222	9.28×10^{-6}	58	233	0.10958	8.56×10^{-6}	35	141	0.10091	7.51×10^{-6}
	x_6	45	190	0.090276	7.14×10^{-6}	50	201	0.21521	7.65×10^{-6}	32	129	0.08054	8.55×10^{-6}
10,000	x_1	211	853	0.60357	9.65×10^{-6}	25	95	0.076427	5.4×10^{-6}	17	67	0.06816	8.81×10^{-6}
	x_2	22	90	0.08012	4.98×10^{-6}	25	95	0.098461	8.9×10^{-6}	17	67	0.08833	7.80×10^{-6}
	x_3	25	103	0.089269	5.89×10^{-6}	25	98	0.07495	8.64×10^{-6}	37	148	0.14732	6.36×10^{-6}
	x_4	25	104	0.11781	5.54×10^{-6}	55	221	0.19048	9.11×10^{-6}	37	149	0.14293	8.25×10^{-6}
	x_5	40	165	0.15859	7.43×10^{-6}	60	241	0.19751	9.01×10^{-6}	36	145	0.14719	8.23×10^{-6}
	x_6	46	194	0.1728	8.62×10^{-6}	51	205	0.28882	9.62×10^{-6}	74	298	0.26456	7.79×10^{-6}
50,000	x_1	225	909	2.1373	9.93×10^{-6}	26	99	0.34575	6.75×10^{-6}	42	169	0.58113	7.78×10^{-6}
	x_2	23	94	0.31098	4.48×10^{-6}	27	103	0.43806	5.16×10^{-6}	42	169	0.58456	7.13×10^{-6}
	x_3	26	107	0.36293	6.83×10^{-6}	27	106	0.4815	5.28×10^{-6}	41	165	0.58717	8.87×10^{-6}
	x_4	26	108	0.32427	9.72×10^{-6}	60	241	0.90868	8.66×10^{-6}	40	161	0.56431	7.17×10^{-6}
	x_5	43	177	0.48938	9.47×10^{-6}	65	261	0.7924	9.05×10^{-6}	82	330	1.08920	8.44×10^{-6}
	x_6	50	210	0.69117	8.12×10^{-6}	56	225	0.72334	8.19×10^{-6}	80	322	1.06670	7.82×10^{-6}
100,000	x_1	231	933	4.2588	9.85×10^{-6}	26	99	0.71242	9.73×10^{-6}	43	173	1.09620	8.47×10^{-6}
	x_2	139	564	2.7266	9.96×10^{-6}	27	103	0.62746	7.39×10^{-6}	43	173	1.10040	7.77×10^{-6}
	x_3	26	107	0.57505	9.92×10^{-6}	27	106	0.82989	7.77×10^{-6}	42	169	1.08330	9.66×10^{-6}
	x_4	27	112	0.62227	8.52×10^{-6}	62	249	1.5474	9×10^{-6}	85	342	2.11880	9.22×10^{-6}
	x_5	45	185	0.8992	7.79×10^{-6}	67	269	1.6692	9.5×10^{-6}	84	338	2.10640	9.78×10^{-6}
	x_6	52	218	1.4318	7.37×10^{-6}	58	233	1.4333	8.32×10^{-6}	167	671	4.06200	9.90×10^{-6}

Table 6. Numerical Results for **Algorithm 1 (DCG)**, **PCG** and **PDY** for Problem 6 with given initial points and dimensions.

		Algorithm 1				PCG				PDY			
DIMENSION	INITIAL POINT	ITER	FVAL	TIME	NORM	ITER	FVAL	TIME	NORM	ITER	FVAL	TIME	NORM
1000	x_1	13	55	1.38	5.68×10^{-6}	23	92	0.4038	9.28×10^{-6}	15	60	0.01671	4.35×10^{-6}
	x_2	13	55	0.013339	5.47×10^{-6}	23	92	0.016325	8.92×10^{-6}	15	60	0.01346	4.18×10^{-6}
	x_3	13	55	0.066142	4.81×10^{-6}	23	92	0.023045	7.86×10^{-6}	15	60	0.01630	3.68×10^{-6}
	x_4	13	55	0.026838	3.3×10^{-6}	23	92	0.016172	5.38×10^{-6}	14	56	0.01339	7.48×10^{-6}
	x_5	12	51	0.009864	9.45×10^{-6}	22	88	0.03785	8.62×10^{-6}	14	56	0.01267	6.01×10^{-6}
	x_1	12	51	0.009881	5.57×10^{-6}	22	88	0.015013	5.08×10^{-6}	14	56	0.01685	3.54×10^{-6}
5000	x_1	14	59	0.042533	3.56×10^{-6}	25	100	0.061642	5.22×10^{-6}	15	60	0.05038	9.73×10^{-6}
	x_2	14	59	0.036648	3.43×10^{-6}	25	100	0.092952	5.02×10^{-6}	15	60	0.04775	9.36×10^{-6}
	x_3	14	59	0.043452	3.02×10^{-6}	24	96	0.068141	8.82×10^{-6}	15	60	0.04923	8.25×10^{-6}
	x_4	13	55	0.032579	7.38×10^{-6}	24	96	0.084625	6.04×10^{-6}	15	60	0.05793	5.64×10^{-6}
	x_5	13	55	0.03295	5.92×10^{-6}	23	92	0.086122	9.67×10^{-6}	15	60	0.04597	4.53×10^{-6}
	x_6	13	55	0.033062	3.49×10^{-6}	23	92	0.093318	5.7×10^{-6}	14	56	0.05070	7.93×10^{-6}
10,000	x_1	14	59	0.064917	5.04×10^{-6}	25	100	0.21424	7.38×10^{-6}	68	274	0.40724	9.06×10^{-6}
	x_2	14	59	0.069913	4.84×10^{-6}	25	100	0.13978	7.09×10^{-6}	68	274	0.41818	8.72×10^{-6}
	x_3	14	59	0.08473	4.27×10^{-6}	25	100	0.1731	6.25×10^{-6}	34	137	0.21905	6.22×10^{-6}
	x_4	14	59	0.075847	2.92×10^{-6}	24	96	0.14744	8.54×10^{-6}	15	60	0.10076	7.98×10^{-6}
	x_5	13	55	0.07974	8.38×10^{-6}	24	96	0.14169	6.85×10^{-6}	15	60	0.12680	6.40×10^{-6}
	x_6	13	55	0.063129	4.94×10^{-6}	23	92	0.15294	8.06×10^{-6}	15	60	0.11984	3.78×10^{-6}
50,000	x_1	15	63	0.25329	3.15×10^{-6}	26	104	0.64669	8.26×10^{-6}	143	575	3.09120	9.42×10^{-6}
	x_2	15	63	0.36394	3.03×10^{-6}	26	104	0.67717	7.95×10^{-6}	143	575	3.06200	9.06×10^{-6}
	x_3	14	59	0.2413	9.54×10^{-6}	26	104	0.5562	7×10^{-6}	142	571	3.04950	9.04×10^{-6}
	x_4	14	59	0.27502	6.53×10^{-6}	25	100	0.56171	9.56×10^{-6}	69	278	1.53920	9.14×10^{-6}
	x_5	14	59	0.36404	5.24×10^{-6}	25	100	0.57982	7.67×10^{-6}	68	274	1.49490	9.43×10^{-6}
	x_6	14	59	0.2506	3.09×10^{-6}	24	96	0.58645	9.03×10^{-6}	15	60	0.38177	8.44×10^{-6}
100,000	x_1	15	63	0.84781	4.45×10^{-6}	27	108	1.3215	5.86×10^{-6}	292	1172	13.59530	9.53×10^{-6}
	x_2	15	63	0.66663	4.28×10^{-6}	27	108	1.5062	5.63×10^{-6}	290	1164	13.30930	9.75×10^{-6}
	x_3	15	63	0.66683	3.77×10^{-6}	26	104	1.166	9.9×10^{-6}	144	579	6.68150	9.96×10^{-6}
	x_4	14	59	0.62697	9.24×10^{-6}	26	104	1.3961	6.78×10^{-6}	141	567	6.50800	9.92×10^{-6}
	x_5	14	59	0.62891	7.41×10^{-6}	26	104	1.2711	5.44×10^{-6}	70	282	3.30510	8.07×10^{-6}
	x_6	14	59	0.62422	4.37×10^{-6}	25	100	1.1685	6.4×10^{-6}	34	137	1.64510	6.37×10^{-6}

Table 7. Numerical Results for **Algorithm 1 (DCG)**, **PCG** and **PDY** for Problem 7 with given initial points and dimensions.

		Algorithm 1				PCG				PDY			
DIMENSION	INITIAL POINT	ITER	FVAL	TIME	NORM	ITER	FVAL	TIME	NORM	ITER	FVAL	TIME	NORM
1000	x_1	6	28	0.25689	2×10^{-6}	17	69	1.2275	6.98×10^{-6}	14	57	0.00953	5.28×10^{-6}
	x_2	6	28	0.008469	1.26×10^{-6}	15	61	0.23396	9.89×10^{-6}	13	53	0.00896	9.05×10^{-6}
	x_3	4	20	0.003619	9.25×10^{-6}	16	65	0.008095	5.79×10^{-6}	3	12	0.00426	8.47×10^{-6}
	x_4	5	24	0.004345	5.7×10^{-6}	16	65	0.010077	5.21×10^{-6}	15	61	0.01169	6.73×10^{-6}
	x_5	6	28	0.007146	4.42×10^{-6}	19	77	0.05354	4.95×10^{-6}	31	126	0.03646	9.03×10^{-6}
	x_6	6	27	0.004299	4.43×10^{-6}	18	72	0.025677	8.93×10^{-6}	15	60	0.01082	3.99×10^{-6}
5000	x_1	6	28	0.012915	4.47×10^{-6}	18	73	0.17722	7.6×10^{-6}	15	61	0.03215	4.25×10^{-6}
	x_2	6	28	0.012272	2.81×10^{-6}	17	69	0.027729	5.25×10^{-6}	14	57	0.02942	7.40×10^{-6}
	x_3	5	24	0.014669	1.16×10^{-6}	17	69	0.02985	6.31×10^{-6}	4	16	0.01107	1.01×10^{-7}
	x_4	6	28	0.012765	7.14×10^{-7}	17	69	0.028176	5.68×10^{-6}	16	65	0.04331	5.43×10^{-6}
	x_5	6	28	0.01331	9.89×10^{-6}	20	81	0.032213	5.39×10^{-6}	33	134	0.09379	7.78×10^{-6}
	x_6	6	27	0.015828	9.91×10^{-6}	19	76	0.044328	9.73×10^{-6}	15	60	0.04077	8.92×10^{-6}
10,000	x_1	6	28	0.022346	6.32×10^{-6}	19	77	0.17863	5.23×10^{-6}	15	61	0.06484	6.01×10^{-6}
	x_2	6	28	0.022669	3.97×10^{-6}	17	69	0.049242	7.42×10^{-6}	15	61	0.07734	3.77×10^{-6}
	x_3	5	24	0.039342	1.64×10^{-6}	17	69	0.048238	8.92×10^{-6}	4	16	0.02707	1.42×10^{-7}
	x_4	6	28	0.021017	1.01×10^{-6}	17	69	0.04807	8.03×10^{-6}	16	65	0.07941	7.69×10^{-6}
	x_5	7	32	0.031654	7.83×10^{-7}	20	81	0.063156	7.62×10^{-6}	34	138	0.14942	6.83×10^{-6}
	x_6	7	31	0.023456	7.85×10^{-7}	20	80	0.059438	6.7×10^{-6}	34	138	0.15224	8.81×10^{-6}
50,000	x_1	7	32	0.092452	7.91×10^{-7}	20	81	1.0808	5.7×10^{-6}	16	65	0.25995	4.89×10^{-6}
	x_2	6	28	0.1068	8.88×10^{-6}	18	73	0.32804	8.08×10^{-6}	15	61	0.24674	8.42×10^{-6}
	x_3	5	24	0.065684	3.66×10^{-6}	18	73	0.2189	9.71×10^{-6}	4	16	0.09405	3.18×10^{-7}
	x_4	6	28	0.10193	2.26×10^{-6}	18	73	0.3497	8.75×10^{-6}	36	146	0.55207	6.39×10^{-6}
	x_5	7	32	0.095676	1.75×10^{-6}	21	85	0.22595	8.3×10^{-6}	35	142	0.54679	9.05×10^{-6}
	x_6	7	31	0.092855	1.76×10^{-6}	21	84	0.22374	7.3×10^{-6}	36	146	0.55764	7.59×10^{-6}
100,000	x_1	7	32	0.17597	1.12×10^{-6}	20	81	2.1675	8.06×10^{-6}	17	69	0.52595	5.68×10^{-6}
	x_2	7	32	0.1741	7.03×10^{-7}	19	77	0.45553	5.57×10^{-6}	16	65	0.52102	4.34×10^{-6}
	x_3	5	24	0.17522	5.18×10^{-6}	19	77	0.43219	6.69×10^{-6}	4	16	0.14864	4.50×10^{-7}
	x_4	6	28	0.20785	3.19×10^{-6}	19	77	0.52259	6.03×10^{-6}	36	146	1.05360	9.04×10^{-6}
	x_5	7	32	0.23979	2.48×10^{-6}	22	89	0.6171	5.72×10^{-6}	74	299	2.10730	8.55×10^{-6}
	x_6	7	31	0.23128	2.48×10^{-6}	22	88	0.57384	5.03×10^{-6}	37	150	1.08240	6.66×10^{-6}

Table 8. Numerical Results for **Algorithm 1 (DCG)**, **PCG** and **PDY** for Problem 8 with given initial points and dimensions.

		Algorithm 1				PCG				PDY			
DIMENSION	INITIAL POINT	ITER	FVAL	TIME	NORM	ITER	FVAL	TIME	NORM	ITER	FVAL	TIME	NORM
1000	x_1	7	28	0.11495	3.03×10^{-6}	9	32	0.85797	7.6×10^{-6}	69	279	0.05538	8.95×10^{-6}
	x_2	7	28	0.005034	3.03×10^{-6}	9	32	0.034675	7.6×10^{-6}	270	1085	0.18798	9.72×10^{-6}
	x_3	7	28	0.006743	3.03×10^{-6}	9	32	0.005985	7.6×10^{-6}	24	52	0.02439	6.57×10^{-6}
	x_4	7	28	0.005856	3.03×10^{-6}	9	32	0.004808	7.6×10^{-6}	27	58	0.01520	7.59×10^{-6}
	x_5	7	28	0.004635	3.03×10^{-6}	9	32	0.015026	7.6×10^{-6}	28	61	0.04330	9.21×10^{-6}
	x_6	7	28	0.006487	3.03×10^{-6}	9	32	0.15778	7.6×10^{-6}	40	85	0.02116	8.45×10^{-6}
5000	x_1	5	22	0.009068	4.52×10^{-6}	7	26	0.67239	1.3×10^{-6}	658	2639	1.13030	9.98×10^{-6}
	x_2	5	22	0.009369	4.52×10^{-6}	7	26	0.010651	1.3×10^{-6}	27	58	0.05101	7.59×10^{-6}
	x_3	5	22	0.010895	4.52×10^{-6}	7	26	0.015758	1.3×10^{-6}	49	104	0.08035	8.11×10^{-6}
	x_4	5	22	0.014958	4.52×10^{-6}	7	26	0.014935	1.3×10^{-6}	40	85	0.07979	8.45×10^{-6}
	x_5	5	22	0.01507	4.52×10^{-6}	7	26	0.01524	1.3×10^{-6}	18	40	0.09128	9.14×10^{-6}
	x_6	5	22	0.008716	4.52×10^{-6}	7	26	0.1999	1.3×10^{-6}	17	38	0.18528	8.98×10^{-6}
10,000	x_1	6	27	0.031198	3.81×10^{-6}	5	19	0.044387	5.06×10^{-6}	49	104	0.20443	7.62×10^{-6}
	x_2	6	27	0.02098	3.81×10^{-6}	5	19	0.0223	5.06×10^{-6}	40	85	0.15801	8.45×10^{-6}
	x_3	6	27	0.01991	3.81×10^{-6}	5	19	0.018209	5.06×10^{-6}	19	42	0.37880	7.66×10^{-6}
	x_4	6	27	0.025402	3.81×10^{-6}	5	19	0.021654	5.06×10^{-6}	90	187	1.25802	9.7×10^{-6}
	x_5	6	27	0.025816	3.81×10^{-6}	5	19	0.017353	5.06×10^{-6}	988	1988	12.68259	9.93×10^{-6}
	x_6	6	27	0.025065	3.81×10^{-6}	5	19	0.019763	5.06×10^{-6}	27	58	0.32859	7.59×10^{-6}
50,000	x_1	4	21	0.083641	2.34×10^{-7}	8	33	0.42902	5.15×10^{-6}	19	42	0.52291	6.42×10^{-6}
	x_2	4	21	0.074156	2.34×10^{-7}	8	33	0.11525	5.15×10^{-6}	148	304	3.93063	9.92×10^{-6}
	x_3	4	21	0.078596	2.34×10^{-7}	8	33	0.14432	5.15×10^{-6}	937	1886	22.97097	9.87×10^{-6}
	x_4	4	21	0.078289	2.34×10^{-7}	8	33	0.11562	5.15×10^{-6}	27	58	0.68467	7.59×10^{-6}
	x_5	4	21	0.073535	2.34×10^{-7}	8	33	0.11674	5.15×10^{-6}	346	702	8.45043	9.79×10^{-6}
	x_6	4	21	0.081909	2.34×10^{-7}	8	33	0.10486	5.15×10^{-6}	40	85	0.99230	8.45×10^{-6}
100,000	x_1	4	22	0.1663	6.25×10^{-6}	6	25	1.2922	6.81×10^{-7}	-	-	-	-
	x_2	4	22	0.15147	6.25×10^{-6}	6	25	0.18839	6.81×10^{-7}	-	-	-	-
	x_3	4	22	0.15582	6.25×10^{-6}	6	25	0.16153	6.81×10^{-7}	-	-	-	-
	x_4	4	22	0.15465	6.25×10^{-6}	6	25	0.17397	6.81×10^{-7}	-	-	-	-
	x_5	4	22	0.16744	6.25×10^{-6}	6	25	0.18586	6.81×10^{-7}	-	-	-	-
	x_6	4	22	0.1687	6.25×10^{-6}	6	25	0.17938	6.81×10^{-7}	-	-	-	-

Table 9. Numerical Results for **Algorithm 1 (DCG)**, **PCG** and **PDY** for Problem 9 with given initial points and dimensions.

		Algorithm 1				PCG				PDY			
DIMENSION	INITIAL POINT	ITER	FVAL	TIME	NORM	ITER	FVAL	TIME	NORM	ITER	FVAL	TIME	NORM
4	x_1	51	215	0.23665	9.01×10^{-6}	79	321	0.5978	9.76×10^{-6}	59	241	0.71268	9.36×10^{-6}
	x_2	51	215	0.04968	9.99×10^{-6}	77	313	0.016326	9.85×10^{-6}	58	237	0.045441	9.73×10^{-6}
	x_3	53	223	0.017211	9.46×10^{-6}	80	325	0.16529	9.38×10^{-6}	59	241	0.019552	9.9×10^{-6}
	x_4	53	223	0.019004	9.68×10^{-6}	83	337	0.041713	9.57×10^{-6}	62	253	0.022007	8.07×10^{-6}
	x_5	57	239	0.023447	8.87×10^{-6}	81	329	0.11972	9.04×10^{-6}	61	249	0.040117	8.36×10^{-6}
	x_6	54	227	0.020832	9.31×10^{-6}	82	333	0.016127	9.3×10^{-6}	61	249	0.017374	9.18×10^{-6}

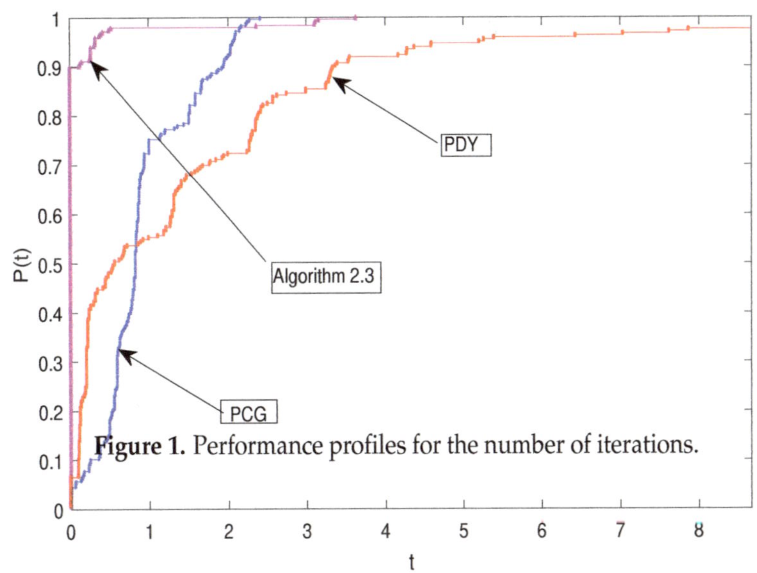

Figure 1. Performance profiles for the number of iterations.

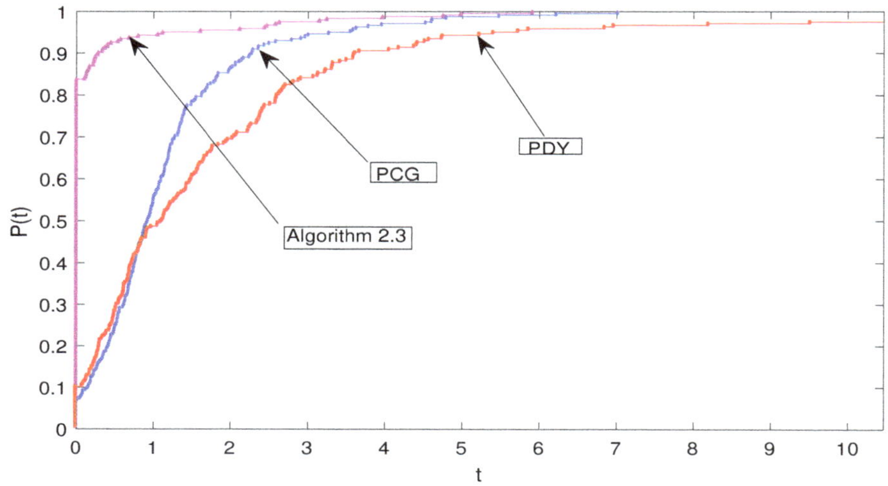

Figure 2. Performance profiles for the CPU time (in seconds).

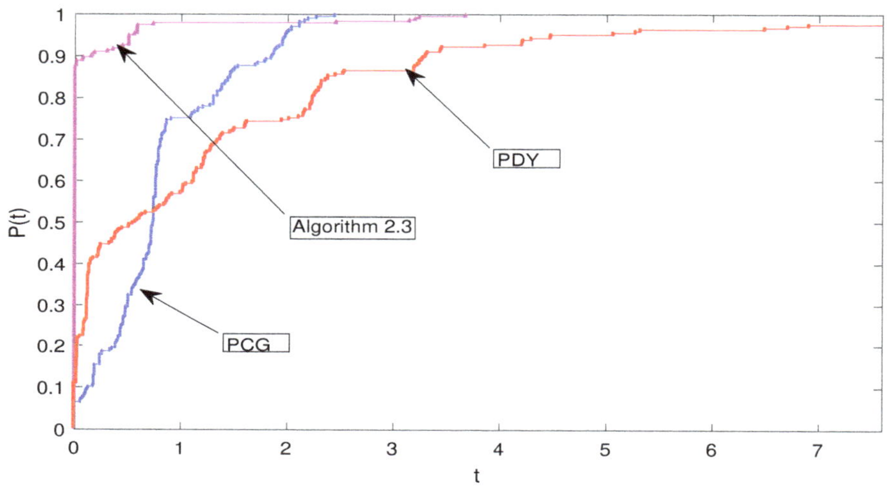

Figure 3. Performance profiles for the number of function evaluations.

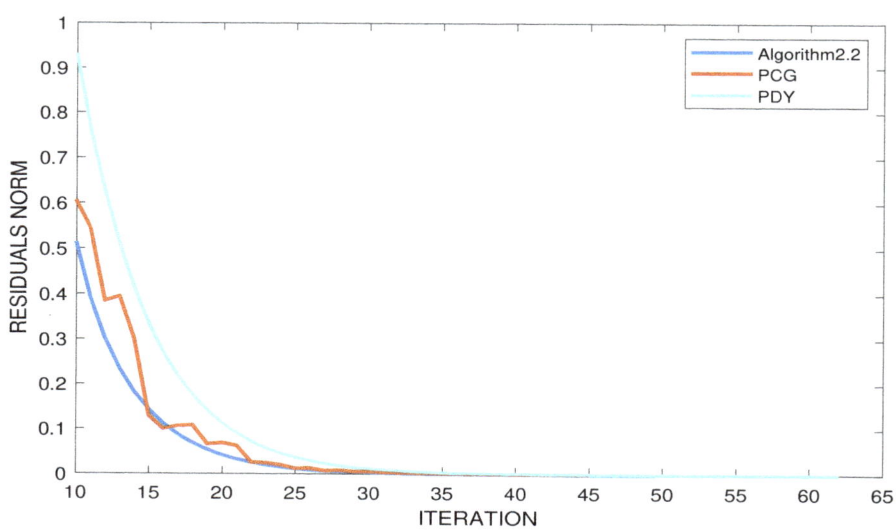

Figure 4. Convergence histories of **Algorithm** 1, **PCG** and **PDY** on Problem 9.

Applications in Compressive Sensing

There are many problems in signal processing and statistical inference involving finding sparse solutions to ill-conditioned linear systems of equations. Among popular approach is minimizing an objective function which contains quadratic (ℓ_2) error term and a sparse ℓ_1-regularization term, that is,

$$\min_x \frac{1}{2}\|y - Bx\|_2^2 + \eta\|x\|_1, \tag{24}$$

where $x \in R^n$, $y \in R^k$ is an observation, $B \in R^{k\times n}$ ($k << n$) is a linear operator, η is a non-negative parameter, $\|x\|_2$ denotes the Euclidean norm of x and $\|x\|_1 = \sum_{i=1}^n |x_i|$ is the ℓ_1-norm of x. It is easy to see that problem (24) is a convex unconstrained minimization problem. Due to the fact that if the original signal is sparse or approximately sparse in some orthogonal basis, problem (24) frequently appears in compressive sensing and hence an exact restoration can be produced by solving (24).

Iterative methods for solving (24) have been presented in many papers (see References [5,31–35]). The most popular method among these methods is the gradient based method and the earliest gradient projection method for sparse reconstruction (GPRS) was proposed by Figueiredo et al. [5]. The first step of the GPRS method is to express (24) as a quadratic problem using the following process. Let $x \in \mathbb{R}^n$ and splitting it into its positive and negative parts. Then x can be formulated as

$$x = u - v, \qquad u \geq 0, \quad v \geq 0,$$

where $u_i = (x_i)_+, v_i = (-x_i)_+$ for all $i = 1, 2, ..., n$ and $(.)_+ = \max\{0, .\}$. By definition of ℓ_1-norm, we have $\|x\|_1 = e_n^T u + e_n^T v$, where $e_n = (1, 1, ..., 1)^T \in R^n$. Now (24) can be written as

$$\min_{u,v} \frac{1}{2}\|y - B(u - v)\|_2^2 + \eta e_n^T u + \eta e_n^T v, \qquad u \geq 0, \quad v \geq 0, \tag{25}$$

which is a bound-constrained quadratic program. However, from Reference [5], Equation (25) can be written in standard form as

$$\min_z \frac{1}{2} z^T D z + c^T z, \qquad \text{such that} \quad z \geq 0, \tag{26}$$

where $z = \begin{pmatrix} u \\ v \end{pmatrix}$, $c = \omega e_{2n} + \begin{pmatrix} -b \\ b \end{pmatrix}$, $b = B^T y$, $D = \begin{pmatrix} B^T B & -B^T B \\ -B^T B & B^T B \end{pmatrix}$.

Clearly, D is a positive semi-definite matrix, which implies that Equation (26) is a convex quadratic problem.

Xiao et al. [19] translated (26) into a linear variable inequality problem which is equivalent to a linear complementarity problem. Furthermore, it was noted that z is a solution of the linear complementarity problem if and only if it is a solution of the nonlinear equation:

$$F(z) = \min\{z, Dz + c\} = 0. \tag{27}$$

The function F is a vector-valued function and the "min" is interpreted as component-wise minimum. It was proved in References [36,37] that $F(z)$ is continuous and monotone. Therefore problem (24) can be translated into problem (1) and thus Algorithm 1 (DCG) can be applied to solve it.

In this experiment, we consider a simple compressive sensing possible situation, where our goal is to restore a blurred image. We use the following well-known gray test images; (P1) Cameraman, (P2) Lena, (P3) House and (P4) Peppers for the experiments. We use 4 different Gaussian blur kernals with standard deviation σ to compare the robustness of DCG method with CGD method proposed in Reference [19]. CGD method is an extension of the well-known conjugate gradient method for unconstrained optimization CG-DESCENT [20] to solve the ℓ_1-norm regularized problems.

To access the performance of each algorithm tested with respect to metrics that indicate a better quality of restoration, in Table 10 we reported the number of iterations, the objective function (ObjFun) value at the approximate solution, the mean of squared error (MSE) to the original image $\tilde{x}$,

$$MSE = \frac{1}{n}\|\tilde{x} - x_*\|^2,$$

where x_* is the reconstructed image and the signal-to-noise-ratio (SNR) which is defined as

$$\text{SNR} = 20 \times \log_{10}(\frac{\|\bar{x}\|}{\|x - \bar{x}\|}).$$

We also reported the structural similarity (SSIM) index that measure the similarity between the original image and the restored image [38]. The MATLAB implementation of the SSIM index can be obtained at http://www.cns.nyu.edu/~lcv/ssim/.

Table 10. Efficiency comparison based on the value of the number of iterations (Iter), objective function (ObjFun) value, mean-square-error (MSE) and signal-to-noise-ratio (SNR) under different Pi (σ).

Image	Iter		ObjFun		MSE		SNR	
	DCG	CGD	DCG	CGD	DCG	CGD	DCG	CGD
P1(1E-8)	8	9	4.397×10^3	4.398×10^3	3.136×10^{-2}	3.157×10^{-2}	9.42	9.39
P1(1E-1)	8	9	4.399×10^3	4.401×10^3	3.147×10^{-2}	3.163×10^{-2}	9.40	9.38
P1(0.11)	11	8	4.428×10^3	4.432×10^3	3.229×10^{-2}	3.232×10^{-2}	9.29	9.29
P1(0.25)	12	8	4.468×10^3	4.473×10^3	3.365×10^{-2}	3.289×10^{-2}	9.11	9.21
P1(1E-8)	9	9	4.555×10^3	4.556×10^3	3.287×10^{-2}	3.3412×10^{-2}	9.14	9.07
P1(1E-1)	9	9	4.558×10^3	4.559×10^3	3.298×10^{-2}	3.348×10^{-2}	9.12	9.06
P1(0.11)	12	12	4.588×10^3	4.591×10^3	3.416×10^{-2}	3.446×10^{-2}	8.97	8.93
P1(0.25)	7	8	4.628×10^3	4.630×10^3	3.621×10^{-2}	3.500×10^{-2}	8.72	8.86
P1(1E-8)	9	9	5.179×10^3	5.179×10^3	3.209×10^{-2}	3.3259×10^{-2}	10.03	9.96
P1(1E-1)	9	9	5.182×10^3	5.182×10^3	3.231×10^{-2}	3.267×10^{-2}	10.00	9.95
P1(0.11)	7	9	5.209×10^3	5.209×10^3	3.436×10^{-2}	3.344×10^{-2}	9.73	9.85
P1(0.25)	10	8	5.250×10^3	5.254×10^3	3.557×10^{-2}	3.438×10^{-2}	9.58	9.73
P1(1E-8)	9	9	4.388×10^3	4.389×10^3	3.299×10^{-2}	3.335×10^{-2}	9.03	8.99
P1(1E-1)	9	9	4.391×10^3	4.393×10^3	3.308×10^{-2}	3.340×10^{-2}	9.02	8.98
P1(0.11)	12	8	4.421×10^3	4.424×10^3	3.425×10^{-2}	3.411×10^{-2}	8.87	8.89
P1(0.25)	7	8	4.461×10^3	4.463×10^3	3.621×10^{-2}	3.483×10^{-2}	8.63	8.80

The original, blurred and restored images by each of the algorithm are given in Figures 5–8. The figures demonstrate that both the two tested algorithm can restored the blurred images. It can be observed from Table 10 and Figures 5–8 that Algorithm 1 (DCG) compete with the CGD algorithm, therefore, it can be used as an alternative to CGD for restoring blurred image.

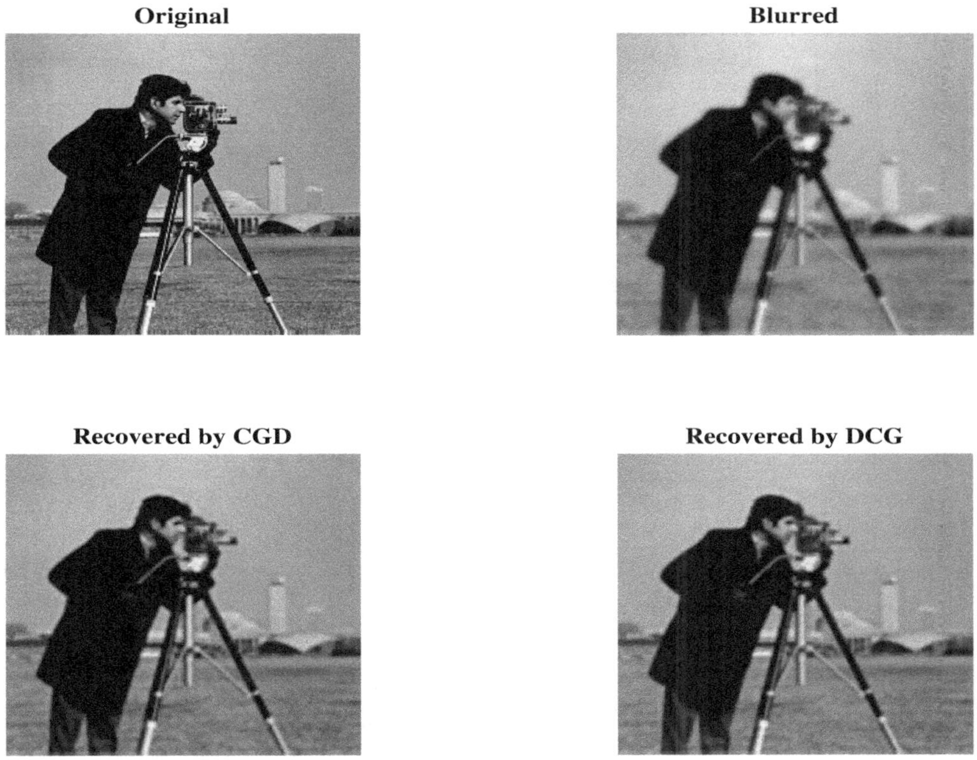

Figure 5. The original image (**top left**), the blurred image (**top right**), the restored image by CGD (**bottom left**) with SNR = 20.05, SSIM = 0.83 and by DCG (**bottom right**) with SNR = 20.12, SSIM = 0.83.

Figure 6. The original image (**top left**), the blurred image (**top right**), the restored image by CGD (**bottom left**) with SNR = 22.93, SSIM = 0.87 and by DCG (**bottom right**) with SNR = 24.36, SSIM = 0.90.

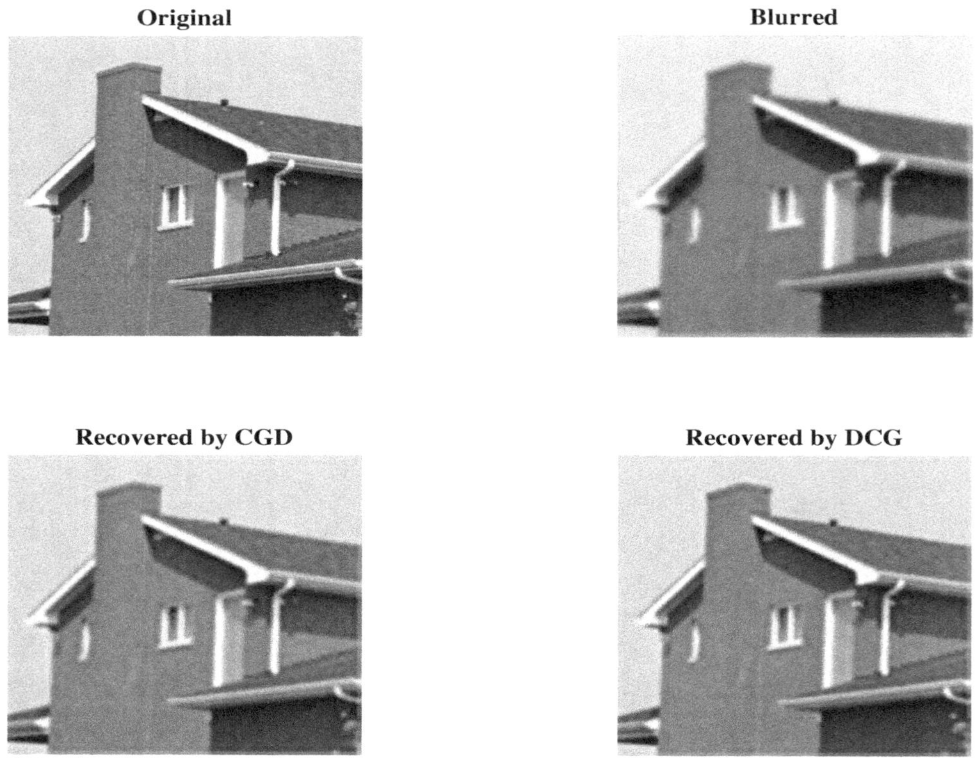

Figure 7. The original image (**top left**), the blurred image (**top right**), the restored image by CGD (**bottom left**) with SNR = 25.65, SSIM = 0.86 and by DCG (**bottom right**) with SNR = 26.37, SSIM = 0.87.

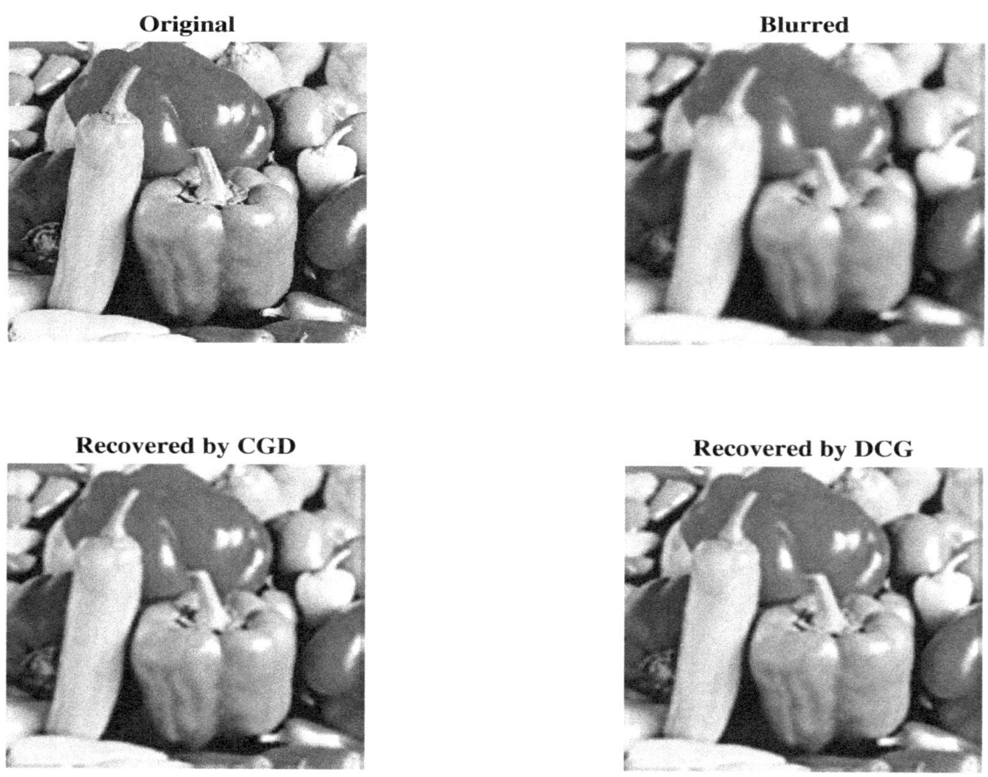

Figure 8. The original image (**top left**), the blurred image (**top right**), the restored image by CGD (**bottom left**) with SNR = 21.50, SSIM = 0.84 and by DCG (**bottom right**) with SNR = 21.81, SSIM = 0.85.

4. Conclusions

In this research article, we present a CG method which possesses the sufficient descent property for solving constrained nonlinear monotone equations. The proposed method has the ability to solve non-smooth equations as it does not require matrix storage and Jacobian information of the nonlinear equation under consideration. The sequence of iterates generated converge the solution under appropriate assumptions. Finally, we give some numerical examples to display the efficiency of the proposed method in terms of number of iterations, CPU time and number of function evaluations compared with some related methods for solving convex constrained nonlinear monotone equations and its application in image restoration problems.

Author Contributions: conceptualization, A.B.A.; methodology, A.B.A.; software, H.M.; validation, P.K. and A.M.A.; formal analysis, P.K. and H.M.; investigation, P.K. and A.M.A.; resources, P.K.; data curation, A.B.A. and H.M.; writing—original draft preparation, A.B.A.; writing—review and editing, H.M.; visualization, A.M.A.; supervision, P.K.; project administration, P.K.; funding acquisition, P.K.

Acknowledgments: We thank Associate Professor Jin Kiu Liu for providing us with the access of the CGD-CS MATLAB codes. The authors acknowledge the financial support provided by King Mongkut's University of Technology Thonburi through the "KMUTT 55th Anniversary Commemorative Fund". This project is supported by the theoretical and computational science (TaCS) center under computational and applied science for smart research innovation (CLASSIC), Faculty of Science, KMUTT. The first author was supported by the "Petchra Pra Jom Klao Ph.D. Research Scholarship from King Mongkut's University of Technology Thonburi".

References

1. Gu, B.; Sheng, V.S.; Tay, K.Y.; Romano, W.; Li, S. Incremental support vector learning for ordinal regression. *IEEE Trans. Neural Netw. Learn. Syst.* **2015**, *26*, 1403–1416. [CrossRef] [PubMed]
2. Li, J.; Li, X.; Yang, B.; Sun, X. Segmentation-based image copy-move forgery detection scheme. *IEEE Trans. Inf. Forensics Secur.* **2015**, *10*, 507–518.
3. Wen, X.; Shao, L.; Xue, Y.; Fang, W. A rapid learning algorithm for vehicle classification. *Inf. Sci.* **2015**, *295*, 395–406. [CrossRef]
4. Michael, S.V.; Alfredo, I.N. Newton-type methods with generalized distances for constrained optimization. *Optimization* **1997**, *41*, 257–278.
5. Figueiredo, M.A.T.; Nowak, R.D.; Wright, S.J. Gradient projection for sparse reconstruction: Application to compressed sensing and other inverse problems. *IEEE J. Sel. Top. Signal Process.* **2007**, *1*, 586–597. [CrossRef]
6. Magnanti, T.L.; Perakis, G. Solving variational inequality and fixed point problems by line searches and potential optimization. *Math. Program.* **2004**, *101*, 435–461. [CrossRef]
7. Pan, Z.; Zhang, Y.; Kwong, S. Efficient motion and disparity estimation optimization for low complexity multiview video coding. *IEEE Trans. Broadcast.* **2015**, *61*, 166–176.
8. Xia, Z.; Wang, X.; Sun, X.; Wang, Q. A secure and dynamic multi-keyword ranked search scheme over encrypted cloud data. *IEEE Trans. Parallel Distrib. Syst.* **2016**, *27*, 340–352. [CrossRef]
9. Zheng, Y.; Jeon, B.; Xu, D.; Wu, Q.M.; Zhang, H. Image segmentation by generalized hierarchical fuzzy c-means algorithm. *J. Intell. Fuzzy Syst.* **2015**, *28*, 961–973.
10. Solodov, M.V.; Svaiter, B.F. A globally convergent inexact newton method for systems of monotone equations. In *Reformulation: Nonsmooth, Piecewise Smooth, Semismooth and Smoothing Methods*; Springer: Dordrecht, The Netherlands, 1998; pp. 355–369.
11. Mohammad, H.; Abubakar, A.B. A positive spectral gradient-like method for nonlinear monotone equations. *Bull. Comput. Appl. Math.* **2017**, *5*, 99–115.
12. Zhang, L.; Zhou, W. Spectral gradient projection method for solving nonlinear monotone equations. *J. Comput. Appl. Math.* **2006**, *196*, 478–484. [CrossRef]
13. Zhou, W.J.; Li, D.H. A globally convergent BFGS method for nonlinear monotone equations without any merit functions. *Math. Comput.* **2008**, *77*, 2231–2240. [CrossRef]
14. Zhou, W.; Li, D. Limited memory BFGS method for nonlinear monotone equations. *J. Comput. Math.* **2007**, *25*, 89–96.
15. Abubakar, A.B.; Waziria, M.Y. A matrix-free approach for solving systems of nonlinear equations. *J. Mod. Methods Numer. Math.* **2016**, *7*, 1–9. [CrossRef]

16. Abubakar, A.B.; Kumam, P. An improved three-term derivative-free method for solving nonlinear equations. *Comput. Appl. Math.* **2018**, *37*, 6760–6773. [CrossRef]
17. Abubakar, A.B.; Kumam, P.; Awwal, A.M. A descent dai-liao projection method for convex constrained nonlinear monotone equations with applications. *Thai J. Math.* **2018**, *17*, 128–152.
18. Wang, C.; Wang, Y.; Xu, C. A projection method for a system of nonlinear monotone equations with convex constraints. *Math. Methods Oper. Res.* **2007**, *66*, 33–46. [CrossRef]
19. Xiao, Y.; Zhu, H. A conjugate gradient method to solve convex constrained monotone equations with applications in compressive sensing. *J. Math. Anal. Appl.* **2013**, *405*, 310–319. [CrossRef]
20. Hager, W.; Zhang, H. A new conjugate gradient method with guaranteed descent and an efficient line search. *SIAM J. Optim.* **2005**, *16*, 170–192. [CrossRef]
21. Liu, S.-Y.; Huang, Y.-Y.; Jiao, H.-W. Sufficient descent conjugate gradient methods for solving convex constrained nonlinear monotone equations. *Abstr. Appl. Anal.* **2014**, *2014*, 305643. [CrossRef]
22. Liu, J.K.; Li, S.J. A projection method for convex constrained monotone nonlinear equations with applications. *Comput. Math. Appl.* **2015**, *70*, 2442–2453. [CrossRef]
23. Ding, Y.; Xiao, Y.; Li, J. A class of conjugate gradient methods for convex constrained monotone equations. *Optimization* **2017**, *66*, 2309–2328. [CrossRef]
24. Liu, J.; Feng, Y. A derivative-free iterative method for nonlinear monotone equations with convex constraints. *Numer. Algorithms* **2018**, 1–18. [CrossRef]
25. Muhammed, A.A.; Kumam, P.; Abubakar, A.B.; Wakili, A.; Pakkaranang, N. A new hybrid spectral gradient projection method for monotone system of nonlinear equations with convex constraints. *Thai J. Math.* **2018**, *16*, 125–147.
26. La Cruz, W.; Martínez, J.; Raydan, M. Spectral residual method without gradient information for solving large-scale nonlinear systems of equations. *Math. Comput.* **2006**, *75*, 1429–1448. [CrossRef]
27. Bing, Y.; Lin, G. An efficient implementation of Merrill's method for sparse or partially separable systems of nonlinear equations. *SIAM J. Optim.* **1991**, *1*, 206–221. [CrossRef]
28. Yu, Z.; Lin, J.; Sun, J.; Xiao, Y.H.; Liu, L.Y.; Li, Z.H. Spectral gradient projection method for monotone nonlinear equations with convex constraints. *Appl. Numer. Math.* **2009**, *59*, 2416–2423. [CrossRef]
29. Yamashita, N.; Fukushima, M. Modified Newton methods for solving a semismooth reformulation of monotone complementarity problems. *Math. Program.* **1997**, *76*, 469–491. [CrossRef]
30. Dolan, E.D.; Moré, J.J. Benchmarking optimization software with performance profiles. *Math. Program.* **2002**, *91*, 201–213. [CrossRef]
31. Figueiredo, M.A.T.; Nowak, R.D. An EM algorithm for wavelet-based image restoration. *IEEE Trans. Image Process.* **2003**, *12*, 906–916. [CrossRef]
32. Hale, E.T.; Yin, W.; Zhang, Y. *A Fixed-Point Continuation Method for ℓ_1-Regularized Minimization with Applications to Compressed Sensing*; CAAM TR07-07; Rice University: Houston, TX, USA, 2007; pp. 43–44.
33. Beck, A.; Teboulle, M. A fast iterative shrinkage-thresholding algorithm for linear inverse problems. *SIAM J. Imaging Sci.* **2009**, *2*, 183–202. [CrossRef]
34. Van Den Berg, E.; Friedlander, M.P. Probing the pareto frontier for basis pursuit solutions. *SIAM J. Sci. Comput.* **2008**, *31*, 890–912. [CrossRef]
35. Birgin, E.G.; Martínez, J.M.; Raydan, M. Nonmonotone spectral projected gradient methods on convex sets. *SIAM J. Optim.* **2000**, *10*, 1196–1211. [CrossRef]
36. Xiao, Y.; Wang, Q.; Hu, Q. Non-smooth equations based method for ℓ_1-norm problems with applications to compressed sensing. *Nonlinear Anal. Theory Methods Appl.* **2011**, *74*, 3570–3577. [CrossRef]
37. Pang, J.-S. Inexact Newton methods for the nonlinear complementarity problem. *Math. Program.* **1986**, *36*, 54–71. [CrossRef]
38. Wang, Z.; Bovik, A.C.; Sheikh, H.R.; Simoncelli, E.P. Image quality assessment: From error visibility to structural similarity. *IEEE Trans. Image Process.* **2004**, *13*, 600–612. [CrossRef]

11

How to Obtain Global Convergence Domains via Newton's Method for Nonlinear Integral Equations

José Antonio Ezquerro * and Miguel Ángel Hernández-Verón

Department of Mathematics and Computation, University of La Rioja, Calle Madre de Dios, 53, 26006 Logrono, Spain; mahernan@unirioja.es
* Correspondence: jezquer@unirioja.es

Abstract: We use the theoretical significance of Newton's method to draw conclusions about the existence and uniqueness of solution of a particular type of nonlinear integral equations of Fredholm. In addition, we obtain a domain of global convergence for Newton's method.

Keywords: Fredholm integral equation; Newton's method; global convergence

---

1. Introduction

Integral equations are very common in physics and engineering, since a lot of problems of these disciplines can be reduced to solve an integral equation. In general, we cannot solve integral equations exactly and are forced to obtain approximate solutions. For this, different numerical methods can be used. So, for example, iterative schemes based on the homotopy analysis method in [1], adapted Newton-Kantorovich schemes in [2] and schemes based on a combination of the Newton-Kantorovich method and quadrature methods in [3]. Besides, techniques based on using iterative methods are also interesting, since the theoretical significance of the methods allows drawing conclusions about the existence and uniqueness of solution of the equations. The use of an iterative method allows approximating a solution and, by analysing the convergence, proving the existence of solution, locating a solution and even separating such solution from other possible solutions by means of results of uniqueness. The theory of fixed point plays an important role in the development of iterative methods for approximating, in general, a solution of an equation and, in particular, for approximating a solution of an integral equation.

In this work, we pay attention to the study of nonlinear Fredholm integral equations with nonlinear Nemytskii operators of type

$$x(s) = \ell(s) + \lambda \int_a^b \mathcal{K}(s,t)\mathcal{H}(x)(t)\,dt, \quad s \in [a,b], \quad \lambda \in \mathbb{R}, \tag{1}$$

where $\ell(s) \in \mathcal{C}[a,b]$, kernel $\mathcal{K}(s,t)$ of integral equation is a known function in $[a,b] \times [a,b]$, $\mathcal{H}$ is a Nemytskii operator [4] given by $\mathcal{H} : \Omega \subseteq \mathcal{C}[a,b] \to \mathcal{C}[a,b]$, such that $\mathcal{H}(x)(t) = H(x(t))$ and $H : \mathbb{R} \to \mathbb{R}$ is a derivable real function, and $x(s) \in \mathcal{C}[a,b]$ is the unknown function to find.

It is common to use the Banach Fixed Point Theorem [5–7] to prove the existence of a unique fixed point of an operator and approximate it by the method of successive approximations. Moreover, global convergence for the method is obtained in the full space. For this, we use that the operator involved is a contraction.

Our main aim of this work is to do a study of integral Equation (1) from Newton's method,

$$x_{n+1} = x_n - [F'(x_n)]^{-1}F(x_n), \quad n \geq 0, \quad \text{with } x_0 \text{ given,}$$

that has quadratic convergence, superior to the convergence of the method of successive approximations, which is linear. This study is similar to that of the Fixed Point Theorem for the method of successive approximations. In addition, we obtain a domain of global convergence, $B(\widetilde{x}, R) = \{x \in \mathcal{C}[a,b] : \|x - \widetilde{x}\| < R\}$, with $\widetilde{x} \in \mathcal{C}[a,b]$, for Newton's method. Also, we obtain a result of uniqueness of solution that separate the approximate solution from other possible solutions. To carry out this study, we develop a technique based on the use of auxiliary points, which allows obtaining domains of global convergence, locating solutions of (1) and domains of uniqueness of these solutions.

On the other hand, if $\mathcal{H}(x) = x$, integral Equation (1) is linear and well-known, it is a Fredholm integral equation of the second kind, which is connected with the eigenvalue problem represented by the homogeneous equation

$$x(s) = \lambda \int_a^b \mathcal{K}(s,t)x(t)\,dt, \quad s \in [a,b],$$

and has non-trivial solutions $x(s) \not\equiv 0$ for the characteristic values or eigenvalues λ (the latter term is sometimes reserved to the reciprocals $\nu = 1/\lambda$) of kernel $\mathcal{K}(s,t)$ and every non-trivial solution of (1) is called characteristic function or eigenfunction corresponding to characteristic value λ. If Equation (1) is nonlinear, our results allow doing a study of the equation based on the values of parameter λ, which is another important aim of our work.

2. Global Convergence and Uniqueness of Solution

If we are interested in proving the convergence of an iteration, we can usually follow three ways to do it: local convergence, semilocal convergence and global convergence. First, from some conditions on the operator involved, if we require conditions to the solution x^*, we establish a local analysis of convergence and obtain a ball of convergence of the iteration, which, from the initial approximation x_0 lying in the ball, shows the accessibility to x^*. Second, from some conditions on the operator involved, if we require conditions to the initial iterate x_0, we establish a semilocal analysis of convergence and obtain a domain of parameters, which corresponds to the conditions required to the initial iterate, so that the convergence of iteration is guaranteed to x^*. Third, from some conditions on the operator involved, the convergence of iteration to x^* in a domain, and independently of the initial approximation x_0, is established and global convergence is called. Observe that the three studies require conditions on the operator involved and requirement of conditions to the solution, to the initial approximation, or to none of these, is what determines the way of analysis.

The local analysis of the convergence has the disadvantage that it requires conditions on the solution and this is unknown. The global analysis of convergence, as a consequence of the absence of conditions on the initial approximations and the solution, is very specific for the operators involved.

In this paper, we focus our attention on the analysis of the global convergence of Newton's method and, as a consequence, we obtain domains of global convergence for nonlinear integral Equation (1) and also locate a solution. For this, we obtain a ball of convergence, by using an auxiliary point, that contains a solution and guarantees the convergence of Newton's method from any point of the ball.

Solving Equation (1) is equivalent to solving the equation $\mathcal{F}(x) = 0$, where $\mathcal{F} : \Omega \subseteq \mathcal{C}[a,b] \longrightarrow \mathcal{C}[a,b]$ and

$$[\mathcal{F}(x)](s) = x(s) - \ell(s) - \lambda \int_a^b \mathcal{K}(s,t)\mathcal{H}(x)(t)\,dt, \quad s \in [a,b], \quad \lambda \in \mathbb{R}, \quad n \in \mathbb{N}. \tag{2}$$

Then,

$$[\mathcal{F}'(x)y](s) = y(s) - \lambda \int_a^b \mathcal{K}(s,t)[\mathcal{H}'(x)y](t)\,dt = \lambda \int_a^b \mathcal{K}(s,t)H'(x(t))y(t)\,dt.$$

As a consequence,

$$\|\mathcal{F}'(x) - \mathcal{F}'(y)\| \leq K\|x-y\|,$$

where $K = |\lambda|\mathcal{S}L$, L is such that $\|\mathcal{H}'(x) - \mathcal{H}'(y)\| \leq L\|x-y\|$, for all $x, y \in \Omega$, and $\mathcal{S} = \left\| \int_a^b \mathcal{K}(s,t)\,dt \right\|$.

From the Banach lemma on invertible operators, it follows

$$\|\widetilde{\Gamma}\| = \|[\mathcal{F}'(\widetilde{x})]^{-1}\| \leq \frac{1}{1-|\lambda|\mathcal{S}\|\mathcal{H}'(\widetilde{x})\|} = \beta, \qquad \|\widetilde{\Gamma}\mathcal{F}(\widetilde{x})\| \leq \frac{\|\widetilde{x}-u\| + |\lambda|\mathcal{S}\|\mathcal{H}(\widetilde{x})\|}{1-|\lambda|\mathcal{S}\|\mathcal{H}'(\widetilde{x})\|} = \eta.$$

provided that

$$|\lambda|\mathcal{S}\|\mathcal{H}'(\widetilde{x})\| < 1. \tag{3}$$

Next, we give some properties that are used later.

Lemma 1. *For operator* (2), *we have:*

(a) $\widetilde{\Gamma}\mathcal{F}(x) = \widetilde{\Gamma}\mathcal{F}(\widetilde{x}) + (x-\widetilde{x}) + \int_0^1 \widetilde{\Gamma}\left(\mathcal{F}'(\widetilde{x}+t(x-\widetilde{x})) - \mathcal{F}'(\widetilde{x})\right)(x-\widetilde{x})\,dt$, *with* $x \in \Omega$.

(b) $\mathcal{F}(x_n) = \int_0^1 \left(\mathcal{F}'(x_{n-1}+t(x_n-x_{n-1})) - \mathcal{F}'(x_{n-1})\right)(x_n-x_{n-1})\,dt$, *with* $x_{n-1}, x_n \in \Omega$.

As a consequence of item (b) of Lemma 1, it follows, for $x_{n-1}, x_n \in \Omega$,

$$\|\mathcal{F}(x_n)\| \leq \frac{K}{2}\|x_n - x_{n-1}\|^2.$$

From the last result, and taking into account the parameters obtained previously, we analyze the first iteration of Newton's method, what leads us to the convergence of the method.

If $x_0 \in B(\widetilde{x}, R)$, then

$$\|\Gamma_0\| = \|[\mathcal{F}'(x_0)]^{-1}\| \leq \frac{\beta}{1-K\beta R} = \alpha, \qquad \|\Gamma_0\mathcal{F}'(\widetilde{x})\| \leq \frac{1}{1-K\beta R}.$$

provided that

$$K\beta R < 1. \tag{4}$$

Moreover, from item (a) of Lemma 1, it follows

$$\|x_1 - x_0\| \leq \|\Gamma_0\mathcal{F}'(\widetilde{x})\|\|\widetilde{\Gamma}\mathcal{F}(x_0)\| < \frac{\eta + R + K\beta R^2/2}{1-K\beta R} = \delta,$$

and, from item (b) of Lemma 1, we have

$$\|x_1 - \widetilde{x}\| = \|-\Gamma_0\left(\mathcal{F}(x_0) + \mathcal{F}'(x_0)(\widetilde{x}-x_0)\right)\| \leq \|\Gamma_0\mathcal{F}'(\widetilde{x})\|\|\widetilde{\Gamma}\mathcal{F}(\widetilde{x})\| + \frac{K\beta R^2/2}{1-K\beta R} \leq \frac{2\eta + K\beta R^2}{2(1-K\beta R)},$$

so that $x_1 \in B(\widetilde{x}, R)$, provided that

$$\frac{2\eta + K\beta R^2}{2(1-K\beta R)} \leq R. \tag{5}$$

Observe now that condition (5) holds if

$$K\beta\eta \leq 1/6 \qquad \text{and} \qquad R \in [R_-, R_+],$$

where $R_- = \frac{1-\sqrt{1-6K\beta\eta}}{3K\beta}$ and $R_+ = \frac{1+\sqrt{1-6K\beta\eta}}{3K\beta}$ are the two real positive roots of quadratic equation

$$2\eta - 2R + 3K\beta R^2 = 0.$$

After that, if we assume that

$$\|x_n - x_{n-1}\| < \gamma^{2^{n-2}} \|x_{n-1} - x_{n-2}\|, \tag{6}$$
$$\|x_n - \widetilde{x}\| < \frac{2\eta + K\beta R^2}{2(1 - K\beta R)} \le R, \tag{7}$$

where $\gamma = K\alpha\delta/2$, for all $n \ge 2$, and provided that condition (5) holds, it follows in the same way that

$$\|x_{n+1} - x_n\| < \gamma^{2^{n-1}} \|x_n - x_{n-1}\|, \qquad \|x_{n+1} - \widetilde{x}\| < \frac{2\eta + K\beta R^2}{2(1 - K\beta R)} \le R,$$

so that (6) and (7) are true for all positive integers n by mathematical induction.

In addition, $\gamma < 1$ if

$$3(K\beta R)^2 - 10(K\beta R) + 2(2 - K\beta\eta) > 0, \tag{8}$$

which is satisfied provided that

$$K\beta\eta \le 1/6 \qquad \text{and} \qquad R < \frac{5 - \sqrt{13 + 6K\beta\eta}}{3K\beta}.$$

As a consequence, condition (4) holds. More precisely, we can establish the following result.

Lemma 2. *There always exists $R > 0$, such that inequalities* (4), (5) *and* (8) *hold, if*

(a) $K\beta\eta \le 0.1547\ldots$ *and* $R \in \left[R_-, \frac{5-\sqrt{13+6K\beta\eta}}{3K\beta}\right)$,
(b) $K\beta\eta \in [0.1547\ldots, 1/6)$ *and* $R \in [R_-, R_+]$,

where $R_- = \frac{1-\sqrt{1-6K\beta\eta}}{3K\beta}$ *and* $R_+ = \frac{1+\sqrt{1-6K\beta\eta}}{3K\beta}$.

Proof. First, we prove item (a) of Lemma 2. Observe that $R_- < \frac{5-\sqrt{13+6K\beta\eta}}{3K\beta}$, since $K\beta\eta \le 0.1547\ldots$, so that $\left[R_-, \frac{5-\sqrt{13+6K\beta\eta}}{3K\beta}\right) \neq \emptyset$. Moreover, as $K\beta\eta \le 0.1547\ldots$, we have $3(K\beta\eta)^2 + 6(K\beta\eta) - 1 \le 0$ and, as a consequence, $R_+ > \frac{5-\sqrt{13+6K\beta\eta}}{3K\beta}$ and $R \in \left[R_-, \frac{5-\sqrt{13+6K\beta\eta}}{3K\beta}\right) \subset [R_-, R_+]$, so that (5) and (8) hold.

Second, if $K\beta\eta \in [0.1547\ldots, 1/6)$, then $3(K\beta\eta)^2 + 6(K\beta\eta) - 1 \ge 0$ and $R_+ < \frac{5-\sqrt{13+6K\beta\eta}}{3K\beta}$, so that $R \in [R_-, R_+]$. Then, (5) and (8) hold.

Third, in both cases, $K\beta R < 1$ follows immediately, since $R < \frac{5-\sqrt{13+6K\beta\eta}}{3K\beta}$ in items (a) and (b) of Lemma 2. □

2.1. Convergence

Now, we can establish the following result.

Theorem 1. *Suppose that $K\beta\eta \le 1/6$ and consider $R > 0$ satisfying item (a) or item (b) of Lemma 2 and such that $B(\widetilde{x}, R) \subset \Omega$. If condition* (3) *holds, then Newtons's method is well-defined and converges to a solution x^* of $\mathcal{F}(x) = 0$ in $\overline{B(\widetilde{x}, R)}$ from every point $x_0 \in B(\widetilde{x}, R)$.*

Proof. From (6) and $\gamma < 1$, we have $\|x_{n+1} - x_n\| < \|x_n - x_{n-1}\|$, for all $n \in \mathbb{N}$, so that sequence $\{\|x_{n+1} - x_n\|\}$ is strictly decreasing for all $n \in \mathbb{N}$ and, therefore, sequence $\{x_n\}$ is convergent. If $x^* = \lim_{n\to\infty} x_n$, then $\mathcal{F}(x^*) = 0$, by the continuity of $\mathcal{F}$ and $\|\mathcal{F}(x_n)\| \to 0$ when $n \to \infty$. □

From Theorem 1, the convergence of Newton's method to a solution of equation $\mathcal{F}(x) = 0$ is guaranteed. Moreover, the best ball of location of the solution is $B(\widetilde{x}, R_-)$ and the biggest ball of convergence is $B(\widetilde{x}, R_+)$ or $B\left(\widetilde{x}, \frac{5-\sqrt{13+6K\beta\eta}}{3K\beta}\right)$, depending on the value of $K\beta\eta$ is: $K\beta\eta \leq 0.1547\ldots$ for the former and $K\beta\eta \in [0.1547\ldots, 1/6)$ for the latter.

2.2. Uniqueness of Solution

For uniqueness of solution, we establish the following result, where uniqueness of solution is proved in $\overline{B(\widetilde{x}, R)}$.

Theorem 2. *Under conditions of Theorem 1, solution x^* of $\mathcal{F}(x) = 0$ is unique in $\overline{B(\widetilde{x}, R)}$.*

Proof. Assume that w^* is another solution of $\mathcal{F}(x) = 0$ in $\overline{B(\widetilde{x}, R)}$ such that $w^* \neq x^*$. If operator $Q = \int_0^1 \mathcal{F}'(w^* + t(x^* - w^*))\, dt$ is invertible, we have $x^* = w^*$, since $Q(w^* - x^*) = \mathcal{F}(w^*) - \mathcal{F}(x^*)$. Then, as

$$\begin{aligned} \|I - \widetilde{\Gamma}Q\| &\leq \|\widetilde{\Gamma}\| \int_0^1 \|\mathcal{F}'(\widetilde{x}) - \mathcal{F}'(w^* + t(x^* - w^*))\| dt \\ &\leq \beta K \int_0^1 \|\widetilde{x} - (w^* + t(x^* - w^*))\| dt \\ &= \beta K R \\ &< 1, \end{aligned} \tag{9}$$

it follows that Q is invertible by the Banach lemma on invertible operators and uniqueness follows immediately. □

Notice that, from Theorems 1 and 3, the best ball of location of a solution of (1) is $B(\widetilde{x}, R_-)$ and the best ball of uniqueness of solution and the biggest ball of convergence is $B\left(\widetilde{x}, \frac{5-\sqrt{13+6K\beta\eta}}{3K\beta}\right)$ or $B(\widetilde{x}, R_+)$, depending on the value of $K\beta\eta$ lies.

Once given the uniqueness of solution in the domain of existence of solution $B(\widetilde{x}, R)$, we enlarge such domain from the following theorem.

Theorem 3. *Under conditions of Theorem 1, we have that the solution x^* is unique in the domain $B(\widetilde{x}, \varrho) \cap \Omega$, where $\varrho = \frac{2}{K\beta} - R$.*

Proof. Assume that w^* is another solution of $\mathcal{F}(x) = 0$ in $B(\widetilde{x}, \varrho) \cap \Omega$ such that $w^* \neq x^*$. Then, from (9), it follows

$$\|I - \widetilde{\Gamma}Q\| < \beta K \int_0^1 ((1-t)\varrho + tR) dt = 1.$$

and Q is again invertible by the Banach lemma on invertible operators. □

Note that $\varrho > 0$, since $\beta K R < 1$, and uniqueness of solution is obtained in the ball of global convergence given in Theorem 1, since $\varrho = \frac{2}{K\beta} - R \geq \frac{5-\sqrt{13+6K\beta\eta}}{3K\beta}, R_+$.

3. Example

Now, we apply the last result to the following nonlinear integral equation:

$$x(s) = s^3 + \frac{18}{25} \int_0^1 s^3 t^3 x(t)^2 dt, \quad s \in [0, 1]. \tag{10}$$

For Equation (10), we have $\lambda = 18/25$ and $\mathcal{S} = \left\| \int_0^1 s^3 t^3\, dt \right\| = 1/4$ with the max-norm. As $\mathcal{H}(\widetilde{x})(t) = \widetilde{x}(t)^2$, then $L = 2$. If we choose $\widetilde{x}(s) = s^3$, then condition (3) holds, since $|\lambda|\mathcal{S}\|H'(\widetilde{x})\| = 9/25 < 1$. Moreover, $\beta = 25/16$ and $\eta = 9/32$ and $K = |\lambda|\mathcal{S}L = 9/25$, so that $K\beta\eta = 0.1582\ldots \in$

$[R_-, R_+]$, where $R_- = 0.4590\ldots$ and $R_+ = 0.7261\ldots$. Therefore, from Theorem 1, the convergence of Newton's method to a solution of Equation (10) is guaranteed and the best ball of location of the solution is $B(s^3, 0.4590\ldots)$ and the biggest ball of convergence is $B(s^3, 0.7261\ldots)$. Furthermore, from Theorem 3, it follows that the domain of uniqueness of solution is $B(s^3, 3.0965\ldots)$.

Next, we approximate a solution of Equation (10) by Newton's method. After four iterations with stopping criterion $\|x_n - x_{n-1}\|_\infty < 10^{-18}$, $n \in \mathbb{N}$, we obtain solution shown in Table 1, where errors $\|x^* - x_n\|$ and sequence $\|\mathcal{F}(x_n)\|$ are also shown. Observe from the last sequence that solution shown in Table 1 is a good approximation of the solution of Equation (10). Finally, we observe in Figure 1 that solution shown in Table 1 lies within the domain of location of solution found above.

Table 1. Approximated solution $x^*(s)$ of (10), absolute errors and $\{\|\mathcal{F}(x_n)\|\}$.

n	$x_n(s)$	$\|x^* - x_n\|$	$\|\mathcal{F}(x_n)\|$
0	s^3	$8.4715\ldots \times 10^{-2}$	7.2×10^{-2}
1	$(1.0841121495327102\ldots)s^3$	$6.0365\ldots \times 10^{-4}$	$5.0938\ldots \times 10^{-4}$
2	$(1.0847157717628998\ldots)s^3$	$3.1090\ldots \times 10^{-8}$	$2.6233\ldots \times 10^{-8}$
3	$(1.0847158028530592\ldots)s^3$	$8.2478\ldots \times 10^{-17}$	$6.9595\ldots \times 10^{-17}$
4	$(1.0847158028530593\ldots)s^3$		

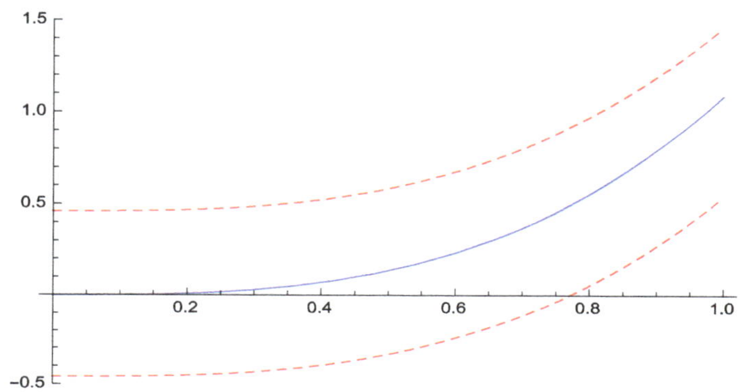

Figure 1. Approximated solution $x^*(s)$ of (10) and domain of location of solution.

4. Study of the Integral Equation from Parameter λ

Next, we study the integral Equation (1) from the values of parameter λ.

First, we observe that $K\beta\eta \leq 1/6$ if

$$6|\lambda|\mathcal{S}L\left(\|\widetilde{x} - \ell\| + |\lambda|\mathcal{S}\|\mathcal{H}(\widetilde{x})\|\right) \leq \left(1 - |\lambda|\mathcal{S}\|\mathcal{H}'(\widetilde{x})\|\right)^2 \tag{11}$$

and condition (3) holds.

Now, we analyze condition (11). Observe that (11) is satisfied if

- $\|\mathcal{H}'(\widetilde{x})\|^2 < 6L\|\mathcal{H}(\widetilde{x})\|$ and $|\lambda| \in [0, \mu_+]$, where

$$\mu_+ = \frac{-(3L\|\widetilde{x} - \ell\| + \|\mathcal{H}'(\widetilde{x})\|) + \sqrt{\Delta}}{\mathcal{S}(6L\|\mathcal{H}(\widetilde{x})\| - \|\mathcal{H}'(\widetilde{x})\|^2)}$$

  and $\Delta = 3L(3L\|\widetilde{x} - \ell\|^2 + 2\|\widetilde{x} - \ell\|\|\mathcal{H}'(\widetilde{x})\| + 2\|\mathcal{H}(\widetilde{x})\|)$.
- $\|\mathcal{H}'(\widetilde{x})\|^2 > 6L\|\mathcal{H}(\widetilde{x})\|$ and $|\lambda| \in [0, \mu_+] \cup [\mu_-, +\infty)$, where

$$\mu_- = \frac{-(3L\|\widetilde{x} - \ell\| + \|\mathcal{H}'(\widetilde{x})\|) - \sqrt{\Delta}}{\mathcal{S}(6L\|\mathcal{H}(\widetilde{x})\| - \|\mathcal{H}'(\widetilde{x})\|^2)}.$$

- $\|\mathcal{H}'(\widetilde{x})\|^2 = 6L\|\mathcal{H}(\widetilde{x})\|$ and $|\lambda| \leq \dfrac{1}{2\mathcal{S}(3L\|\widetilde{x} - \ell\| + \|\mathcal{H}'(\widetilde{x})\|)}$.

Second, once $\widetilde{x}$ is fixed, we have two chances: $K\beta\eta \leq 0.1547\ldots$ or $K\beta\eta \in [0.1547\ldots, 1/6)$. If first holds, then $R \in \left[R_-, \frac{5-\sqrt{13+6K\beta\eta}}{3K\beta}\right)$ and, if second does, then $R \in [R_-, R_+]$.

Finally, as condition (3) is satisfied, then Newtons's method is well-defined and converges to a solution x^* of $\mathcal{F}(x) = 0$ in $\overline{B(\widetilde{x}, R)}$ from every point $x_0 \in B(\widetilde{x}, R)$ by Theorem 1.

5. Application

Now, we apply the last study to the following particular Davis-type integral Equation [8]:

$$x(s) = s + \lambda \int_0^1 G(s,t)x(t)^2 dt, \quad \lambda \in \mathbb{R}, \quad s \in [0,1], \tag{12}$$

where the kernel of (12) is a Green's function defined as follows:

$$G(s,t) = \begin{cases} (1-s)t, & t \leq s, \\ s(1-t), & s \leq t. \end{cases}$$

One can show that the function $x(s)$ that satisfied Equation (12) is any solution of the differential equation

$$x''(s) + \lambda x(s)^2 = 0,$$

that also satisfies the two-point boundary condition: $x(0) = 0$, $x(1) = 1$.

For Equation (12), we have $\mathcal{S} = \left\| \int_0^1 G(s,t)\, dt \right\| = 1/8$ with the max-norm and $\mathcal{H}(\widetilde{x})(t) = \widetilde{x}(t)^2$. Therefore, $L = 2$ and condition (3) is reduced to $|\lambda| < 4/\|\widetilde{x}\|$. In addition,

$$\|\mathcal{H}'(\widetilde{x})\|^2 = 4\|\widetilde{x}\|^2 < 12\|\widetilde{x}\|^2 = 6L\|\mathcal{H}(\widetilde{x})\|$$

and, as a consequence,

$$|\lambda| \leq \mu_+ = \frac{-(6\|\widetilde{x} - s\| + 2\|\widetilde{x}\|) + \sqrt{12\left(3\|\widetilde{x} - s\|^2 + 2\|\widetilde{x} - s\|\|\widetilde{x}\| + \|\widetilde{x}\|^2\right)}}{\|\widetilde{x}\|^2}.$$

After that, we choose $\widetilde{x}(s) = s$ and hence $\mu_+ = 2(-1+\sqrt{3}) = 1.4641\ldots$, so that $|\lambda| \leq 1.4641\ldots$, that satisfies condition (3). In this case, from Theorem 1, we can guarantee the convergence of Newton's method to a solution of Equation (12) with λ such that $|\lambda| \leq 1.4641\ldots$ Moreover, once λ is fixed, depending on the value of $K\beta\eta$, we can obtain the best ball of location of solution and the biggest ball of convergence.

Observe that we cannot apply Newton's method directly, since we do not know the inverse operator that is involved in the algorithm of Newton's method. Then, we use a process of discretization to transform (12) into a finite dimensional problem. For this, we use a Gauss–Legendre quadrature formula to approximate the integral of (12),

$$\int_0^1 \phi(t)\, dt \simeq \sum_{j=1}^{m} w_j \phi(t_j),$$

where the m nodes t_j and weights w_j are known.

Next, we denote the approximations $x(t_i)$ by x_i, with $i = 1, 2, \ldots, m$, so that (12) is equivalent to the nonlinear system given by

$$x_j = t_j + \lambda \sum_{k=1}^{m} a_{jk}\, x_k^2, \quad k = 1, 2, \ldots, m, \tag{13}$$

where

$$a_{jk} = \begin{cases} w_k\,(1-t_j)t_k, & k \le j, \\ w_k\,(1-t_k)t_j, & k > j. \end{cases}$$

After that, we write system (13) compactly in matrix form as

$$F(\mathbf{x}) \equiv \mathbf{x} - \mathbf{v} - \lambda A\,\mathbf{y} = 0, \qquad F : \mathbb{R}^m \longrightarrow \mathbb{R}^m, \tag{14}$$

where

$$\mathbf{x} = (x_1, x_2, \ldots, x_m)^t, \quad \mathbf{v} = (t_1, t_2, \ldots, t_m)^t, \quad A = \left(a_{jk}\right)_{j,k=1}^{m}, \quad \mathbf{y} = \left(x_1^2, x_2^2, \ldots, x_m^2\right)^t.$$

Choose $m = 8$, $\lambda = 7/5$, $\widetilde{\mathbf{x}} = \mathbf{v}$ and hence $K = 0.2471\ldots$, $\beta = 1.2179\ldots$, $\eta = 0.0665\ldots$ and $K\beta\eta = 0.0200\ldots$ As $K\beta\eta < 0.1547\ldots$, it follows, from Theorem 1, that the best ball of location of solution is $\overline{B(\mathbf{v}, 0.0686\ldots)}$ and the biggest ball of convergence is $B(\mathbf{v}, 1.5259\ldots)$.

If the starting point for Newton's method is $x_0 = \mathbf{v}$, the method converges to the solution $x^* = (x_1^*, x_2^*, \ldots, x_8^*)^t$ of system (14), which is shown in Table 2, after four iterations with stopping criterion $\|x_n - x_{n-1}\|_\infty < 10^{-18}$, $n \in \mathbb{N}$.

Table 2. Numerical solution x^* of system (14) with $\lambda = 7/5$.

i	x_i^*	i	x_i^*
1	0.02267000...	5	0.65888692...
2	0.11607746...	6	0.82291926...
3	0.27057507...	7	0.93276524...
4	0.46275932...	8	0.98797444...

Moreover, errors $\|x^* - x_n\|$ and sequence $\{\|F(x_n)\|\}$ are shown in Table 3. Observe then that vector shown in Table 2 is a good approximation of a solution of (14).

Table 3. Absolute errors and $\{\|F(x_n)\|\}$.

n	$\|x^* - x_n\|$	$\|F(x_n)\|$
0	$6.7169\ldots \times 10^{-2}$	$9.6624\ldots \times 10^{-1}$
1	$6.4734\ldots \times 10^{-4}$	$5.3956\ldots \times 10^{-4}$
2	$6.0570\ldots \times 10^{-8}$	$5.0516\ldots \times 10^{-8}$
3	$5.4063\ldots \times 10^{-16}$	$4.5077\ldots \times 10^{-16}$

Furthermore, as a solution of (12) satisfies $x(0) = 0$ and $x(1) = 1$, if values of Table 2 are interpolated, an approximated solution is obtained, which is painted in Figure 2. Notice that this approximated solution lies in the domain of location of solution $\overline{B(\mathbf{v}, 0.0686\ldots)}$ which is obtained from Theorem 1.

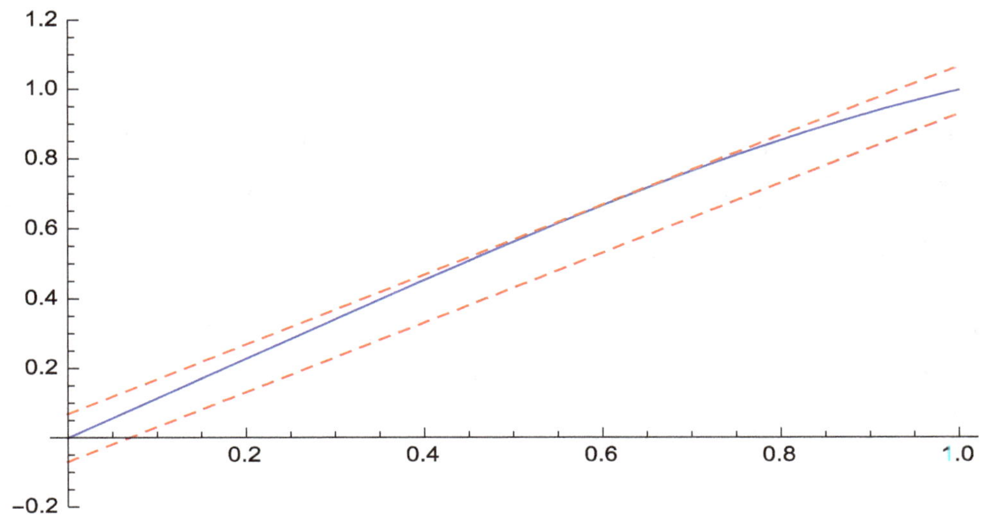

Figure 2. Solution $\mathbf{x}^*$ of system (14) and domain of location of solution.

6. Conclusions

Following the idea of the Fixed Point Theorem for the method of successive approximations, we do an analysis for Newton's method, use the theoretical significance of the method to prove the existence and uniqueness of solution of a particular type of nonlinear integral equations of Fredholm and, in addition, obtain a domain of global convergence for the method that allows locating a solution and separating it from other possible solutions. For this, we use a technique based on using auxiliary points. Moreover, we present a study of the nonlinear equations which is based on the real parameter involved in the equation.

Author Contributions: The contributions of the two authors have been similar. Both authors have worked together to develop the present manuscript.

References

1. Awawdeh, F.; Adawi, A.; Al-Shara, S. A numerical method for solving nonlinear integral equations. *Int. Math. Forum* **2009**, *4*, 805–817.
2. Nadir, M.; Khirani, A. Adapted Newton-Kantorovich method for nonlinear integral equations. *J. Math. Stat.* **2016**, *12*, 176–181. [CrossRef]
3. Saberi-Nadja, J.; Heidari, M. Solving nonlinear integral equations in the Urysohn form by Newton-Kantorovich-quadrature method. *Comput. Math. Appl.* **2010**, *60*, 2018–2065.
4. Matkowski, J. Functional Equations and Nemytskii Operators. *Funkc. Ekvacioj* **1982**, *25*, 127–132.
5. Berinde, V. *Iterative Approximation of Fixed Point*; Springer: New York, NY, USA, 2005.
6. Ragusa, M.A. Local Hölder regularity for solutions of elliptic systems. *Duke Math. J.* **2002**, *13*, 385–397. [CrossRef]
7. Wang, H.; Zhang, L.; Wang, X. Fixed point theorems for a class of nonlinear sum-type operators and application in a fractional differential equation. *Bound. Value Probl.* **2018**, *2018*, 140. [CrossRef]
8. Davis, H.T. *Introduction to Nonlinear Differential and Integral Equations*; Dover: New York, NY, USA, 1962.

12

Extended Local Convergence for the Combined Newton-Kurchatov Method Under the Generalized Lipschitz Conditions

Ioannis K. Argyros [1] and Stepan Shakhno [2,*]

[1] Department of Mathematical Sciences, Cameron University, Lawton, OK 73505, USA; iargyros@cameron.edu
[2] Department of Theory of Optimal Processes, Ivan Franko National University of Lviv, 79000 Lviv, Ukraine
[*] Correspondence: stepan.shakhno@lnu.edu.ua

Abstract: We present a local convergence of the combined Newton-Kurchatov method for solving Banach space valued equations. The convergence criteria involve derivatives until the second and Lipschitz-type conditions are satisfied, as well as a new center-Lipschitz-type condition and the notion of the restricted convergence region. These modifications of earlier conditions result in a tighter convergence analysis and more precise information on the location of the solution. These advantages are obtained under the same computational effort. Using illuminating examples, we further justify the superiority of our new results over earlier ones.

Keywords: nonlinear equation; iterative process; non-differentiable operator; Lipschitz condition

MSC: 65H10; 65J15; 47H17

---

1. Introduction

Consider the nonlinear equation

$$F(x) + Q(x) = 0, \tag{1}$$

where F is a Fréchet-differentiable nonlinear operator on an open convex subset D of a Banach space E_1 with values in a Banach space E_2, and $Q : D \to E_2$ is a continuous nonlinear operator.

Let x, y be two points of D. A linear operator from E_1 into E_2, denoted $Q(x,y)$, which satisfies the condition

$$Q(x,y)(x-y) = Q(x) - Q(y) \tag{2}$$

is called a divided difference of Q at points x and y.

Let x, y, z be three points of D. A operator $Q(x,y,z)$ will be called a divided difference of the second order of the operator Q at the points x, y and z, if it satisfies the condition

$$Q(x,y,z)(y-z) = Q(x,y) - Q(x,z). \tag{3}$$

A well-known simple difference method for solving nonlinear equations $F(x) = 0$ is the Secant method

$$x_{n+1} = x_n - (F(x_{n-1}, x_n))^{-1} F(x_n),\ n = 0,1,2,\ldots, \tag{4}$$

where $F(x_{n-1}, x_n)$ is a divided difference of the first order of $F(x)$ and x_0, x_{-1} are given.

Secant method for solving nonlinear operator equations in a Banach space was explored by the authors [1–6] under the condition that the divided differences of a nonlinear operator F satisfy the Lipschitz (Hölder) condition with constant L of type

$$\|F(x,y) - F(u,v)\| \leq L(\|x-u\| + \|y-v\|).$$

In [7] a one-point iterative Secant-type method with memory was psoposed.

In [8,9] the Kurchatov method under the classical Lipschitz conditions for the divided differences of the first and second order was explored and its quadratic convergence of it was determined. The iterative formula of Kurchatov method has the form [1,8–11]

$$x_{n+1} = x_n - (F(2x_n - x_{n-1}, x_{n-1}))^{-1} F(x_n),\ n = 0,1,2,\ldots. \tag{5}$$

Related articles but with stronger convergence criteria exist; see works of Argyros, Ezquerro, Hernandez, Rubio, Gutierrez, Wang, Li [1,12–15] and references therein.

In [14] which dealt with the study of the Newton method, it was proposed that there are generalized Lipschitz conditions for the nonlinear operator, in which instead of constant L, some positive integrable function is used.

In our work [16], we introduced, for the first time, a similar generalized Lipschitz condition for the operator of the first order divided difference, and under this condition, the convergence of the Secant method was studied and it was found that its convergence order is $(1+\sqrt{5})/2$.

In [17], we introduced a generalized Lipschitz condition for the divided differences of the second order, and we have studied the local convergence of the Kurchatov method (5).

Note that in many papers, such as [3,18–21], the authors investigated the Secant and Secant-type methods under the generalized conditions for the first divided differences of the form

$$\|(F(x,y) - F(u,v)))\| \leq \omega(\|x-y\|, \|u-v\|) \quad \forall x,y,u,v \in D, \tag{6}$$

where $\omega : \mathbf{R}_+ \times \mathbf{R}_+ \longrightarrow \mathbf{R}_+$ is continuous nondecreasing function in their two arguments. Under these same conditions, in the work of Argyros [10], it was proven that there is a semi-local convergence of the Kurchatov method and in [22] of Ren and Argyros the semi-local convergence of a combined Kurchatov method and Secant method was demonstrated. In both cases, only the linear convergence of the methods is received.

We also refer the reader to the intersting paper by Donchev et al. [23], where several other relaxed Lipschitz conditions are used in the setting of fixed points for these conditions. Clearly, our results can be written in this setting too in an analogous way.

In [24], we first proposed and studied the local convergence of the combined Newton-Kurchatov method

$$x_{n+1} = x_n - (F'(x_n) + Q(2x_n - x_{n-1}, x_{n-1}))^{-1}(F(x_n) + Q(x_n)),\ n = 0,1,2,\ldots, \tag{7}$$

where $F'(u)$ is a Fréchet derivative, $Q(u,v)$ is a divided difference of the first order, x_0, x_{-1} are given, which is built on the basis of the mentioned Newton and Kurchatov methods. Semi-local convergence of the method (7) under the classical Lipschitz conditions is studied in the mentioned article, but the convergence only with the order $(1+\sqrt{5})/2$ has been determined.

In [25], we studied the method (7) under relatively weak, generalized Lipschitz conditions for the derivatives and divided differences of nonlinear operators. Setting $Q(x) \equiv 0$, we receive the results for the Newton method [14], and when $F(x) \equiv 0$ we got the known results for Kurchatov method [9,17]. We proved the quadratic order of convergence of the method (7), which is higher than the convergence order $(1+\sqrt{5})/2$ for the Newton–Secant method [1,26–28]

$$x_{n+1} = x_n - (F'(x_n) + Q(x_{n-1}, x_n))^{-1}(F(x_n) + Q(x_n)),\ n = 0,1,2,\ldots,\ . \tag{8}$$

The results of the numerical study of the method (7) and other combined methods on the test problems are provided in our works [24,28].

In this work, we continue to study a combined method (7) for solving nonlinear Equation (1), but with optimization considerations resulting in a tighter analysis than in [25].

The rest of the article is structured as follows: In Section 2, we present the local convergence analysis of the method (7) and the uniquness ball for solution of the equation. Section 3 contains the Corollaries of Theorems from Section 2. In Section 4, we provide the numerical example. The article ends with some conclusions.

2. Local Convergence of Newton-Kurchatov Method (7)

Let us denote $B(x_0, r) = \{x : \|x - x_0\| < r\}$ an open ball of radius $r > 0$ with center at point $x_0 \in D$, $B(x_0, r) \subset D$.

Condition on the divided difference operator $Q(x, y)$

$$\|Q(x,y) - Q(u,v)\| \leq L(\|x-u\| + \|y-v\|) \quad \forall x,y,u,v \in D \tag{9}$$

is called Lipschitz condition in domain D with constant $L > 0$. If the condition is being fulfilled

$$\|Q(x,y) - Q'(x_0)\| \leq L(\|x-x_0\| + \|y-x_0\|) \quad \forall x,y \in B(x_0,r), \tag{10}$$

then we call it the center Lipschitz condition in the ball $B(x_0, r)$ with constant L.

However, L in Lipschitz conditions can be not a constant, and can be a positive integrable function. In this case, if for $x_* \in D$ inverse operator $[F'(x_*)]^{-1}$ exists, then the conditions (9) and (10) for $x_0 = x_*$ can be replaced respectively for

$$\|Q'(x_*)^{-1}(Q(x,y) - Q(u,v))\| \leq \int_0^{\|x-y\|+\|u-v\|} L(t)dt \quad \forall x,y,u,v \in D \tag{11}$$

and

$$\|Q'(x_*)^{-1}(Q(x,y) - Q'(x_*))\| \leq \int_0^{\|x-x_*\|+\|y-x_*\|} L(t)dt \quad \forall x,y \in B(x_*,r). \tag{12}$$

Simultaneously

Lipschitz conditions (11) and (12) are called generalized Lipschitz conditions or Lipschitz conditions with the L average.

Similarly, we introduce the generalized Lipschitz condition for the divided difference of the second order

$$\|Q'(x_*)^{-1}(Q(u,x,y) - Q(v,x,y))\| \leq \int_0^{\|u-v\|} N(t)dt \; \forall x,y,u,v \in B(x_*,r), \tag{13}$$

where N is a positive integrable function.

Remark 1. *Note than the operator F is Fréchet differentiable on D when the Lipschitz conditions (9) or (11) are fulfilled $\forall x,y,u,v \in D$ (the divided differences $F(x,y)$ are Lipschitz continuous on D) and $F(x,x) = F'(x)$ $\forall x \in D$ [29].*

Suppose that equation

$$\int_0^r L_1^0(u)du + \int_0^{2r} L_2^0(u)du + 2r\int_0^{2r} N_0(u)du = 1.$$

has at least one positive solution. Denote by r_0 the smallest such solution. Set $D_0 = D \cap B(x_*, r_0)$

The radius of the convergence ball and the convergence order of the combined Newton–Kurchatov method (7) are determined in next theorem.

Theorem 1. *Let F and Q be continuous nonlinear operators defined in open convex domain D of a Banach space E_1 with values in the Banach space E_2. Let us suppose, that: (1) $H(x) \equiv F(x) + Q(x) = 0$ has a solution $x_* \in D$, for which there exists a Fréchet derivative $H'(x_*)$ and it is invertible; (2) F has the Fréchet derivative of the first order, and Q has divided differences of the first and second order on $B(x_*, 3r) \subset D$, so that for each $x, y, u, v \in D$*

$$\|H'(x_*)^{-1}(F'(x) - F'(x_*))\| \leq \int_0^{\rho(x)} L_1^0(u)du, \tag{14}$$

$$\|H'(x_*)^{-1}(Q(x,y) - Q(x_*, x_*))\| \leq \int_0^{\|x-x_*\|+\|y-x_*\|} L_2^0(t)dt, \tag{15}$$

$$\|H'(x_*)^{-1}(Q(u,x,y) - Q(v,x,y))\| \leq \int_0^{\|u-v\|} N_0(t)dt, \tag{16}$$

and for each $x, y, u, v \in D_0$

$$\|H'(x_*)^{-1}(F'(x) - F'(x^\theta))\| \leq \int_{\theta\rho(x)}^{\rho(x)} L_1(u)du,\ 0 \leq \tau \leq 1, \tag{17}$$

$$\|H'(x_*)^{-1}(Q(x,y) - Q(u,v))\| \leq \int_0^{\|x-u\|+\|y-v\|} L_2(t)dt, \tag{18}$$

$$\|H'(x_*)^{-1}(Q(u,x,y) - Q(v,x,y))\| \leq \int_0^{\|u-v\|} N(t)dt, \tag{19}$$

where $x^\theta = x_* + \theta(x - x_*)$, $\varrho(x) = \|x - x_*\|$, L_1^0, L_2^0, N_0 L_1, L_2 and N are positive nondecreasing integrable functions and $r > 0$ satisfies the equation

$$\frac{\frac{1}{r}\int_0^r L_1(u)udu + \int_0^r L_2(u)du + 2r\int_0^{2r} N(u)du}{1 - \left(\int_0^r L_1^0(u)du + \int_0^{2r} L_2^0(u)du + 2r\int_0^{2r} N_0(u)du\right)} = 1. \tag{20}$$

Then for all $x_0, x_{-1} \in B(x_*, r)$ the iterative method (7) is well defined and the generated by it sequence $\{x_n\}_{n\geq 0}$, which belongs to $B(x_*, r)$, converges to x_* and satisfies the inequality

$$\|x_{n+1} - x_*\| \leq e_n := \frac{\frac{1}{\rho(x_n)}\int_0^{\rho(x_n)} L_1(u)udu + \int_0^{\rho(x_n)} L_2(u)du + \int_0^{\|x_n-x_{n-1}\|} N(u)du\|x_n - x_{n-1}\|}{1 - \left(\int_0^{\rho(x_n)} L_1^0(u)du + \int_0^{2\rho(x_n)} L_2^0(u)du + \int_0^{\|x_n-x_{n-1}\|} N_0(u)du\|x_n - x_{n-1}\|\right)}\|x_n - x_*\|. \tag{21}$$

Proof. First we show that $f(t) = \dfrac{1}{t^2}\displaystyle\int_0^t L_1(u)udu$, $g(t) = \dfrac{1}{t}\displaystyle\int_0^t L_2(u)du$, $h(t) = \dfrac{1}{t}\displaystyle\int_0^t N(u)du$, $f_0(t) = \dfrac{1}{t^2}\displaystyle\int_0^t L_1^0(u)udu$, $g_0(t) = \dfrac{1}{t}\displaystyle\int_0^t L_2^0(u)du$, $h_0(t) = \dfrac{1}{t}\displaystyle\int_0^t N_0(u)du$ monotonically nondecreasing with respect to t. Indeed, under the monotony of L_1, L_2, N we have

$$\left(\frac{1}{t_2^2}\int_0^{t_2} - \frac{1}{t_1^2}\int_0^{t_1}\right)L_1(u)udu = \left(\frac{1}{t_2^2}\int_{t_1}^{t_2} + \left(\frac{1}{t_2^2} - \frac{1}{t_1^2}\right)\int_0^{t_1}\right)L_1(u)udu \geq$$

$$\geq L(t_1)\left(\frac{1}{t_2^2}\int_{t_1}^{t_2} + \left(\frac{1}{t_2^2} - \frac{1}{t_1^2}\right)\int_0^{t_1}\right)udu = L_1(t_1)\left(\frac{1}{t_2^2}\int_0^{t_2} - \frac{1}{t_1^2}\int_0^{t_1}\right)udu = 0,$$

$$\Big(\frac{1}{t_2}\int_0^{t_2}-\frac{1}{t_1}\int_0^{t_1}\Big)L_2(u)du=\Big(\frac{1}{t_2}\int_{t_1}^{t_2}+\Big(\frac{1}{t_2}-\frac{1}{t_1}\Big)\int_0^{t_1}\Big)L_2(u)du\geq$$

$$\geq L_2(t_1)\Big(\frac{1}{t_2}\int_{t_1}^{t_2}+\Big(\frac{1}{t_2}-\frac{1}{t_1}\Big)\int_0^{t_1}\Big)du=L_2(t_1)\Big(\frac{t_2-t_1}{t_2}+t_1\Big(\frac{1}{t_2}-\frac{1}{t_1}\Big)\Big)=0$$

for $0<t_1<t_2$. So, $f(t)$, $g(t)$ are nondecreasing with respect to t. Similarly we get for $h(t)$,$f_0(t)$, $g_0(t)$ and $h_0(t)$.

We denote by A_n linear operator $A_n=F'(x_n)+Q(2x_n-x_{n-1},x_{n-1})$. Easy to see that if $x_n,x_{n-1}\in B(x_*,r)$, then $2x_n-x_{n-1},x_{n-1}\in B(x_*,3r)$. Then A_n is invertible and the inequality holds

$$\begin{aligned}&\|A_n^{-1}H'(x_*)\|=\|[I-(I-H'(x_*)^{-1}A_n)]^{-1}\|\leq\\&\leq\Big(1-\Big(\int_0^{\rho(x_n)}L_1^0(u)du+\int_0^{2\rho(x_n)}L_2^0(u)du+\int_0^{\|x_n-x_{n-1}\|}N_0(u)du\|x_n-x_{n-1}\|\Big)\Big)^{-1}.\end{aligned}\tag{22}$$

Indeed from the formulas (14)–(16) we get

$$\|I-H'(x_*)^{-1}A_n\|=\|H'(x_*)^{-1}(F'(x_*)-F'(x_n)+Q(x_*,x_*)-Q(x_n,x_n)+$$

$$+Q(x_n,x_n)-Q(2x_n-x_{n-1},x_{n-1})\|)\leq\int_0^{\rho(x_n)}L_1^0(u)du+\|H'(x_*)^{-1}(Q(x_*,x_*)-$$

$$-Q(x_n,x_n)+Q(x_n,x_n)-Q(x_n,x_{n-1})+Q(x_n,x_{n-1})-Q(2x_n-x_{n-1},x_{n-1}))\|\leq$$

$$\leq\int_0^{\rho(x_n)}L_1^0(u)du+\int_0^{2\rho(x_n)}L_2^0(u)du+$$

$$+\|H'(x_*)^{-1}(Q(x_n,x_{n-1},x_n)-Q(2x_n-x_{n-1},x_{n-1},x_n))(x_n-x_{n-1})\|\leq$$

$$\leq\int_0^{\rho(x_n)}L_1^0(u)du+\int_0^{2\rho(x_n)}L_2^0(u)du+\int_0^{\|x_n-x_{n-1}\|}N_0(u)du\|x_n-x_{n-1}\|.$$

From the definition r_0 (20), we get

$$\int_0^{r_0}L_1(u)du+\int_0^{2r_0}L_2(u)du+2r\int_0^{2r_0}N(u)du<1,\tag{23}$$

since $r<r_0$.

Using the Banach theorem on inverse operator [30], we get formula (22). Then we can write

$$\|x_{n+1}-x_*\|=\|x_n-x_*-A_n^{-1}(F(x_n)-F(x_*)+Q(x_n)-Q(x_*))\|=$$

$$\begin{aligned}&=\|-A_n^{-1}\Big(\int_0^1(F'(x_n^\tau)-F'(x_n))d\tau+Q(x_n,x_*)-Q(2x_n-x_{n-1},x_{n-1}))(x_n-x_*)\|\leq\\&\leq\|A_n^{-1}H'(x_*)\|(\|H'(x_*)^{-1}\int_0^1\int_{\tau\rho(x_n)}^{\rho(x_n)}L_1(u)dud\tau+\|H'(x_*)^{-1}(+Q(x_n,x_*)-\\&-Q(2x_n-x_{n-1},x_{n-1}))\|)\|x_n-x_*\|.\end{aligned}\tag{24}$$

According to the condition (17)–(19) of the theorem we get

$$\|H'(x_*)^{-1}(\int_0^1 \int_{\tau\rho(x_n)}^{\rho(x_n)} L_1(u)dud\tau + Q(x_n, x_*) - A_n)\| =$$

$$= \frac{1}{\rho(x_n)} \int_0^{\rho(x_n)} L_1(u)udu + \|H'(x_*)^{-1}(Q(x_n, x_*) - Q(x_n, x_n) +$$

$$+ Q(x_n, x_n) - Q(x_n, x_{n-1}) + Q(x_n, x_{n-1}) - Q(2x_n - x_{n-1}, x_{n-1}))\| \leq$$

$$\leq \frac{1}{\rho(x_n)} \int_0^{\rho(x_n)} L_1(u)udu + \|H'(x_*)^{-1}(Q(x_n, x_*) - Q(x_n, x_n))\| +$$

$$+ \|H'(x_*)^{-1}(Q(x_n, x_{n-1}, x_n) - Q(2x_n - x_{n-1}, x_{n-1}, x_n))(x_n - x_{n-1})\| \leq$$

$$\leq \frac{1}{\rho(x_n)} \int_0^{\rho(x_n)} L_1(u)udu + \int_0^{\rho(x_n)} L_2(u)du + \int_0^{\|x_n - x_{n-1}\|} N(u)du\|x_n - x_{n-1}\|.$$

From (22) and (24) shows that fulfills (21). Then from (21) and (20) we get

$$\|x_{n+1} - x_*\| < \|x_n - x_*\| < \ldots < \max\{\|x_0 - x_*\|, \|x_{-1} - x_*\|\} < r.$$

Therefore, the iterative process (5) is correctly defined and the sequence that it generates belongs to $B(x_*, r)$. From the last inequality and estimates (21) we get $\lim_{n\to\infty} \|x_n - x_*\| = 0$. Since the sequence $\{x_n\}_{n\geq 0}$ converges to x_*, then

$$\|x_n - x_{n-1}\| \leq \|x_n - x_*\| + \|x_{n-1} - x_*\| \leq 2\|x_{n-1} - x_*\|$$

and $\lim_{n\to\infty} \|x_n - x_{n-1}\| = 0$. □

Corollary 1. *The order of convergence of the iterative procedure (7) is quadratic.*

Proof. Let us denote $\rho_{\max} = \max\{\rho(x_0), \rho(x_{-1})\}$. Since $g(t)$ and $h(t)$ are monotonically nondecreasing, then with taking into account the expressions

$$\frac{1}{\rho(x_n)} \int_0^{\rho(x_n)} L_1(u)udu = \frac{\int_0^{\rho(x_n)} L_1(u)udu\rho(x_n))}{(\rho(x_n))^2} \leq \frac{\int_0^{\rho_{\max}} L_1(u)udu\rho(x_n)}{(\rho_{\max})^2} =: A_1\rho(x_n),$$

$$\int_0^{\rho(x_n)} L_2(u)du = \frac{\int_0^{\rho(x_n)} L_2(u)du\rho(x_n)}{\rho(x_n)} \leq \frac{\int_0^{\rho_{\max}} L_2(u)du\rho(x_n)}{\rho_{\max}} =: A_2\rho(x_n),$$

$$\int_0^{\|x_n - x_{n-1}\|} N(u)du = \frac{\int_0^{\|x_n - x_{n-1}\|} N(u)du\|x_n - x_{n-1}\|}{\|x_n - x_{n-1}\|} <$$

$$< \frac{\int_0^{\|x_0 - x_{-1}\|} N(u)du\|x_n - x_{n-1}\|}{\|x_0 - x_{-1}\|} =: A_3\|x_n - x_{n-1}\|$$

and

$$\Big(1 - \Big(\int_0^{\rho(x_n)} L_1^0(u)du + 2\int_0^{\rho(x_n)} L_2^0(u)du + \int_0^{\|x_n - x_{n-1}\|} N_0(u)du\|x_n - x_{n-1}\|\Big)\Big)^{-1} <$$

$$< \Big(1 - \Big(\int_0^{\rho_{\max}} L_1^0(u)du + 2\int_0^{\rho_{\max}} L_2^0(u)du + \int_0^{\|x_0 - x_{-1}\|} N_0(u)du\|x_0 - x_{-1}\|\Big)\Big)^{-1} =: A_4,$$

from the inequality (21) follows

$$\|x_{n+1} - x_*\| \leq A_4(A_1\rho(x_n) + A_2\rho(x_n) + A_3\|x_n - x_{n-1}\|^2)\|x_n - x_*\|.$$

or

$$\|x_{n+1} - x_*\| \leq C_3\|x_n - x_*\|^2 + C_4\|x_n - x_{n-1}\|^2\|x_n - x_*\|. \tag{25}$$

Here $A_k, k = 1, ..., 4, C_3, C_4$ are some positive constants.

Assume that the order of convergence of the iterative process (7) is not lower 2, therefore there exist $C_5 \geq 0$ and $N > 0$, that for all $n \geq N$ the inequality holds

$$\|x_n - x_*\| \geq C_5\|x_{n-1} - x_*\|^2.$$

Since

$$\|x_n - x_{n-1}\|^2 \leq (\|x_n - x_*\| + \|x_{n-1} - x_*\|)^2 \leq 4\|x_{n-1} - x_*\|^2,$$

then from (44) we get

$$\begin{aligned} \|x_{n+1} - x_*\| &\leq C_3\|x_n - x_*\|^2 + 4C_4\|x_{n-1} - x_*\|^2\|x_n - x_*\| \\ &\leq (C_3 + 4C_4/C_5)\|x_n - x_*\|^2 = C_6\|x_n - x_*\|^2. \end{aligned} \tag{26}$$

inequality (26) means that the order of convergence is not lower than 2. Thus, the convergence rate of sequence $\{x_n\}_{n\geq 0}$ to x_* is quadratic. □

3. Uniqueness Ball of the Solution

The next theorem determines the ball of uniqueness of the solution x_* of (1) in $B(x_*, r)$.

Theorem 2. *Let us assume that: (1) $H(x) \equiv F(x) + Q(x) = 0$ has a solution $x_* \in D$, in which there exists a Fréchet derivative $H'(x_*)$ and it is invertible; (2) F has a continuous Frećhet derivative in $B(x_*, r)$, F' satisfies the generalized Lipschitz condition*

$$\|H'(x_*)^{-1}(F'(x) - F'(x_*))\| \leq \int_0^{\rho(x)} L_1^0(u)du \quad \forall x \in B(x_*, r),$$

the divided difference $Q(x, y)$ satisfies the generalized Lipschitz condition

$$\|H'(x_*)^{-1}(Q(x, x_*) - G'(x_*))\| \leq \int_0^{\rho(x)} L_2^0(u)du \quad \forall x \in B(x_*, r),$$

where L_1 and L_2 are positive integrable functions. Let $r > 0$ satisfy

$$\frac{1}{r}\int_0^r (r - u)L_1^0(u)du + \int_0^r L_2^0(u)du \leq 1.$$

Then the equation $H(x) = 0$ has a unique solution x_* in $B(x_*, r)$.

Proof analogous to [27,31].

4. Corollaries

In the study of iterative methods, the traditional assumption is that the derivatives and/or the divided differences satisfy the classical Lipschitz conditions. Assuming that L_1, L_2 and N are constants, we get from Theorems 1 and 2 important corollaries, which are of interest.

Corollary 2. *Let us assume that: (1) $H(x) \equiv F(x) + Q(x) = 0$ has a solution $x_* \in D$, in which there exists Fréchet derivative $H'(x_*)$ and it is invertible; (2) F has a continuous Fréchet derivative and Q has divided differences of the first and second order $Q(x,y)$ and $Q(x,y,z)$ in $B(x_*, 3r) \subset D$, which satisfy the Lipschitz conditions for each $x, y, u, v \in D$*

$$\|H'(x_*)^{-1}(F'(x) - F'(x_*)\| \leq L_1^0 \|x - x_*\|,$$

$$\|H'(x_*)^{-1}(Q(x,y) - Q(u,v))\| \leq L_2^0(\|x-u\| + \|y-v\|),$$

for $x, y, u, v \in D_0$

$$\|H'(x_*)^{-1}(Q(u,x,y) - Q(v,x,y))\| \leq N_0\|u - v\|,$$

$$\|H'(x_*)^{-1}(F'(x) - F'(x_* + \tau(x - x_*))\| \leq (1-\tau)L_1\|x - x_*\|,$$

$$\|H'(x_*)^{-1}(Q(x,y) - Q(u,v))\| \leq L_2(\|x-u\| + \|y-v\|),$$

$$\|H'(x_*)^{-1}(Q(u,x,y) - Q(v,x,y))\| \leq N\|u - v\|,$$

where L_1^0, L_2^0, N_0, L_1, L_2 and N are positive numbers,

$$r_0 = \frac{2}{L_1^0 + 2L_2^0 + \sqrt{(L_1^0 + 2L_2^0)^2 + 16N_0}},$$

and r is the positive root of the equation

$$\frac{L_1 r/2 + L_2 r + 4Nr^2}{1 - L_1^0 r - 2L_2^0 r - 4N_0 r^2} = 1.$$

Then Newton-Kurchatov method (5) converges for all $x_{-1}, x_0 \in B(x_*, r)$ and there fulfills

$$\|x_{n+1} - x_*\| \leq \frac{(L_1/2 + L_2)\|x_n - x_*\| + N\|x_n - x_{n-1}\|^2}{1 - \left(L_1^0 + 2L_2^0\|x_n - x_*\| + N_0\|x_n - x_{n-1}\|^2\right)}.$$

Moreover, r is the best of all possible.

Note that value of $r = \dfrac{2}{3L}$ improves $\bar{r} = \dfrac{2}{3L_1^1}$ for Newton method for solving equation $F(x) = 0$ [14,32,33], and with $r = 2/(3L_2 + \sqrt{9L_2^2 + 32N})$ improves $\bar{r} = 2/(3L_2^1 + \sqrt{9(L_2^1)^2 + 32N_1})$ for Kurchatov method for solving the equation $Q(x) = 0$, as derived in [8].

Corollary 3. *Suppose that: (1) $H(x) \equiv F(x) + Q(x) = 0$ has a solution $x_* \in D$, in which there exists the Fréchet derivative $H'(x_*)$ and it is invertible; (2) F has continuous derivative and Q has divided difference $Q(x, x_*)$ in $B(x_*, r) \subset D$, which satisfy the Lipschitz conditions*

$$\|H'(x_*)^{-1}(F'(x) - F'(x_*))\| \leq L_1^0\|x - x_*\|,$$

$$\|H'(x_*)^{-1}(Q(x, x_*) - G'(x_*))\| \leq L_2^0\|x - x_*\|$$

for all $x \in B(x_*, r)$, where L_1^0 and L_2^0 are positive numbers and $r = \dfrac{2}{L_1^0 + 2L_2^0}$. Then x_* is the only solution in $B(x_*, r)$ of $H(x) = 0$, r does not depend on F and Q and is the best choice.

Note that the resulting radius of the uniqueness ball of the solution $r = \frac{2}{L_1}$ improves $\bar{r} = \frac{2}{L_1^1}$ for Newton method for solving the equation $F(x) = 0$ [14] and $r = \frac{1}{L_2}$ improves $\bar{r} = \frac{1}{L_2^1}$ for Kurchatov method for solving the equation $Q(x) = 0$ [8]. (See also the numerical examples).

Remark 2. *We compare the results in [25] with the new results in this article. In order to do this, let us consider the conditions given in [25] corresponding to our conditions (15)–(17):*

For each $x, y, u, v \in D$

$$\|H'(x_*)^{-1}(F'(x) - F'(x^\theta))\| \leq \int_{\theta\rho(x)}^{\rho(x)} L_1^1(u)du, \ 0 \leq \theta \leq 1, \tag{27}$$

$$\|H'(x_*)^{-1}(Q(x,y) - Q(u,v))\| \leq \int_0^{\|x-u\|+\|y-v\|} L_2^1(t)dt, \tag{28}$$

$$\|H'(x_*)^{-1}(Q(u,x,y) - Q(v,x,y))\| \leq \int_0^{\|u-v\|} N^1(t)dt, \tag{29}$$

$$\frac{\frac{1}{\bar{r}}\int_0^{\bar{r}} L_1^1(u)udu + \int_0^{\bar{r}} L_2^1(u)du + 2\bar{r}\int_0^{2\bar{r}} N^1(u)du}{1 - \left(\int_0^{\bar{r}} L_1^1(u)du + \int_0^{2\bar{r}} L_2^1(u)du + 2\bar{r}\int_0^{2\bar{r}} N^1 1(u)du\right)} = 1, \tag{30}$$

$$\|x_{n+1} - x_n\| \leq \bar{e}_n. \tag{31}$$

It follows from (14)–(16), (17)–(19), (27)–(29), that

$$L_1^0(t) \leq L_1^1(t), \tag{32}$$

$$L_1(t) \leq L_1^1(t), \tag{33}$$

$$L_2^0(t) \leq L_2^1(t), \tag{34}$$

$$L_2(t) \leq L_2^1(t), \tag{35}$$

$$N_0(t) \leq N^1(t), \tag{36}$$

$$N(t) \leq N^1(t), \tag{37}$$

leading to

$$\bar{r} \leq r, \tag{38}$$

$$e_n \leq \bar{e}_n, \tag{39}$$

$$A_l \leq \bar{A}_l, \ l = 1,2,3,4, \tag{40}$$

$$C_l \leq \bar{C}_l, \ l = 1,2,3,4,6 \tag{41}$$

and

$$C_5 \geq \bar{C}_5, \tag{42}$$

$$\bar{e}_n := \frac{\frac{1}{\rho(x_n)}\int_0^{\rho(x_n)} L_1^1(u)udu + \int_0^{\rho(x_n)} L_2^1(u)du + \int_0^{\|x_n-x_{n-1}\|} N^1(u)du\|x_n - x_{n-1}\|}{1 - \left(\int_0^{\rho(x_n)} L_1^1(u)du + \int_0^{2\rho(x_n)} L_2^1(u)du + \int_0^{\|x_n-x_{n-1}\|} N^1(u)du\|x_n - x_{n-1}\|\right)}\|x_n - x_*\|, \tag{43}$$

$$\|x_{n+1} - x_*\| \leq \bar{A}_4(\bar{A}_1\rho(x_n) + \bar{A}_2\rho(x_n) + \bar{A}_3\|x_n - x_{n-1}\|^2)\|x_n - x_*\|,$$

or

$$\|x_{n+1}-x_*\| \leq \bar{C}_3\|x_n-x_*\|^2+\bar{C}_4\|x_n-x_{n-1}\|^2\|x_n-x_*\|, \tag{44}$$

or

$$\|x_{n+1}-x_*\| \leq \bar{C}_6\|x_n-x_*\|^2$$

with

$$\bar{C}_6=(\bar{C}_3+4\bar{C}_4/\bar{C}_5)$$

for some

$$\|x_n-x_*\| \geq \bar{C}_5\|x_{n-1}-x_*\|^2.$$

Hence, we obtain the impovements:

(1) At least as many initial choices x_{-1}, x_0 as before.
(2) At least as few iterations than before to obtain a predetermined error accuracy.
(3) At least as precice information on the location of the solution as before.

Moreover, if any of (32)–(37) holds as a strict inequality, then so do (38)–(42). Furthermore, we notice that these improvements are found using the same information, since the functions $L_1^0, L_2^0, N_0, L_1, L_2, N$ are special cases of functions L_1^1, L_2^1, N^1 used in [25]. Finally, if $G=0$ or $F=0$, we obtain the results for Newton's method or the Kurchatov method as special cases. Clearly, the results for these methods are also improved. Our technique can also be used to improve the results of other iterative methods in an analogous way.

5. Numerical Examples

Example 1. *Let $E_1=E_2=R^3$ and $\Omega=S(x_*,1)$. Define functions F and Q for $v=(v_1,v_2,v_3)^T$ on Ω by*

$$\begin{aligned} F(v) &= \left(e^{v_1}-1,\ \tfrac{e-1}{2}v_2^2+v_2,\ v_3\right)^T, \\ Q(v) &= \left(|v_1|,\ |v_2|,\ |v_2|, |\sin(v_3)|\right)^T \end{aligned} \tag{45}$$

$$\begin{aligned} F'(v) &= \operatorname{diag}\left(e^{v_1},\ (e-1)v_2+1, 1\right), \\ Q(v,\bar{v}) &= \operatorname{diag}\left(\tfrac{|\bar{v}_1|-|v_1|}{\bar{v}_1-v_1},\ \tfrac{|\bar{v}_2|-|v_2|}{\bar{v}_2-v_2},\ \tfrac{|\sin(\bar{v}_3)|-|\sin(v_3)|}{\bar{v}_3-v_3}\right) \end{aligned} \tag{46}$$

Choose:

$$H(x)=F(x)+Q(x),$$

$$\|H'(x_*)^{-1}\|=1, L_1^0=e-1, L_2^0=1, N_0=\frac{1}{2},$$

$$L_1=e^{\frac{1}{e-1}}, L_2=1, N=\frac{1}{2},$$

$$L_1^1=e, L_2^1=1, N^1=\frac{1}{2}.$$

Then compute:
r using (20), $r=0.1599$;
$\bar{r}$ using (30), $\bar{r}=0.1315$.
Also, $\bar{r}<r$.
Notice that $L_1^0<L_1<L_1^1$, so the improve ments stated in Remark 1 hold.

6. Conclusions

In [1,8,34], we studied the local convergence of Secant and Kurchatov methods in the case of fulfilment of Lipschitz conditions for the divided differences, which hold for some Lipschitz constants. In [14], the convergence of the Newton method is shown for the generalized Lipschitz conditions for the Fréchet derivative of the first order. We explored the local convergence of the Newton-Kurchatov method under the generalized Lipschitz conditions for Fréchet derivative of a differentiable part of the operator and the divided differences of the nondifferentiable part. Our results contain known parts as partial cases.

By using our idea of restricted convergence regions, we find tighter Lipschitz constants leading to a finer local convergence analysis of method (7) and its special cases compared to in [25].

Author Contributions: All authors contributed equally and significantly to writing this article. All authors read and approved the final manuscript.

Acknowledgments: The authors would like to express their sincere gratitude to the referees for their valuable comments which have significantly improved the presentation of this paper.

References

1. Argyros, I.K. *Convergence and Applications of Newton-Type Iterations*; Springer: New York, NY, USA, 2008.
2. Hernandez, M.A.; Rubio, M.J. The Secant method and divided differences Hölder continuous. *Appl. Math. Comput.* **2001**, *124*, 139–149. [CrossRef]
3. Hernandez, M.A.; Rubio, M.J. The Secant method for nondifferentiable operators. *Appl. Math. Lett.* **2002**, *15*, 395–399. [CrossRef]
4. Ortega, J.M.; Rheinboldt, W.C. *Iterative Solution of Nonlinear Equations in Several Variables*; Academic Press: New York, NY, USA, 1970.
5. Traub, J.F. *Iterative Methods for the Solution of Equations*; Prentice-Hall, Inc.: Englewood Cliffs, NJ, USA, 1964.
6. Shakhno, S.M. Application of nonlinear majorants for intestigation of the secant method for solving nonlinear equations. *Matematychni Studii* **2004**, *22*, 79–86.
7. Ezquerro, J.A.; Grau-Sánchez, M.; Hernández, M.A. Solving non-differentiable equations by a new one-point iterative method with memory. *J. Complex.* **2012**, *28*, 48–58. [CrossRef]
8. Shakhno, S.M. On a Kurchatov's method of linear interpolation for solving nonlinear equations. *Proc. Appl. Math. Mech.* **2004**, *4*, 650–651. [CrossRef]
9. Shakhno, S.M. About the difference method with quadratic convergence for solving nonlinear operator equations. *Matematychni Studii* **2006**, *26*, 105–110. (In Ukrainian)
10. Argyros, I.K. A Kantorovich-type analysis for a fast iterative method for solving nonlinear equations. *J. Math. Anal. Appl.* **2007**, *332*, 97–108. [CrossRef]
11. Kurchatov, V.A. On a method of linear interpolation for the solution of functional equations. *Dokl. Akad. Nauk SSSR* **1971**, *198*, 524–526. (In Russian); translation in *Soviet Math. Dokl.* **1971**, *12*, 835–838.
12. Ezquerro, J.A.; Hernández, M. Generalized differentiability conditions for Newton's method. *IMA J. Numer. Anal.* **2002**, *22*, 187–205. [CrossRef]
13. Gutiérrez, J.M.; Hernández, M.A. Newton's method under weak Kantorovich conditions. *IMA J. Numer. Anal.* **2000**, *20*, 521–532. [CrossRef]
14. Wang, X.H. Convergence of Newton's method and uniqueness of the solution of equations in Banach space. *IMA J. Numer. Anal.* **2000**, *20*, 123–134. [CrossRef]
15. Wang, X.H.; Li, C. Local and global behavior for algorithms of solving equations. *Chin. Sci. Bull.* **2001**, *46*, 444–451. [CrossRef]
16. Shakhno, S.M. On the secant method under generalized Lipschitz conditions for the divided difference operator. *Proc. Appl. Math. Mech.* **2007**, *7*, 2060083–2060084. [CrossRef]
17. Shakhno, S.M. Method of linear interpolation of Kurchatov under generalized Lipschitz conditions for divided differences of first and second order. *Visnyk Lviv. Univ. Ser. Mech. Math.* **2012**, *77*, 235–242. (In Ukrainian)
18. Amat, S. On the local convergence of Secant-type methods. *Intern. J. Comput. Math.* **2004**, *81*, 1153–1161. [CrossRef]

19. Amat, S.; Busquier, S. On a higher order Secant method. *Appl. Math. Comput.* **2003**, *141*, 321–329. [CrossRef]
20. Argyros, I.K.; Ezquerro, J.A.; Gutiérrez, J.M.; Hernández, M.A.; Hilout, S. Chebyshev-Secant type methods for non-differentiable operator. *Milan J. Math.* **2013**, *81*, 25–35.
21. Ren, H. New sufficient convergence conditions of the Secant method nondifferentiable operators. *Appl. Math. Comput.* **2006**, *182*, 1255–1259. [CrossRef]
22. Ren, H.; Argyros, I.K. A new semilocal convergence theorem with nondifferentiable operators. *J. Appl. Math. Comput.* **2010**, *34*, 39–46. [CrossRef]
23. Donchev, T.; Farkhi, E.; Reich, S. Fixed set iterations for relaxed Lipschitz multimaps. *Nonlinear Anal.* **2003**, *53*, 997–1015. [CrossRef]
24. Shakhno, S.M.; Yarmola, H.P. Two-point method for solving nonlinear equation with nondifferentiable operator. *Matematychni Studii.* **2011**, *36*, 213–220. (In Ukrainian)
25. Shakhno, S.M. Combined Newton-Kurchatov method under the generalized Lipschitz conditions for the derivatives and divided differences. *J. Numer. Appl. Math.* **2015**, *2*, 78–89.
26. Catinas, E. On some iterative methods for solving nonlinear equations. *Revue d'Analyse Numérique et de Théorie de l'Approximation* **1994**, *23*, 47–53.
27. Shakhno, S. Convergence of combined Newton-Secant method and uniqueness of the solution of nonlinear equations. *Visnyk Ternopil Nat. Tech. Univ.* **2013**, *69*, 242–252. (In Ukrainian)
28. Shakhno, S.M.; Mel'nyk, I.V.; Yarmola, H.P. Analysis of convergence of a combined method for the solution of nonlinear equations. *J. Math. Sci.* **2014**, *201*, 32–43. [CrossRef]
29. Argyros, I. K. On the secant method. *Publ. Math. Debr.* **1993**, *43*, 233–238.
30. Kantorovich, L.V.; Akilov, G.P. *Functional Analysis*; Pergamon Press: Oxford, UK, 1982.
31. Shakhno, S.M. Convergence of the two-step combined method and uniqueness of the solution of nonlinear operator equations. *J. Comput. Appl. Math.* **2014**, *261*, 378–386. [CrossRef]
32. Potra, F.A. On an iterative algorithm of order 1.839... for solving nonlinear operator equations. *Numer. Funct. Anal. Optim.* **1985**, *7*, 75–106. [CrossRef]
33. Traub, J.F.; Woźniakowski, H. Convergence and complexity of Newton iteration for operator equations. *J. Assoc. Comput. Mach.* **1979**, *26*, 250–258. [CrossRef]
34. Hernandez, M.A.; Rubio, M.J. A uniparametric family of iterative processes for solving nondifferentiable equations. *J. Math. Anal. Appl.* **2002**, *275*, 821–834. [CrossRef]

13

A Generic Family of Optimal Sixteenth-Order Multiple-Root Finders and their Dynamics Underlying Purely Imaginary Extraneous Fixed Points

Min-Young Lee [1], Young Ik Kim [1,*] and Beny Neta [2]

[1] Department of Applied Mathematics, Dankook University, Cheonan 330-714, Korea; leemy@dankook.ac.kr
[2] Naval Postgraduate School, Department of Applied Mathematics, Monterey, CA 93943, USA; bneta@nps.edu
* Correspondence: yikbell79@yahoo.com.

Abstract: A generic family of optimal sixteenth-order multiple-root finders are theoretically developed from general settings of weight functions under the known multiplicity. Special cases of rational weight functions are considered and relevant coefficient relations are derived in such a way that all the extraneous fixed points are purely imaginary. A number of schemes are constructed based on the selection of desired free parameters among the coefficient relations. Numerical and dynamical aspects on the convergence of such schemes are explored with tabulated computational results and illustrated attractor basins. Overall conclusion is drawn along with future work on a different family of optimal root-finders.

Keywords: sixteenth-order optimal convergence; multiple-root finder; asymptotic error constant; weight function; purely imaginary extraneous fixed point; attractor basin

MSC: 65H05; 65H99

1. Introduction

Many nonlinear equations governing real-world natural phenomena cannot be solved exactly by virtue of their intrinsic complexities. It would be certainly an important matter to discuss methods for approximating such solutions of the nonlinear equations. The most widely accepted method under general circumstances is Newton's method, which has quadratic convergence for a simple-root and linear convergence for a multiple-root. Other higher-order root-finders have been developed by many researchers [1–9] with optimal convergence satisfying Kung–Traub's conjecture [10]. Several authors [10–14] have proposed optimal sixteenth-order simple-root finders, although their applications to real-life problems are limited due to the high degree of their algebraic complexities. Optimal sixteenth-order multiple-root finders are hardly found in the literature to the best of our knowledge at the time of writing this paper. It is not too much to emphasize the theoretical importance of developing optimal sixteenth-order multiple root-finders as well as to apply them to numerically solve real-world nonlinear problems.

In order to develop an optimal sixteenth-order multiple-root finders, we pursue a family of iterative methods equipped with generic weight functions of the form:

$$\begin{cases} y_n = x_n - m\frac{f(x_n)}{f'(x_n)}, \\ z_n = y_n - mQ_f(s)\frac{f(y_n)}{f'(x_n)} = x_n - m\big[1+sQ_f(s)\big]\frac{f(x_n)}{f'(x_n)}, \\ w_n = z_n - mK_f(s,u)\frac{f(z_n)}{f'(x_n)} = x_n - m\big[1+sQ_f(s)+suK_f(s,u)\big]\frac{f(x_n)}{f'(x_n)}, \\ x_{n+1} = w_n - mJ_f(s,u,v)\frac{f(w_n)}{f'(x_n)} = x_n - m\big[1+sQ_f(s)+suK_f(s,u)+suvJ_f(s,u,v)\big]\frac{f(x_n)}{f'(x_n)}, \end{cases} \quad (1)$$

where $s = \left(\frac{f(y_n)}{f(x_n)}\right)^{1/m}$, $u = \left(\frac{f(z_n)}{f(y_n)}\right)^{1/m}$, $v = \left(\frac{f(w_n)}{f(z_n)}\right)^{1/m}$; $Q_f : \mathbb{C} \to \mathbb{C}$ is analytic [15] in a neighborhood of 0, $K_f : \mathbb{C}^2 \to \mathbb{C}$ holomorphic [16,17] in a neighborhood of $(0,0)$, and $J_f : \mathbb{C}^3 \to \mathbb{C}$ holomorphic in a neighborhood of $(0,0,0)$. Since s, u and v are respectively one-to-m multiple-valued functions, their principal analytic branches [15] are considered. Hence, for instance, it is convenient to treat s as a principal root given by $s = \exp[\frac{1}{m}\mathrm{Log}(\frac{f(y_n)}{f(x_n)})]$, with $\mathrm{Log}(\frac{f(y_n)}{f(x_n)}) = \mathrm{Log}|\frac{f(y_n)}{f(x_n)}| + i\,\mathrm{Arg}(\frac{f(y_n)}{f(x_n)})$ for $-\pi < \mathrm{Arg}(\frac{f(y_n)}{f(x_n)}) \leq \pi$; this convention of Arg(z) for $z \in \mathbb{C}$ agrees with that of Log[z] command of Mathematica [18] to be employed later in numerical experiments.

The case for $m = 1$ has been recently developed by Geum–Kim–Neta [19]. Many other existing cases for $m = 1$ are special cases of (1) with appropriate forms of weight functions Q_f, K_f, and J_f; for example, the case developed in [10] uses the following weight functions:

$$\begin{cases} Q_f(s) = \frac{1}{(1-s)^2}, \\ K_f(s,u) = \frac{1+(1-u)s^2}{(1-s)^2(1-u)(1-su)^2}, \\ J_f(s,u,v) = \frac{-1+2su^2(v-1)+s^4(u-1)u^2(v-1)(uv-1)+s^2[uv-1-u^3(v^2-1)]}{(1-s)^2(u-1)(su-1)^2(v-1)(uv-1)(suv-1)^2}. \end{cases} \tag{2}$$

One goal of this paper is to construct a family of optimal sixteenth-order multiple-root finders by characterizing the generic forms of weight functions $Q_f(s)$, $K_f(s,u)$, and $J_f(s,u,v)$. The other goal is to investigate the convergence behavior by exploring their numerical behavior and dynamics through basins of attractions [20] underlying the extraneous fixed points [21] when $f(z) = (z-a)^m(z-b)^m$ is applied. In view of the right side of final substep of (1), we can conveniently locate extraneous fixed points from the roots of the weight function $m[1 + sQ_f(s) + suK_f(s,u) + suvJ_f(s,u,v)]$.

A motivation undertaking this research is to investigate the local and global characters on the convergence of proposed family of methods (1). The local convergence of an iterative method for solving nonlinear equations is usually guaranteed with an initial guess taken in a sufficiently close neighborhood of the sought zero. On the other hand, effective information on its global convergence is hardly achieved under general circumstances. We can obtain useful information on the global convergence from attractor basins through which relevant dynamics is worth exploring. Especially the dynamics underlying the extraneous fixed points (to be described in Section 3) would influence the dynamical behavior of the iterative methods by the presence of possible attractive, indifferent, repulsive, and other chaotic orbits. One way of reducing such influence is to control the location of the extraneous fixed points. We prefer the location to be the imaginary axis that divides the entire complex plane into two symmetrical half-planes. The dynamics underlying the extraneous fixed points on the imaginary axis would be less influenced by the presence of the possible periodic or chaotic attractors.

The main theorem is presented in Section 2 with required constraints on weight functions, Q_f, K_f, and J_f to achieve the convergence order of 16. Section 2 discusses special cases of rational weight functions. Section 3 extensively investigates the purely imaginary extraneous fixed points and investigates their stabilities. Section 4 presents numerical experiments as well as the relevant dynamics, while Section 5 states the overall conclusions along with the short description of future work.

2. Methods and Special Cases

A main theorem on the convergence of (1) is established here with the error equation and relationships among generic weight functions $Q_f(s)$, $K_f(s,u)$, and $J_f(s,u,v)$:

Theorem 1. *Suppose that $f : \mathbb{C} \to \mathbb{C}$ has a multiple root α of multiplicity $m \geq 1$ and is analytic in a neighborhood of α. Let $c_j = \frac{m!}{(m-1+j)!}\frac{f^{(m-1+j)}(\alpha)}{f^{(m)}(\alpha)}$ for $j = 2,3,\cdots$. Let x_0 be an initial guess selected in a sufficiently small region containing α. Assume $L_f : \mathbb{C} \to \mathbb{C}$ is analytic in a neighborhood of 0. Let $Q_i = \frac{1}{i!}\frac{d^i}{ds^i}Q_f(s)|_{(s=0)}$ for $0 \leq i \leq 6$. Let $K_f : \mathbb{C}^2 \to \mathbb{C}$ be holomorphic in a neighborhood of $(0,0)$. Let $J_f : \mathbb{C}^3 \to \mathbb{C}$ be holomorphic in a neighborhood of $(0,0,0)$. Let $K_{ij} = \frac{1}{i!j!}\frac{\partial^{i+j}}{\partial s^i \partial u^j}K_f(s,u)|_{(s=0,u=0)}$ for*

$0 \le i \le 12$ and $0 \le j \le 6$. Let $J_{ijk} = \frac{1}{i!j!k!}\frac{\partial^{i+j+k}}{\partial s^i \partial u^j \partial v^j} J_f(s,u,v)\big|_{(s=0,u=0,v=0)}$ for $0 \le i \le 8$, $0 \le j \le 4$ and $0 \le k \le 2$. If $Q_0 = 1, Q_1 = 2, K_{00} = 1, K_{10} = 2, K_{01} = 1, K_{20} = 1 + Q_2, K_{11} = 4, K_{30} = -4 + 2Q_2 + Q_3, J_{000} = 1, J_{100} = 2, J_{200} = 1 + Q_2, J_{010} = 1, J_{110} = 4, J_{300} = -4 + 2Q_2 + Q_3, J_{001} = 1, J_{020} = K_{02}, J_{210} = 1 + K_{21}, J_{400} = K_{40}, J_{101} = 2, J_{120} = 2 + K_{12}, J_{310} = -4 + K_{31} + 2Q_2, J_{500} = K_{50}, J_{011} = 2, J_{201} = 1 + Q_2, J_{030} = -1 + K_{02} + K_{03}, J_{220} = 1 + K_{21} + K_{22} - Q_2, J_{410} = -3 + K_{40} + K_{41} + Q_2 - Q_4, J_{600} = K_{60}, J_{111} = 8, J_{301} = -4 + 2Q_2 + Q_3, J_{130} = -4 + 2K_{02} + K_{12} + K_{13}, J_{320} = -6 + 2K_{21} + K_{31} + K_{32} - 2Q_2 - Q_3, J_{510} = 6 + 2K_{40} + K_{50} + K_{51} - 3Q_3 - 2Q_4 - Q_5, J_{700} = K_{70}$ are fulfilled, then Scheme (1) leads to an optimal class of sixteenth-order multiple-root finders possessing the following error equation: with $e_n = x_n - \alpha$ for $n = 0, 1, 2, \cdots$,

$$e_{n+1} = \frac{1}{3456m^{15}} c_2(\rho c_2^2 - 2mc_3)\left[\beta_0 c_2^4 + \beta_1 c_2^2 c_3 + 12m^2(K_{02}-1)c_3^2 - 12m^2 c_2 c_4\right]\Psi\, e_n^{16} + O(e_n^{17}), \quad (3)$$

where $\rho = 9 + m - 2Q_2$, $\beta_0 = (-431 + 12K_{40} - 7m^2 + 6m(-17 + Q_2) + 102Q_2 - 24Q_3 - 12Q_4 + 6K_{21}\rho + 3K_{02}\rho^2)$, $\beta_1 = -12m(-17 + K_{21} - 2m + Q_2 + K_{02}\rho)$, $\Psi = \Delta_1 c_2^8 + \Delta_2 c_2^6 c_3 + \Delta_3 c_2^5 c_4 + \Delta_4 c_2^3 c_3 c_4 + \Delta_5 c_2^4 + \Delta_6 c_2^2 + \Delta_7 c_3^4 + \Delta_8 c_2 c_3^2 c_4$,

$\Delta_1 = (-255124 + 144J_{800} - 144K_{80} - 122577m - 23941m^2 - 2199m^3 - 79m^4 + 24K_{40}(472 + 93m + 5m^2) - 72(17+m)Q_2^3 - 576Q_3^2 + Q_3(48(-566 + 6K_{40} - 117m - 7m^2) - 576Q_4) + 24(-485 + 6K_{40} - 108m - 7m^2)Q_4 - 144Q_4^2 + Q_2^2(36(-87 + 14m + m^2) + 288Q_3 + 144Q_4) + Q_2(18(5300 + 1529m + 172m^2 + 7m^3 - 8K_{40}(18+m)) + 144(35+m)Q_3 + 72(29+m)Q_4) + 18\rho^3\sigma + 6\rho(12J_{610} - 12(2K_{50} + K_{60} + K_{61} - 2Q_5 - Q_6) + J_{211}(-12K_{40} + \sigma_2) + 2K_{21}(-J_{002}\sigma_2 + \sigma_3 + 6\eta_0)) + \rho^2(36J_{420} - 36J_{211}K_{21} + 36J_{002}K_{21}^2 - 72K_{31} - 36J_{021}K_{40} - 36K_{41} - 36K_{42} + 3J_{021}\sigma_2 + 6K_{02}(-6J_{401} + 12J_{002}K_{40} - J_{002}\sigma_2 + \sigma_3)) + 12J_{401}\sigma_7 + J_{002}\sigma_7^2 + 9\rho^4\tau)$,

$\Delta_2 = m(144(Q_2^3 - 2Q_5 - Q_6) + 288Q_3(-K_{21} + (39+4m) - K_{02}\rho) + 144Q_2^2(-2K_{21} - (7+m) - K_{02}\rho) + 144Q_4(-K_{21} + 4(9+m) - K_{02}\rho) + Q_2(144K_{21}(58+5m) - 36(1529 - 8K_{40} + m(302 + 17m)) - 288Q_3 - 144Q_4 + 144K_{02}(38+3m)\rho) - 108\rho^2\sigma + 6(40859 - 24J_{610} + 48K_{50} + 24K_{60} + 24K_{61} + 24K_{40}(-31 + J_{211} - 3m) + m(14864 + m(1933 + 88m)) - 2J_{211}\sigma_8) - 72\rho^3\tau - 24\rho_2(J_{002}\sigma_2 - 6\eta_0) - 24K_{21}(1309 + m(267 + 14m) - J_{002}\sigma_8 + 6\eta_0) + \rho(144J_{211}K_{21} - 144J_{002}K_{21}^2 + 12(-12J_{420} + 12(2K_{31} + J_{021}K_{40} + K_{41} + K_{42}) - J_{021}\sigma_4) - 12K_{02}(1781 + m(360 + 19m) - 2J_{002}\sigma_4 + 12\eta_0)))$,

$\Delta_3 = 12m^2(6(63+5m)Q_2 - 48Q_3 - 24Q_4 + 6(J_{211} + K_{21} - 2J_{002}K_{21})\rho - (1645 - 12J_{401} + 12(-1 + 2J_{002})K_{40} + 372m + 23m^2 - 2J_{002}\sigma_2) + 3\rho^2\sigma_5)$,

$\Delta_4 = 144m^3(-3Q_2 + (53 - J_{211} + (-1 + 2J_{002})K_{21} + 6m - 2J_{002}\rho_2) - \rho\sigma_5)$,

$\Delta_5 = 72\rho m^3 c_5 + 12m^2 c_3^2((-12K_{21}(J_{211} - 4(10+m)) + K_{02}(4778 - 12J_{401} + m(990 + 52m)) + 3(-1929 + 4J_{401} + 4J_{420} - 8K_{31} + 8K_{40} - 4K_{41} - 4K_{42} - 476m - 31m^2 + 2J_{211}(43+5m) + J_{021}(115 + 22m + m^2))) - 6(-88 + 121J_{021} + 4J_{211} + 232K_{02} + 8K_{21} + (-10 + 13J_{021} + 22K_{02})m + 2J_{002}(51 - 4K_{21} + 5m))Q_2 + 12(1 + J_{002} + 3J_{021} + 6K_{02})Q_2^2 + 18\rho\sigma + 18\rho^2\tau + Q_4(12(-2 + K_{02}) + 12\eta_1) + Q_3(24(-2 + K_{02}) + 24\eta_1) - J_{021}\eta_2 + \eta_1((2165 - 12K_{40} + m(510 + 31m)) + \eta_2))$,

$\Delta_6 = 72m^3(2(-1 + J_{002})mc_4^2 - 2mc_3c_5 + c_3^3(2Q_2(-1 + 6K_{02}) - 2K_{02}(49 - J_{211} + 5m + 2J_{002}\rho_3) + (85 - 2J_{211} - 2J_{230} + 4K_{12} - 4K_{21} + 4J_{002}(K_{21} - \rho_2) + 2K_{22} + 2K_{23} + 11m + 2J_{021}\rho_3) - 4\rho\tau))$,

$\Delta_7 = 144m^4(-1 + J_{002} + J_{021} + J_{040} - K_{03} - K_{04} + (2 - \eta_1)K_{02} + J_{002}K_{02}^2)$, $\Delta_8 = 144m^4(-3 + \eta_1 + (1 - 2J_{002})K_{02})$,

$\tau = J_{040} - J_{021}K_{02} + J_{002}K_{02}^2 - K_{03} - K_{04}$, $\rho_2 = 17 + 2m - Q_2$, $\rho_3 = -26 + K_{21} - 3m + 3Q_2$, $\sigma = J_{230} - J_{211}K_{02} - 2K_{12} - J_{021}K_{21} + 2J_{002}K_{02}K_{21} - K_{22} - K_{23}$, $\sigma_2 = 431 + 7m^2 - 6m(-17 + Q_2) - 102Q_2 + 24Q_3 + 12Q_4$, $\sigma_3 = 472 + 5m^2 + m(93 - 6Q_2) - 108Q_2 + 12Q_3 + 6Q_4$, $\sigma_4 = 890 + 13m^2 - 231Q_2 + 6Q_2^2 - 3m(-69 + 7Q_2) + 24Q_3 + 12Q_4$, $\sigma_5 = J_{021} + K_{02} - 2J_{002}K_{02}$, $\sigma_6 = -1255 + 6K_{40} - 288m - 17m^2 + 363Q_2 + 39mQ_2 - 18Q_2^2 - 12Q_3 - 6Q_4$, $\sigma_7 = 431 - 12K_{40} + 7m^2 - 6m(-17 + Q_2) - 102Q_2 + 24Q_3 + 12Q_4$, $\sigma_8 = 1349 + 19m^2 + m(312 - 36Q_2) - 360Q_2 + 12Q_2^2 + 24Q_3 + 12Q_4$, $\eta_0 = -J_{401} + 2J_{002}K_{40}$, $\eta_1 = 2J_{002} + J_{021}$, $\eta_2 = 6K_{21}^2 - 6K_{21}(43 + 5m) + 2\sigma_6K_{02}$.

Proof. Since Scheme (1) employs five functional evaluations, namely, $f'(x_n)$, $f(x_n)$, $f(y_n)$, $f(z_n)$, and $f(w_n)$, optimality can be achieved if the corresponding convergence order is 16. In order to induce the desired order of convergence, we begin by the 16th-order Taylor series expansion of $f(x_n)$ about α:

$$f(x_n) = \frac{f'(\alpha)}{m!} e_n^m \{1 + \sum_{i=2}^{17} c_i\, e_n^{i-1} + O(e_n^{17})\}. \tag{4}$$

It follows that

$$f'(x_n) = \frac{f'(\alpha)}{(m-1)!} e_n^{m-1} \{1 + \sum_{i=2}^{16} i\, \frac{m+i-1}{m} c_i\, e_n^{i-1} + O(e_n^{16})\}. \tag{5}$$

For brevity of notation, we abbreviate e_n as e. Using Mathematica [18], we find:

$$y_n = x_n - m\frac{f(x_n)}{f'(x_n)} = \alpha + \frac{c_2}{m}e^2 + \frac{(-(m+1)c_2^2 + 2mc_3)}{m^2}e^3 + \frac{Y_4}{m^3}e^4 + \sum_{i=5}^{16} \frac{Y_i}{m^{i-1}}\, e_n^i + O(e^{17}), \tag{6}$$

where $Y_4 = (1+m)^2 c_2^3 - m(4+3m)c_2c_3 + 3m^2c_4$ and $Y_i = Y_i(c_2, c_3, \cdots, c_{16})$ for $5 \le i \le 16$.

After a lengthy computation using the fact that $f(y_n) = f(x_n)|_{e_n \to (y_n - \alpha)}$, we get:

$$s = \left(\frac{f(y_n)}{f(x_n)}\right)^{1/m} = \frac{c_2}{m}e + \frac{(-(m+2)c_2^2 + 2mc_3)}{m^2}e^2 + \frac{\gamma_3}{2m^3}e^3 + \sum_{i=4}^{15} E_i\, e^i + O(e^{16}), \tag{7}$$

where $\gamma_3 = (7 + 7m + 2m^2)c_2^3 - 2m(7+3m)c_2c_3 + 6m^2c_4$, $E_i = E_i(c_2, c_3, \cdots, c_{16})$ for $4 \le i \le 15$.

In the third substep of Scheme (1), $w_n = O(e^8)$ can be achieved based on Kung–Traub's conjecture. To reflect the effect on w_n from z_n in the second substep, we need to expand z_n up to eighth-order terms; hence, we carry out a sixth-order Taylor expansion of $Q_f(s)$ about 0 by noting that $s = O(e)$ and $\frac{f(y_n)}{f'(x_n)} = O(e^2)$:

$$Q_f(s) = Q_0 + Q_1 s + Q_2 s^2 + Q_3 s^3 + Q_4 s^4 + Q_5 s^5 + Q_6 s^6 + O(e^7), \tag{8}$$

where $Q_j = \frac{1}{j!}\frac{d^j}{ds^j} Q_f(s)$ for $0 \le j \le 6$. As a result, we come up with:

$$z_n = x_n - mQ_f(s)\frac{f(y_n)}{f'(x_n)} = \alpha + \frac{(1-Q_0)}{m}e^2 + \frac{\mu_3}{m^2}e^3 + \sum_{i=4}^{16} W_i\, e^i + O(e^{17}),$$

where $\mu_3 = (-1 + m(Q_0 - 1) + 3Q_0 - Q_1)c_2^2 - 2m(Q_0 - 1)c_3$ and $W_i = W_i(c_2, c_3, \cdots, c_{16}, Q_0, \cdots, Q_6)$ for $4 \le i \le 16$. Selecting $Q_0 = 1$ and $Q_1 = 2$ leads us to an expression:

$$z_n = \alpha + \frac{c_2(\rho c_2^2 - 2mc_3)}{m^2}e^4 + \sum_{i=5}^{16} W_i\, e^i + O(e^{17}). \tag{9}$$

By a lengthy computation using the fact that $f(z_n) = f(x_n)|_{e_n \to (z_n - \alpha)}$, we deduce:

$$u = \left(\frac{f(z_n)}{f(y_n)}\right)^{1/m} = \frac{(\rho c_2^2 - 2mc_3)}{2m^2}e^2 + \frac{\delta_3}{3m^3}e^3 + \sum_{i=4}^{16} G_i\, e^i + O(e^{17}), \tag{10}$$

where $\delta_3 = (49 + 2m^2 + m(27 - 6Q_2) - 18Q_2 + 3Q_3)c_2^3 - 6m\rho c_2c_3 + 6m^2c_4$ and $G_i = G_i(c_2, c_3, \cdots, c_{16}, Q_2, \cdots, Q_6)$ for $4 \le i \le 16$.

In the last substep of Scheme (1), $x_{n+1} = O(e^{16})$ can be achieved based on Kung-Traub's conjecture. To reflect the effect on x_{n+1} from w_n in the third substep, we need to expand w_n up to sixteenth-order terms; hence, we carry out a 12th-order Taylor expansion of $K_f(s,u)$ about $(0,0)$ by noting that: $s = O(e)$, $u = O(e^2)$ and $\frac{f(z_n)}{f'(x_n)} = O(e^4)$ with $K_{ij} = 0$ satisfying $i + 2j > 12$ for all $0 \le i \le 12, 0 \le j \le 6$:

$$\begin{aligned} K_f(s,u) = {} & K_{00} + K_{10}s + K_{20}s^2 + K_{30}s^3 + K_{40}s^4 + K_{50}s^5 + K_{60}s^6 + K_{70}s^7 + K_{80}s^8 + K_{90}s^9 + K_{100}s^{10} + K_{110}s^{11} + \\ & K_{120}s^{12} + (K_{01} + K_{11}s + K_{21}s^2 + K_{31}s^3 + K_{41}s^4 + K_{51}s^5 + K_{61}s^6 + K_{71}s^7 + K_{81}s^8 + K_{91}s^9 + K_{101}s^{10})u + \\ & (K_{02} + K_{12}s + K_{22}s^2 + K_{32}s^3 + K_{42}s^4 + K_{52}s^5 + K_{62}s^6 + K_{72}s^7 + K_{82}s^8)u^2 + \\ & (K_{03} + K_{13}s + K_{23}s^2 + K_{33}s^3 + K_{43}s^4 + K_{53}s^5 + K_{63}s^6)u^3 + \\ & (K_{04} + K_{14}s + K_{24}s^2 + K_{34}s^3 + K_{44}s^4)u^4 + (K_{05} + K_{15}s + K_{25}s^2)u^5 + K_{06}u^6 + O(e^{13}). \end{aligned} \tag{11}$$

Substituting $z_n, f(x_n), f(y_n), f(z_n), f'(x_n)$, and $K_f(s,u)$ into the third substep of (1) leads us to:

$$w_n = z_n - mK_f(s,u) \cdot \frac{f(z_n)}{f'(x_n)} = \alpha + \frac{(1 - K_{00})c_2(\rho c_2^2 - 2mc_3)}{2m^3}e^4 + \sum_{i=5}^{16} \Gamma_i\, e^i + O(e^{17}), \tag{12}$$

where $\Gamma_i = \Gamma_i(c_2, c_3, \cdots, c_{16}, Q_2, \cdots, Q_6, K_{j\ell})$, for $5 \le i \le 16, 0 \le j \le 12$ and $0 \le \ell \le 6$. Thus $K_{00} = 1$ immediately annihilates the fourth-order term. Substituting $K_{00} = 1$ into $\Gamma_5 = 0$ and solving for K_{10}, we find:

$$K_{10} = 2. \tag{13}$$

Continuing the algebraic operations in this manner at the i-th $(6 \le i \le 7)$ stage with known values of $K_{j\ell}$, we solve $\Gamma_i = 0$ for remaining $K_{j\ell}$ to find:

$$K_{20} = 1 + Q_2,\ K_{01} = 1. \tag{14}$$

Substituting $K_{00} = 1, K_{10} = 2, K_{20} = 1 + Q_2, K_{01} = 1$ into (12) and simplifying we find:

$$v = \left(\frac{f(w_n)}{f(z_n)}\right)^{1/m} = -\frac{\left[\beta_0 c_2^4 + \beta_1 c_2^2 c_3 + 12m^2(K_{02} - 1)c_3^2 - 12m^2 c_2 c_4\right]}{12m^4}e^4 + \sum_{i=5}^{16} T_i\, e^i + O(e^{17}), \tag{15}$$

where β_0 and β_1 are described in (3) and $T_i = T_i(c_2, c_3, \cdots, c_{16}, Q_2, \cdots, Q_6)$ for $5 \le i \le 16$.

To compute the last substep of Scheme (1), it is necessary to have an eighth-order Taylor expansion of $J_f(s,u,v)$ about $(0,0,0)$ due to the fact that $\frac{f(w_n)}{f'(x_n)} = O(e^8)$. It suffices to expand J_f up to eighth-, fourth-, and second-order terms in s, u, v in order, by noting that $s = O(e)$, $u = O(e^2), v = O(e^4)$ with $J_{ijk} = 0$ satisfying $i + 2j + 4k > 8$ for all $0 \le i \le 8, 0 \le j \le 4, 0 \le k \le 2$:

$$\begin{aligned} J_f(s,u,v) = {} & J_{000} + J_{100}s + J_{200}s^2 + J_{300}s^3 + J_{400}s^4 + J_{500}s^5 + J_{600}s^6 + J_{700}s^7 + J_{800}s^8 + (J_{010} + J_{110}s + J_{210}s^2 + \\ & J_{310}s^3 + J_{410}s^4 + J_{510}s^5 + J_{610}s^6)u + (J_{020} + J_{120}s + J_{220}s^2 + J_{320}s^3 + J_{420}s^4)u^2 + (J_{030} + J_{130}s + J_{230}s^2)u^3 + \\ & J_{040}u^4 + (J_{001} + J_{101}s + J_{201}s^2 + J_{301}s^3 + J_{401}s^4 + (J_{011} + J_{111}s + J_{211}s^2)u + J_{021}u^2)v + J_{002}v^2. \end{aligned} \tag{16}$$

Substituting $w_n, f(x_n), f(y_n), f(z_n), f(w_n), f'(x_n)$ and $J_f(s,u,v)$ in (1), we arrive at:

$$x_{n+1} = w_n - mJ_f(s,u,v) \cdot \frac{f(w_n)}{f'(x_n)} = \alpha + \phi e^8 + \sum_{i=9}^{16} \Omega_i\, e^i + O(e^{17}), \tag{17}$$

where $\phi = \frac{1}{24m^7}(1 - J_{000})c_2(\rho c_2^2 - 2mc_3)\left[\beta_0 c_2^4 + \beta_1 c_2^2 c_3 + 12m^2(K_{02} - 1)c_3^2 - 12m^2 c_2 c_4\right]$ and $\Omega_i = \Omega_i(c_2,\ c_3,\ \cdots, c_{16}, Q_2, \cdots, Q_6, K_{\delta\theta}, J_{jk\ell})$, for $9 \le i \le 16$, $0 \le \delta \le 12$, $0 \le \theta \le 6$, $0 \le j \le 8$, $0 \le k \le 4$, $0 \le \ell \le 2$.

Since $J_{000} = 1$ makes $\phi = 0$, we substitute $J_{000} = 1$ into $\Omega_9 = 0$ and solve for J_{100} to find:

$$J_{100} = 2. \tag{18}$$

Continuing the algebraic operations in the same manner at the i-th ($10 \le i \le 15$) stage with known values of $J_{jk\ell}$, we solve $\Omega_i = 0$ for remaining $J_{jk\ell}$ to find:

$$\begin{cases} J_{200} = 1 + Q_2, J_{010} = 1, J_{110} = 4, J_{300} = -4 + 2Q_2 + Q_3, J_{001} = 1, J_{020} = K_{02}, J_{210} = 1 + K_{21}, \\ J_{400} = K_{40}, J_{101} = 2, J_{120} = 2 + K_{12}, J_{310} = -4 + K_{31} + 2Q_2, J_{500} = K_{50}, J_{011} = 2, J_{201} = 1 + Q_2, \\ J_{111} = 8, J_{030} = -1 + K_{02} + K_{03}, J_{220} = 1 + K_{21} + K_{22} - Q_2, J_{410} = -3 + K_{40} + K_{41} + Q_2 - Q_4, \\ J_{301} = -4 + 2Q_2 + Q_3, J_{130} = -4 + 2K_{02} + K_{12} + K_{13}, J_{320} = -6 + 2K_{21} + K_{31} + K_{32} - 2Q_2 - Q_3, \\ J_{600} = K_{60}, J_{510} = 6 + 2K_{40} + K_{50} + K_{51} - 3Q_3 - 2Q_4 - Q_5, \ J_{700} = K_{70}. \end{cases} \tag{19}$$

Upon substituting Relation (19) into Ω_{16}, we finally obtain:

$$\Omega_{16} = \frac{1}{3456m^{15}} c_2(\rho c_2^2 - 2mc_3)\left[\beta_0 c_2^4 + \beta_1 c_2^2 c_3 + 12m^2(K_{02} - 1)c_3^2 - 12m^2 c_2 c_4\right]\Psi, \tag{20}$$

where ρ, β_0, β_1, and Ψ as described in (3). This completes the proof. □

Remark 1. *Theorem 1 clearly reflects the case for $m = 1$ with the same constraints on weight functions Q_f, K_f, J_f studied in [19].*

Special Cases of Weight Functions

Theorem 1 enables us to obtain $Q_f(s)$, $K_f(s,u)$, and $J_f(s,u,v)$ by means of Taylor polynomials:

$$\begin{cases} Q_f(s) = 1 + 2s + Q_2 s^2 + Q_3 s^3 + Q_4 s^4 + Q_5 s^5 + Q_6 s^6 + O(e^7), \\ K_f(s,u) = 1 + 2s + (1 + Q_2)s^2 + (2Q_2 + Q_3 - 4)s^3 + K_{40}s^4 + K_{50}s^5 + K_{60}s^6 + K_{70}s^7 + K_{80}s^8 \\ +K_{90}s^9 + K_{100}s^{10} + K_{110}s^{11} + K_{120}s^{12} + (1 + 4s + K_{21}s^2 + K_{31}s^3 + K_{41}s^4 + K_{51}s^5 + K_{61}s^6 \\ +K_{71}s^7 + K_{81}s^8 + K_{91}s^9 + K_{101}s^{10})u + (K_{02} + K_{12}s + K_{22}s^2 + K_{32}s^3 + K_{42}s^4 + K_{52}s^5 \\ +K_{6}s^6 + K_{72}s^7 + K_{82}s^8)u^2 + (K_{03} + K_{13}s + K_{23}s^2 + K_{33}s^3 + K_{43}s^4 + K_{53}s^5 + K_{63}s^6)u^3 \\ +(K_{04} + K_{14}s + K_{24}s^2 + K_{34}s^3 + K_{44}s^4)u^4 + (K_{05} + K_{15}s + K_{25}s^2)u^5 + K_{06}u^6 + O(e^{13}), \\ J_f(s,u,v) = 1 + 2s + (1 + Q_2)s^2 + (2Q_2 + Q_3 - 4)s^3 + K_{40}s^4 + K_{50}s^5 + K_{60}s^6 + K_{70}s^7 + J_{800}s^8 \\ +(1 + 4s + (1 + K_{21})s^2 + (K_{31} + 2Q_2 - 4)s^3 + (K_{40} + K_{41} - 3 + Q_2 - Q_4)s^4 + (2K_{40} + K_{50} + K_{51} + 6 \\ -3Q_3 - 2Q_4 - Q_5)s^5 + J_{610}s^6)u + (K_{02} + (2 + K_{12})s + (K_{21} + K_{22} - Q_2 + 1)s^2 + (2K_{21} + K_{31} + K_{32} - 6 \\ -2Q_2 - Q_3)s^3 + J_{420}s^4)u^2 + (K_{02} + K_{03} - 1 + (2K_{02} + K_{12} + K_{13} - 4)s + J_{230}s^2)u^3 + J_{040}u^4 \\ +(1 + 2s + (1 + Q_2)s^2 + (2Q_2 + Q_3 - 4)s^3 + J_{401}s^4 + (2 + 8s + J_{211}s^2)u + J_{021}u^2)v + J_{002}v^2 + O(e^9), \end{cases} \tag{21}$$

where parameters Q_2–Q_6, $K_{40}, K_{50}, K_{60}, K_{70}, K_{80}, K_{90}, K_{100}, K_{110}, K_{120}, K_{21}, K_{31}, K_{41}, K_{51}, K_{61}, K_{71}, K_{81}, K_{91}, K_{101}, K_{02}, K_{12}, K_{22}, K_{32}, K_{42}, K_{52}, K_{62}, K_{72}, K_{82}, K_{03}, K_{13}, K_{23}, K_{33}, K_{43}, K_{53}, K_{63}, K_{04}, K_{14}, K_{24}, K_{34}, K_{44}, K_{05}, K_{15}, K_{25}, K_{06}$ and $J_{040}, J_{002}, J_{021}, J_{211}, J_{230}, J_{401}, J_{420}, J_{610}, J_{800}$ may be free.

Although various forms of weight functions $Q_f(s)$, $K_f(s,u)$ and $J_f(s,u,v)$ are available, in the current study we limit ourselves to all three weight functions in the form of rational functions, leading us to possible purely imaginary extraneous fixed points when $f(z) = (z^2 - 1)^m$ is employed. In the current study, we will consider two special cases described below:

The first case below will represent the best scheme, W3G7, studied in [19] only for $m = 1$.

Case 1:

$$\begin{cases} Q_f(s) & = \frac{1}{1-2s}, \\ K_f(s,u) & = Q_f(s) \cdot \frac{(s-1)^2}{1-2s-u+2s^2u}, \\ J_f(s,u,v) & = K_f(s,u) \cdot \frac{1+\sum_{i=1}^{3} q_i s^i + u\sum_{i=4}^{8} q_i s^{i-4} + u^2 \sum_{i=9}^{14} q_i s^{i-9} + u^3 \sum_{i=15}^{21} q_i s^{i-15}}{\mathcal{A}_1(s,u) + v\cdot(\sum_{i=22}^{25} r_i s^{i-22} + u(r_{26} + r_{27}s + \lambda s^2)}, \end{cases} \tag{22}$$

where

$$\begin{cases} q_9 = q_{12} = q_{13} = q_{17} = q_{18} = q_{19} = q_{20} = r_9 = r_{10} = r_{19} = r_{20} = 0, \\ q_1 = \frac{-3055820263252-76497245\lambda}{142682111242}, q_2 = \frac{56884034112404+44614515451\lambda}{285364222484}, \\ q_3 = \frac{-45802209949332-44308526471\lambda}{142682111242}, q_4 = -\frac{3(17778426888128+67929066997\lambda)}{1426821112420}, \\ q_5 = \frac{2(21034820227211+132665343294\lambda)}{356705278105}, q_6 = \frac{-1589080655012451+134087681464\lambda}{142682111242}, \\ q_7 = \frac{2(-780300304419180+71852971399\lambda)}{71341055621}, q_8 = \frac{12288(-727219117761+128167952\lambda)}{71341055621}, \\ q_{10} = 2, q_{11} = \frac{2(-741727036224277+126275739062\lambda)}{71341055621}, \\ q_{14} = -\frac{8192(-3964538065856+615849113\lambda)}{71341055621}, q_{15} = \frac{8(-226231159891830+34083208621\lambda)}{71341055621}, \\ q_{16} = -\frac{24(-908116719056544+136634733499\lambda)}{356705278105}, q_{21} = \frac{131072(-918470889768+136352293\lambda)}{356705278105}, \\ r_1 = q_1, r_2 = q_2, r_3 = q_3, r_4 = q_4, r_5 = q_5, r_6 = q_6 - 1, r_7 = q_7 - q_1 - 2, r_8 = q_8 + \frac{q_3}{2}, \\ r_{11} = \frac{-29558910226378916+5256346708371\lambda}{1426821112420}, r_{12} = \frac{-55018830261476-109759858153\lambda}{142682111242}, \\ r_{13} = \frac{25(-75694849962572+11301475999\lambda)}{71341055621}, r_{14} = -\frac{4096(-1500792372416+228734011\lambda)}{15508925135}, \\ r_{15} = q_{15}, r_{16} = \frac{43641510974266076-6354680006961\lambda}{713410556210}, r_{17} = -\frac{2(-1060205894022116+202907726307\lambda)}{71341055621}, \\ r_{18} = \frac{2(-2870055173156756+475573395275\lambda)}{71341055621}, r_{21} = \frac{q_{21}}{2}, r_{22} = -1, r_{23} = -q_1, r_{24} = -q_2, \\ r_{25} = -q_3, r_{26} = -1 - q_4, r_{27} = -2 - q_1 - q_5, \lambda = \frac{1353974063793787}{212746858830}, \end{cases} \tag{23}$$

and $\mathcal{A}_1(s,u) = 1 + \sum_{i=1}^{3} r_i s^i + u \sum_{i=4}^{8} r_i s^{i-4} + u^2 \sum_{i=9}^{14} r_i s^{i-9} + u^3 \sum_{i=15}^{21} r_i s^{i-15}$.

As a second case, we will consider the following set of weight functions:

Case 2:

$$\begin{cases} Q_f(s) &= \frac{1}{1-2s}, \\ K_f(s,u) &= Q_f(s) \cdot \frac{(s-1)^2}{1-2s-u+2s^2u}, \\ J_f(s,u,v) &= \frac{1+\sum_{i=1}^{3} q_i s^i + u \sum_{i=4}^{8} q_i s^{i-4} + u^2 \sum_{i=9}^{14} q_i s^{i-9} + u^3 \sum_{i=15}^{19} q_i s^{i-15}}{\mathcal{A}_0(s,u) + v \cdot (\sum_{i=20}^{23} r_i s^{i-20} + u \sum_{i=24}^{28} r_i s^{i-24} + r_{29} u^2)}, \end{cases} \tag{24}$$

where $\mathcal{A}_0(s,u) = 1 + \sum_{i=1}^{3} r_i s^i + u \sum_{i=4}^{8} r_i s^{i-4} + u^2 \sum_{i=9}^{14} r_i s^{i-9} + u^3 \sum_{i=15}^{19} r_i s^{i-15}$ and determination of the 48 coefficients q_i, r_i of J_f is described below. Relationships were sought among all free parameters of $J_f(s,u,v)$, giving us a simple governing equation for extraneous fixed points of the proposed family of methods (1).

To this end, we first express s, u and v for $f(z) = (z^2-1)^m$ as follows with $t = z^2$:

$$s = \frac{1}{4}\left(1 - \frac{1}{t}\right), \quad u = \frac{1}{4} \cdot \frac{(t-1)^2}{(t+1)^2}, \quad v = \frac{(t-1)^4}{4(1+6t+t^2)^2}. \tag{25}$$

In order to obtain a simple form of $J_f(s,u,v)$, we needed to closely inspect how it is connected with $K_f(s,u)$. When applying to $f(z) = (z^2-1)^m$, we find $K_f(s,u)$ with $t = z^2$ as shown below:

$$K_f(s,u) = \frac{4t(1+t)}{t^2+6t+1}. \tag{26}$$

Using the two selected weight functions Q_f, K_f, we continue to determine coefficients q_i, r_i of J_f yielding a simple governing equation for extraneous fixed points of the proposed methods when $f(z) = (z^2-1)^m$ is applied. As a result of tedious algebraic operations reflecting the 25 constraints (with possible rank deficiency) given by (18) and (19), we find only 23 effective relations, as follows:

$$\begin{cases} q_1 = \frac{1}{4}(-8 - r_{23}), q_2 = -3 - 2q_1, q_3 = 2 + q_1, q_4 = -r_{24}, q_5 = 2q_4 - r_{25}, \\ q_6 = 5 + \frac{7q_3}{4} + \frac{13q_4}{4} + \frac{9q_5}{8} - \frac{5q_9}{4} - \frac{25q_{10}}{8} + \frac{5q_{12}}{8} + \frac{r_8}{4} - \frac{5r_{11}}{4} - \frac{5r_{12}}{8}, \\ q_7 = -\frac{4q_3}{5} + \frac{4q_4}{5} + \frac{q_5}{5} - \frac{2q_6}{5} - \frac{2r_8}{5}, q_8 = q_4 + \frac{q_5}{2} - \frac{q_7}{2}, q_9 = q_{15} - r_{15}, \\ q_{10} = -2q_4 - 2q_{15} + q_{16} - r_{16}, q_{11} = -6 + \frac{q_5}{2} - \frac{5q_7}{2} + q_9 + 2q_{10} - r_8 + r_{11}, \\ r_1 = -2 + q_1, r_2 = 2(1 + q_2), r_3 = 4q_3, r_4 = -1 + q_4, r_5 = 2 - q_3 - 2q_4 + q_5, \\ r_6 = 1 + 2q_3 - q_4 - 2q_5 + q_6, r_7 = -2 + 5q_3 - 4q_4 - 2q_5 + 6q_7 + 2r_8, r_9 = -q_4 + q_9, \\ r_{10} = -2 - q_5 - 2q_9 + q_{10}, r_{20} = -1, r_{21} = 4 - q_3, r_{22} = -4(1 - q_3). \end{cases} \tag{27}$$

The three relations, $J_{500} = K_{50}, J_{600} = K_{60}$, and $J_{700} = K_{70}$ give one relation $r_{22} = -4(1 - q_3)$.

Due to 23 constraints in Relation (27), we find that 18 free parameters among 48 coefficients of J_f in (24) are available. We seek relationships among the free parameters yielding purely imaginary extraneous fixed points of the proposed family of methods when $f(z) = (z^2 - 1)^m$ is applied.

To this end, after substituting the 23 effective relations given by (27) into J_f in (24) and by applying to $f(z) = (z^2 - 1)^m$, we can construct $H(z) = 1 + sQ_f(s) + suK_f(s,u) + suvJ_f(s,u,v)$ in (1) and seek its roots for extraneous fixed points with $t = z^2$:

$$H(z) = \frac{\mathcal{A} \cdot G(t)}{t(1+t)^2(1+6t+t^2) \cdot W(t)}, \tag{28}$$

where $\mathcal{A}$ is a constant factor, $G(t) = \sum_{i=0}^{20} g_i t^i$, with $g_0 = -q_{14}, g_1 = -16 - 2q_{12} - 4q_{13} - 8q_{14} - 16q_{15} + 10q_{16} - 4r_8 + 4r_{11} + 2r_{12} - 16_{14} - 4r_{15} - 10r_{16} + 3r_{20} + 60r_{21} + 10r_{22}, g_i = g_i(q_{12}, r_{12}, \cdots, r_{25})$, for $2 \le i \le 20$ and $W(t) = \sum_{i=0}^{15} w_i t^i$, with $w_0 = -r_{14}$, $w_1 = 16r_8 + 4r_{13} - 5r_{14} + 4r_{25}$, $w_i = w_i(q_{12}, r_{12}, \cdots, r_{25})$, for $2 \le i \le 15$. The coefficients of both polynomials, $G(t)$ and $W(t)$, contain at most 18 free parameters.

We first observe that partial expressions of $H(z)$ with $t = z^2$, namely, $1 + sQ_f(s) = \frac{1+3t}{2(1+t)}$, $1 + sQ_f(s) + suK_f(s,u) = \frac{1+21t+35t^2+7t^3}{4(1+t)(1+6t+t^2)}$ and the denominator of (28) contain factors $t, (1+3t), (1+t), (1+6t+t^2), (1+21t+35t^2+7t^3)$ when $f(z) = (z^2-1)^m$ is applied. With an observation of presence of such factors, we seek a special subcase in which $G(t)$ may contain all the interested factors as follows:

$$G(t) = t(1+3t)(1+t)^{\lambda}(1+6t+t^2)^{\beta}(1+21t+35t^2+7t^3) \cdot (1+10t+5t^2)(1+92t+134t^2+28t^3+t^4) \cdot \Phi(t), \tag{29}$$

where $\Phi(t)$ is a polynomial of degree $(9 - (\lambda + 2\beta))$, with $\lambda \in \{0,1,2\}, \beta \in \{1,2,3\}$ and $1 \le \lambda + \beta \le 3$; two polynomial factors $(1 + 10t + 5t^2)$ and $(1 + 92t + 134t^2 + 28t^3 + t^4)$ were found in Case 3G of the previous study done by Geum–Kim–Neta [19]. Notice that factors $(1 + 6t + t^2)$, $(1 + 21t + 35t^2 + 7t^3)$, $(1 + 10t + 5t^2)$ and $(1 + 92t + 134t^2 + 28t^3 + t^4)$ of $G(t)$ are all negative, i.e., the corresponding extraneous fixed points are all purely imaginary.

In fact, the degree of $\Phi(t)$ will be decreased by annihilating the relevant coefficients containing free parameters to make all its roots negative. We take the 6 pairs of $(\lambda, \beta) \in \{(0,1),(0,2),(0,3),(1,1),(1,2),(2,1)\}$ to form 6 subcases named as Case 2A–2F in order. The lengthy algebraic process eventually leads us to additional constraints to each subcase described below:

Case 2A: $(\lambda, \beta) = (0,1)$

$$\begin{pmatrix} q_{12} \\ q_{15} \\ r_8 \\ r_{11} \\ r_{12} \\ r_{13} \\ r_{14} \\ r_{15} \\ r_{16} \\ r_{18} \\ r_{23} \\ r_{24} \end{pmatrix} = \begin{bmatrix} -3 & 9 & 9 & \frac{13}{2} & \frac{7}{2} & -13 & 0 & 0 & -12 & -\frac{23}{2} & -13 & -134 \\ 0 & -\frac{1}{2} & -\frac{1}{4} & -\frac{1}{8} & -\frac{1}{16} & 0 & 0 & 0 & 0 & 0 & 0 & 0 \\ -\frac{7}{4} & \frac{17}{4} & 5 & \frac{59}{16} & 2 & -\frac{1}{4} & \frac{1}{16} & -\frac{181}{28} & -\frac{699}{112} & -\frac{99}{16} & -\frac{1807}{224} & -\frac{1163}{16} \\ -\frac{11}{4} & 34 & \frac{171}{4} & \frac{125}{4} & \frac{159}{8} & -3 & \frac{3}{4} & -\frac{291}{7} & -\frac{274}{7} & -\frac{75}{2} & -\frac{629}{14} & -454 \\ -\frac{11}{2} & 14 & \frac{23}{2} & \frac{35}{4} & \frac{9}{2} & 1 & -\frac{1}{4} & -\frac{255}{14} & -\frac{425}{28} & -14 & -\frac{821}{56} & -\frac{693}{4} \\ 7 & -30 & -33 & -\frac{49}{2} & -\frac{29}{2} & 1 & -\frac{1}{2} & \frac{297}{7} & \frac{537}{14} & 37 & \frac{1199}{28} & \frac{937}{2} \\ 0 & 1 & 1 & \frac{3}{4} & \frac{1}{2} & 0 & 0 & -\frac{9}{7} & -\frac{15}{14} & -1 & -\frac{27}{28} & -\frac{27}{2} \\ \frac{1}{2} & -\frac{21}{4} & -\frac{23}{4} & -\frac{67}{16} & -\frac{21}{8} & \frac{1}{4} & -\frac{1}{16} & \frac{165}{28} & \frac{599}{112} & \frac{79}{16} & \frac{1291}{224} & \frac{967}{16} \\ -\frac{3}{2} & 16 & \frac{35}{2} & \frac{51}{4} & 8 & -1 & \frac{1}{4} & -\frac{241}{14} & -\frac{439}{28} & -\frac{29}{2} & -\frac{947}{56} & -\frac{707}{4} \\ 2 & -21 & -23 & -\frac{67}{4} & -\frac{21}{2} & 0 & -1 & \frac{152}{7} & \frac{279}{14} & \frac{37}{2} & \frac{603}{28} & \frac{447}{2} \\ 0 & 0 & 0 & 0 & 0 & 0 & 0 & -\frac{2}{7} & -\frac{4}{7} & -1 & -\frac{12}{7} & 0 \\ 0 & 0 & 0 & 0 & 0 & 0 & 0 & -\frac{2}{7} & -\frac{1}{14} & 0 & \frac{1}{28} & -\frac{1}{2} \end{bmatrix} \begin{pmatrix} q_{13} \\ q_{16} \\ q_{17} \\ q_{18} \\ q_{19} \\ r_{17} \\ r_{19} \\ r_{25} \\ r_{26} \\ r_{27} \\ r_{28} \\ r_{29} \end{pmatrix} + \begin{pmatrix} -74 \\ 0 \\ -\frac{2207}{56} \\ -\frac{1900}{7} \\ -\frac{1355}{14} \\ \frac{1929}{7} \\ -\frac{61}{7} \\ \frac{1987}{56} \\ -\frac{1453}{14} \\ \frac{919}{7} \\ \frac{16}{7} \\ \frac{2}{7} \end{pmatrix} \tag{30}$$

and $q_{14} = 0$. These 12 additional constraints $q_{12}, q_{15}, r_8, r_{11}, r_{12}, r_{13}, r_{14}, r_{15}, r_{16}, r_{18}, r_{23}, r_{24}$ are expressed in terms of 12 parameters $q_{13}, q_{16}, q_{17}, q_{18}, q_{19}, r_{17}, r_{19}, r_{25}, r_{26}, r_{27}, r_{28}, r_{29}$ that are arbitrarily free for the purely imaginary extraneous fixed points. Those 12 free parameters are chosen at our disposal. Then, using Relations (27) and (30), the desired form of $J_f(s,u,v)$ in (24) can be constructed.

Case 2B: $(\lambda,\beta)=(0,2)$

$$\begin{pmatrix} q_{12}\\ q_{15}\\ r_8\\ r_{11}\\ r_{12}\\ r_{13}\\ r_{14}\\ r_{15}\\ r_{16}\\ r_{18}\\ r_{23}\\ r_{24}\\ r_{25}\\ r_{26} \end{pmatrix} = \begin{bmatrix} -3 & 9 & 9 & \frac{13}{2} & \frac{7}{2} & 0 & 0 & \frac{1}{6} & -\frac{1}{3} & -\frac{178}{3}\\ 0 & -\frac{1}{2} & -\frac{1}{4} & -\frac{1}{8} & -\frac{1}{16} & 0 & 0 & 0 & 0 & 0\\ -\frac{7}{4} & \frac{17}{4} & 5 & \frac{59}{16} & 2 & -\frac{1}{4} & \frac{1}{16} & -\frac{1}{48} & -\frac{161}{96} & -\frac{917}{24}\\ -\frac{11}{4} & 34 & \frac{171}{4} & \frac{125}{4} & \frac{159}{8} & -3 & \frac{3}{4} & \frac{5}{6} & -\frac{25}{6} & -\frac{668}{3}\\ -\frac{11}{2} & 14 & \frac{23}{2} & \frac{35}{4} & \frac{9}{2} & 1 & -\frac{1}{4} & \frac{1}{6} & \frac{61}{24} & -\frac{317}{6}\\ 7 & -30 & -33 & -\frac{49}{2} & -\frac{29}{2} & 1 & -\frac{1}{2} & 0 & \frac{7}{4} & 217\\ 0 & 1 & 1 & \frac{3}{4} & \frac{1}{2} & 0 & 0 & 0 & \frac{1}{4} & -5\\ \frac{1}{2} & -\frac{21}{4} & -\frac{23}{4} & -\frac{67}{16} & -\frac{21}{8} & \frac{1}{4} & -\frac{1}{16} & -\frac{11}{48} & \frac{5}{96} & \frac{617}{24}\\ -\frac{3}{2} & 16 & \frac{35}{2} & \frac{51}{4} & 8 & -1 & \frac{1}{4} & \frac{2}{3} & -\frac{5}{24} & -\frac{455}{6}\\ 2 & -21 & -23 & -\frac{67}{4} & -\frac{21}{2} & 0 & -1 & -\frac{5}{6} & \frac{5}{12} & \frac{293}{3}\\ 0 & 0 & 0 & 0 & 0 & 0 & 0 & -\frac{1}{3} & -\frac{4}{3} & -\frac{4}{3}\\ 0 & 0 & 0 & 0 & 0 & 0 & 0 & 0 & \frac{1}{4} & 3\\ 0 & 0 & 0 & 0 & 0 & 0 & 0 & \frac{1}{3} & -\frac{2}{3} & -\frac{44}{3}\\ 0 & 0 & 0 & 0 & 0 & 0 & 0 & -\frac{4}{3} & -\frac{1}{3} & \frac{29}{3} \end{bmatrix} \begin{pmatrix} q_{13}\\ q_{16}\\ q_{17}\\ q_{18}\\ q_{19}\\ r_{17}\\ r_{19}\\ r_{27}\\ r_{28}\\ r_{29} \end{pmatrix} + \begin{pmatrix} -10\\ 0\\ -\frac{87}{8}\\ -76\\ \frac{25}{2}\\ 57\\ -1\\ \frac{43}{8}\\ -\frac{33}{2}\\ 23\\ 0\\ 4\\ -16\\ 12 \end{pmatrix} \tag{31}$$

and $q_{14}=0$. These 14 additional constraints are expressed in terms of 10 parameters $q_{13}, q_{16}, q_{17}, q_{18}, q_{19}, r_{17}, r_{19}, r_{27}, r_{28}, r_{29}$ that are arbitrarily free for the purely imaginary extraneous fixed points. Those 10 free parameters are chosen at our disposal. Then, using Relations (27) and (31), the desired form of $J_f(s,u,v)$ in (24) can be constructed.

Case 2C: $(\lambda,\beta)=(0,3)$

$$\begin{pmatrix} q_{12}\\ q_{15}\\ q_{16}\\ r_8\\ r_{11}\\ r_{12}\\ r_{13}\\ r_{14}\\ r_{15}\\ r_{16}\\ r_{18}\\ r_{23}\\ r_{24}\\ r_{25}\\ r_{26}\\ r_{27} \end{pmatrix} = \begin{bmatrix} -3 & 0 & -\frac{1}{4} & -1 & 0 & 0 & -\frac{23}{16} & -\frac{29}{4}\\ 0 & \frac{1}{4} & \frac{1}{4} & \frac{3}{16} & 0 & 0 & \frac{1}{32} & -\frac{21}{8}\\ 0 & -1 & -\frac{3}{4} & -\frac{1}{2} & 0 & 0 & -\frac{1}{16} & \frac{21}{4}\\ -\frac{7}{4} & \frac{3}{4} & \frac{1}{2} & -\frac{1}{8} & -\frac{1}{4} & \frac{1}{16} & -\frac{15}{8} & -\frac{33}{2}\\ -\frac{11}{4} & \frac{35}{4} & \frac{23}{4} & \frac{23}{8} & -3 & \frac{3}{4} & -9 & -20\\ -\frac{11}{2} & -\frac{5}{2} & -\frac{7}{4} & -\frac{5}{2} & 1 & -\frac{1}{4} & \frac{9}{8} & \frac{51}{2}\\ 7 & -3 & -2 & \frac{1}{2} & 1 & -\frac{1}{2} & \frac{29}{8} & \frac{119}{2}\\ 0 & 0 & 0 & 0 & 0 & 0 & \frac{3}{16} & \frac{1}{4}\\ \frac{1}{2} & -\frac{1}{2} & -\frac{1}{4} & 0 & \frac{1}{4} & -\frac{1}{16} & \frac{9}{8} & -\frac{17}{2}\\ -\frac{3}{2} & \frac{3}{2} & \frac{3}{4} & 0 & -1 & \frac{1}{4} & -\frac{27}{8} & \frac{55}{2}\\ 2 & -2 & -1 & 0 & 0 & -1 & \frac{71}{16} & -\frac{147}{4}\\ 0 & 0 & 0 & 0 & 0 & 0 & -\frac{1}{4} & -11\\ 0 & 0 & 0 & 0 & 0 & 0 & \frac{1}{4} & 3\\ 0 & 0 & 0 & 0 & 0 & 0 & -\frac{7}{4} & -5\\ 0 & 0 & 0 & 0 & 0 & 0 & 4 & -29\\ 0 & 0 & 0 & 0 & 0 & 0 & -\frac{13}{4} & 29 \end{bmatrix} \begin{pmatrix} q_{13}\\ q_{17}\\ q_{18}\\ q_{19}\\ r_{17}\\ r_{19}\\ r_{28}\\ r_{29} \end{pmatrix} + \begin{pmatrix} -\frac{57}{4}\\ -\frac{1}{8}\\ \frac{1}{4}\\ -9\\ -100\\ \frac{19}{2}\\ \frac{99}{2}\\ -\frac{3}{4}\\ 13\\ -\frac{77}{2}\\ \frac{201}{4}\\ 13\\ 4\\ -29\\ 64\\ -39 \end{pmatrix} \tag{32}$$

and $q_{14}=0$. These 16 additional constraints are expressed in terms of 8 parameters $q_{17}, q_{18}, q_{19}, r_{17}, r_{18}, r_{19}, r_{28}, r_{29}$ that are arbitrarily free for the purely imaginary extraneous fixed points. Those 8 free parameters are chosen at our disposal. Then, using Relations (27) and (32), the desired form of $J_f(s,u,v)$ in (24) can be constructed.

Case 2D: $(\lambda,\beta)=(1,1)$, being identical with Case 2A.

Case 2E: $(\lambda,\beta)=(1,2)$, being identical with Case 2B.

Case 2F: $(\lambda, \beta) = (2, 1)$,

$$\begin{pmatrix} q_{12} \\ q_{15} \\ q_{16} \\ r_8 \\ r_{11} \\ r_{12} \\ r_{13} \\ r_{14} \\ r_{15} \\ r_{16} \\ r_{18} \\ r_{23} \\ r_{24} \end{pmatrix} = \begin{bmatrix} -3 & 0 & -\frac{1}{4} & -1 & 0 & 0 & -13 & -12 & -\frac{23}{2} & -13 & -134 \\ 0 & \frac{1}{4} & \frac{1}{4} & \frac{3}{16} & 0 & 0 & 0 & 0 & 0 & 0 & 0 \\ 0 & -1 & -\frac{3}{4} & -\frac{1}{2} & 0 & 0 & 0 & 0 & 0 & 0 & 0 \\ -\frac{7}{4} & \frac{3}{4} & \frac{1}{2} & -\frac{1}{8} & -\frac{1}{4} & \frac{1}{16} & -\frac{181}{28} & -\frac{699}{112} & -\frac{99}{16} & -\frac{1807}{224} & -\frac{1163}{16} \\ -\frac{11}{4} & \frac{35}{4} & \frac{23}{4} & \frac{23}{8} & -3 & \frac{3}{4} & -\frac{291}{7} & -\frac{274}{7} & -\frac{75}{2} & -\frac{629}{14} & -454 \\ -\frac{11}{2} & -\frac{5}{2} & -\frac{7}{4} & -\frac{5}{2} & 1 & -\frac{1}{4} & -\frac{255}{14} & -\frac{425}{28} & -14 & -\frac{821}{56} & -\frac{693}{4} \\ 7 & -3 & -2 & \frac{1}{2} & 1 & -\frac{1}{2} & \frac{297}{7} & \frac{537}{14} & 37 & \frac{1199}{28} & \frac{937}{2} \\ 0 & 0 & 0 & 0 & 0 & 0 & -\frac{9}{7} & -\frac{15}{14} & -1 & -\frac{27}{28} & -\frac{27}{2} \\ \frac{1}{2} & -\frac{1}{2} & -\frac{1}{4} & 0 & \frac{1}{4} & -\frac{1}{16} & \frac{165}{28} & \frac{599}{112} & \frac{79}{16} & \frac{1291}{224} & \frac{967}{16} \\ -\frac{3}{2} & \frac{3}{2} & \frac{3}{4} & 0 & -1 & \frac{1}{4} & -\frac{241}{14} & -\frac{439}{28} & -\frac{29}{2} & -\frac{947}{56} & -\frac{707}{4} \\ 2 & -2 & -1 & 0 & 0 & -1 & \frac{152}{7} & \frac{279}{14} & \frac{37}{2} & \frac{603}{28} & \frac{447}{2} \\ 0 & 0 & 0 & 0 & 0 & 0 & -\frac{2}{7} & -\frac{4}{7} & -1 & -\frac{12}{7} & 0 \\ 0 & 0 & 0 & 0 & 0 & 0 & -\frac{2}{7} & -\frac{1}{14} & 0 & \frac{1}{28} & -\frac{1}{2} \end{bmatrix} \begin{pmatrix} q_{13} \\ q_{17} \\ q_{18} \\ q_{19} \\ r_{17} \\ r_{19} \\ r_{25} \\ r_{26} \\ r_{27} \\ r_{28} \\ r_{29} \end{pmatrix} + \begin{pmatrix} -74 \\ 0 \\ 0 \\ -\frac{2207}{56} \\ -\frac{1900}{7} \\ -\frac{1355}{14} \\ \frac{1929}{7} \\ -\frac{61}{7} \\ \frac{1987}{56} \\ -\frac{1453}{14} \\ \frac{919}{7} \\ \frac{16}{7} \\ \frac{2}{7} \end{pmatrix} \tag{33}$$

and $q_{14} = 0$. These 13 additional constraints are expressed in terms of 11 parameters $q_{13}, q_{17}, q_{18}, q_{19}, r_{17}, r_{19}, r_{25}, r_{26}, r_{27}, r_{28}, r_{29}$ that are arbitrarily free for the purely imaginary extraneous fixed points. Those 11 free parameters are chosen at our disposal. Then, using Relations (27) and (33), the desired form of $J_f(s, u, v)$ in (24) can be constructed. After a process of careful factorization, we find the expression for $H(z)$ in (28) stated in the following lemma.

Proposition 1. *The expression $H(z)$ in* (28) *is identical in each subcase of 2A–2F and given by a unique relation below:*

$$H(z) = \frac{(1+3t)(1+10t+5t^2)(1+92t+134t^2+28t^3+t^4)}{8(1+t)(1+6t+t^2)(1+28t+70t^2+28t^3+t^4)}, \ t = z^2, \tag{34}$$

despite the possibility of different coefficients in each subcase.

Proof. Let us write $G(t)$ in (28) as $G(t) = t(1+3t) \cdot \psi_1(t) \cdot \psi_2(t) \cdot \Phi(t) \cdot (1+t)^{\lambda}(1+6t+t^2)^{\beta-1}$ with $\psi_1(t) = (1+6t+t^2)(1+21t+35t^2+7t^3)$ and $\psi_2(t) = (1+10t+5t^2)(1+92t+134t^2+28t^3+t^4)$. Then after a lengthy process of a series of factorizations with the aid of Mathematica symbolic ability, we find $\Phi(t)$ and $W(t)$ in each subcase as follows.

(1) Case 2A: with $\lambda = 0$ and $\beta = 1$, we get

$$\begin{cases} \Phi(t) = -\frac{2}{7}(1+t) \cdot \Gamma_1(t), \\ W(t) = -\frac{16}{7}\psi_1(t)(1+28t+70t^2+28t^3+t^4)\Gamma_1(t), \end{cases} \tag{35}$$

where $\Gamma_1(t) = -244 + 28q_{16} + 28q_{17} + 21q_{18} + 14q_{19} - 36r_{25} - 30r_{26} - 28r_{27} - 27r_{28} - 378r_{29} + 14t(-72 + 4q_{16} + 4q_{17} + 3q_{18} + 2q_{19} - 4r_{25} + 2r_{27} + 4r_{28} - 72r_{29}) + t^2(1692 - 476q_{16} - 476q_{17} - 357q_{18} - 238q_{19} + 548r_{25} + 578r_{26} + 672r_{27} + 957r_{28} + 6006r_{29}) + 4t^3(-2288 + 196q_{16} + 196q_{17} + 147q_{18} + 98q_{19} - 148r_{25} - 100r_{26} - 42r_{27} - 6r_{28} - 1540r_{29}) - 7t^4(1636 + 68q_{16} + 68q_{17} + 51q_{18} + 34q_{19} + 4r_{25} + 22r_{26} + 36r_{27} + 55r_{28} + 386r_{29}) + t^5(-4176 + 56q_{16} + 56q_{17} + 42q_{18} + 28q_{19} + 648r_{25} + 400r_{26} + 140r_{27} - 32r_{28} + 7168r_{29}) + t^6(-4332 + 28q_{16} + 28q_{17} + 21q_{18} + 14q_{19} - 484r_{25} - 394r_{26} - 392r_{27} - 545r_{28} - 2926r_{29})$.

(2) Case 2B: with $\lambda = 0$ and $\beta = 2$, we get

$$\begin{cases} \Phi(t) = -\frac{2}{3}(1+t) \cdot \Gamma_2(t), \\ W(t) = -\frac{16}{3}(1+6t+t^2)\psi_1(t)(1+28t+70t^2+28t^3+t^4)\Gamma_2(t), \end{cases} \tag{36}$$

where $\Gamma_2(t) = (1+6t+t^2)(3(-4 + 4q_{16} + 4q_{17} + 3q_{18} + 2q_{19} + r_{28} - 20r_{29}) + t(24 - 48q_{16} - 48q_{17} - 36q_{18} - 24q_{19} + 4r_{27} + 22r_{28} + 280r_{29}) + 6t^2(-32 + 12q_{16} + 12q_{17} + 9q_{18} + 6q_{19} + 2r_{27} + 6r_{28} - 16r_{29}) - 2t^3(396 + 24q_{16} + 24q_{17} + 18q_{18} + 12q_{19} + 2r_{27} + 11r_{28} + 140r_{29}) + 3t^4(-188 + 4q_{16} + 4q_{17} + 3q_{18} + 2q_{19} - 4r_{27} - 13r_{28} + 52r_{29}))$.

(3) Case 2C: with $\lambda = 0$ and $\beta = 3$, we get

$$\begin{cases} \Phi(t) = \frac{1}{2}(1+t) \cdot \Gamma_3(t), \\ W(t) = 4(1+6t+t^2)^2\psi_1(t)(1+28t+70t^2+28t^3+t^4)\Gamma_3(t), \end{cases} \tag{37}$$

where $\Gamma_3(t) = (1+6t+t^2)^2(12-3r_{28}+2t(60+r_{28}-84r_{29})-4r_{29}+t^2(124+r_{28}+172r_{29}))$.
(4) Case 2D: with $\lambda = 1$ and $\beta = 1$, we get

$$\begin{cases} \Phi(t) = -\frac{2}{7} \cdot \Gamma_1(t), \\ W(t) = -\frac{16}{7}\psi_1(t)(1+28t+70t^2+28t^3+t^4)\Gamma_1(t). \end{cases} \tag{38}$$

(5) Case 2E: with $\lambda = 1$ and $\beta = 2$, we get

$$\begin{cases} \Phi(t) = -\frac{2}{3} \cdot \Gamma_2(t), \\ W(t) = -\frac{16}{3}(1+6t+t^2)\psi_1(t)(1+28t+70t^2+28t^3+t^4)\Gamma_2(t), \end{cases} \tag{39}$$

(6) Case 2F: with $\lambda = 2$ and $\beta = 1$, we get

$$\begin{cases} \Phi(t) = \frac{2}{7} \cdot \Gamma_4(t), \\ W(t) = \frac{2}{7}(1+t)\psi_1(t)(1+28t+70t^2+28t^3+t^4)\Gamma_4(t), \end{cases} \tag{40}$$

where $\Gamma_4(t) = 244+36r_{25}+30r_{26}+28r_{27}+27r_{28}+378r_{29}+t(764+20r_{25}-30r_{26}-56r_{27}-83r_{28}+630r_{29})-2t^2(1228+284r_{25}+274r_{26}+308r_{27}+437r_{28}+3318r_{29})+2t^3(5804+580r_{25}+474r_{26}+392r_{27}+449r_{28}+6398r_{29})-t^4(156+1132r_{25}+794r_{26}+532r_{27}+513r_{28}+10094r_{29})+t^5(4332+484r_{25}+394r_{26}+392r_{27}+545r_{28}+2926r_{29})$.

Substituting each pair of $(\Phi(t), W(t))$ into (28) yields an identical Relation (34) as desired. □

Remark 2. *The factorization process in the above proposition yields the additional constraints given by* (30)–(33) *for subcases 2A–2F, after a lengthy computation. Case 2D and Case 2E are found to be identical with Case 2A and Case 2B, respectively, by direct computation.*

In Table 1, we list free parameters selected for typical subcases of 2A–2F. Combining these selected free parameters with Relations (27) and (30)–(33), we can construct special iterative schemes named as W2A1, W2A2, $\cdots$, W2F3, W2F4. Such schemes together with W3G7 for Case 1 shall be used in Section 4 to display results on their numerical and dynamical aspects.

Table 1. Free parameters selected for typical subcases of 2A1–2F4.

SCN	q_{13}	q_{16}	q_{17}	q_{18}	q_{19}	r_{17}	r_{19}	r_{25}	r_{26}	r_{27}	r_{28}	r_{29}
2A1	0	0	0	0	0	0	0	0	0	0	0	0
2A2	20	0	0	0	0	0	1012	4	2	-8	0	0
2A3	$-\frac{89}{26}$	0	0	0	0	0	0	$-\frac{149}{26}$	0	0	0	0
2A4	0	0	$\frac{711}{26}$	$-\frac{622}{13}$	$\frac{222}{13}$	0	0	$-\frac{149}{26}$	0	0	0	0
2B1	0	0	0	0	0	0	0	-	-	0	0	0
2B2	0	0	0	0	0	0	96	-	-	-31	-1	$-\frac{1}{4}$
2B3	-19	0	0	0	0	0	0	-	-	-18	0	0
2B4	0	0	45	-52	-12	0	0	-	-	-18	0	0
2C1	0	-	0	0	0	0	0	-	-	-	0	0
2C2	-34	-	0	0	0	174	0	-	-	-	4	0
2C3	0	-	0	0	0	0	280	-	-	-	4	0
2C4	$-\frac{39}{4}$	-	$\frac{375}{4}$	$-\frac{627}{4}$	0	0	0	-	-	-	0	0
2F1	0	-	0	0	0	-40	0	$-\frac{122}{3}$	$\frac{290}{3}$	$-\frac{184}{3}$	0	0
2F2	16	-	0	0	0	0	0	1	0	-10	0	0
2F3	$\frac{138}{7}$	-	$-\frac{356}{7}$	$\frac{3963}{14}$	0	0	0	0	0	0	0	0
2F4	0	-	0	0	0	-32	0	-33	78	-46	-4	0

3. The Dynamics behind the Extraneous Fixed Points

The dynamics behind the extraneous fixed points [21] of iterative map (1) have been investigated by Stewart [20], Amat et al. [22], Argyros–Magreñan [23], Chun et al. [24], Chicharro et al. [25], Chun–Neta [26], Cordero et al. [27], Geum et al. [14,19,28–30], Rhee at al. [9], Magreñan [31], Neta et al. [32,33], and Scott et al. [34].

We locate a root α of a given function $f(x)$ as a fixed point ξ of the iterative map R_f:

$$x_{n+1} = R_f(x_n), n = 0, 1, \cdots, \tag{41}$$

where R_f is the iteration function associated with f. Typically, R_f is written in the form: $R_f(x_n) = x_n - \frac{f(x_n)}{f'(x_n)} H_f(x_n)$, where H_f is a weight function whose zeros are other fixed points $\xi \neq \alpha$ called extraneous fixed points of R_f. The dynamics of R_f might be influenced by presence of possible attractive, indifferent, or repulsive, and other periodic or chaotic orbits underlying the extraneous fixed points. For ease of analysis, we rewrite the iterative map (41) in a more specific form:

$$x_{n+1} = R_f(x_n) = x_n - m\frac{f(x_n)}{f'(x_n)} H_f(x_n), \tag{42}$$

where $H_f(x_n) = 1 + sQ_f(s) + suK_f(s,u) + suvJ_f(s,u,v)$ can be regarded as a weight function in the classical modified Newton's method for a multiple root of integer multiplicity m. Notice that α is a fixed point of R_f, while $\xi \neq \alpha$ for which $H_f(\xi) = 0$ are extraneous fixed points of R_f.

The influence of extraneous fixed points on the convergence behavior was well demonstrated for simple zeros via König functions and Schröder functions [21] applied to a class of functions $\{f_k(x) = x^k - 1, k \geq 2\}$. The basins of attraction may be altered due to the trapped sequence $\{x_n\}$ by the attractive extraneous fixed points of R_f. An initial guess x_0 chosen near a desired root may converge to another unwanted remote root when repulsive or indifferent extraneous fixed points are present. These aspects of the Schröder functions were observed when applied to the same class of functions $\{f_k(x) = x^k - 1, k \geq 2\}$.

To simply treat dynamics underlying the extraneous fixed points of iterative map (42), we select a member $f(z) = (z^2-1)^m$. By a similar approach made by Chun et al. [35] and Neta et al. [33,36], we construct $H_f(x_n) = s \cdot Q_f(s) + s \cdot u \cdot K_f(s,u) + s \cdot u \cdot v \cdot J_f(s,u,v)$ in (42). Applying $f(z) = (z^2-1)^m$ to H_f, we find a rational function $H(z)$ with $t = z^2$:

$$H(z) = \frac{\mathcal{N}(t)}{\mathcal{D}(t)}, \tag{43}$$

where both $\mathcal{D}(t)$ and $\mathcal{N}(t)$ are co-prime polynomial functions of t. The underlying dynamics of the iterative map (42) can be favorably investigated on the Riemann sphere [37] with possible fixed points "0(zero)" and "∞". As can be seen in Section 5, the relevant dynamics will be illustrated in a 6×6 square region centered at the origin.

Indeed, the roots t of $\mathcal{N}(t)$ provide the extraneous fixed points ξ of R_f in Map (42) by the relation:

$$\xi = \begin{cases} t^{\frac{1}{2}}, & \text{if } t \neq 0, \\ 0(\text{double root}), & \text{if } t = 0. \end{cases} \tag{44}$$

Extraneous Fixed Points and their Stability

The following proposition describes the stability of the extraneous fixed points of (42).

Proposition 2. *Let $f(z) = (z^2 - 1)^m$. Then the extraneous fixed points ξ for Case 2 discussed earlier are all found to be repulsive.*

Proof. By direct computation of $R'_f(z)$ with $f(z) = (z^2 - 1)^m$, we write it as with $t = z^2$:

$$R'_f(z) = \frac{\Psi_n(t)}{\Psi_d(t)},$$

where $\Psi_n(t) = (-1+t)^{15}$ and $\Psi_d(t) = 16t(1+t)^2(1+6t+t^2)^2(1+28t+70t^2+28t^3+t^4)^2$. With the help of Mathematica, we are able to express $\Psi_n(t) = \frac{1}{61509375}(1+3t)(1+10t+5t^2)(1+92t+134t^2+28t^3+t^4)\cdot Q_n(t) - 2097152\cdot R_n(t)$ and $\Psi_d(t) = -\frac{1}{61509375}16(1+3t)(1+10t+5t^2)(1+92t+134t^2+28t^3+t^4)\cdot Q_d(t) - 131072\cdot R_d(t)$, with $Q_n(t)$ and $Q_d(t)$ as six- and eight-degree polynomials, while $R_n(t) = (327,923,929,643 + 34,417,198,067,010t + 446,061,306,116,505t^2 + 1621107643125740t^3 + 2,036,953,856,667,405t^4 + 892,731,761,917,554t^5 + 108,873,731,877,775t^6)$ and $R_d(t) = (327,923,929,643 + 34417198067010t + 446,061,306,116,505t^2 + 1621107643125740t^3 + 2,036,953,856,667,405t^4 + 892,731,761,917,554t^5 + 108,873,731,877,775t^6)$. Further, we express $R_n(t) = (1+10t+5t^2)Q_\nu(t) + R_\nu(t)$ and $R_d(t) = (1+10t+5t^2)Q_\delta(t) + R_\delta(t)$, with $R_\nu(t) = -\frac{10,077,696}{25}(36+341t) = R_\delta(t)$. Now let $t = \xi^2$, then

$$R'_f(\xi) = 16$$

using the fact that $(1+3t)(1+10t+5t^2)(1+92t+134t^2+28t^3+t^4) = 0$. Hence ξ for Case 2 are all found to be repulsive. □

Remark 3. *Although not described here in detail due to limited space, by means of a similar proof as shown in Proposition 2, extraneous fixed points ξ for Case 1 was found to be indifferent in [19].*

If $f(z) = p(z)$ is a generic polynomial other than $(z^2-1)^m$, then theoretical analysis of the relevant dynamics may not be feasible as a result of the highly increased algebraic complexity. Nevertheless, we explore the dynamics of the iterative map (42) applied to $f(z) = p(z)$, which is denoted by R_p as follows:

$$z_{n+1} = R_p(z_n) = z_n - m\frac{p(z_n)}{p'(z_n)}H_p(z_n). \tag{45}$$

Basins of attraction for various polynomials are illustrated in Section 5 to observe the complicated dynamics behind the fixed points or the extraneous fixed points. The letter W was conveniently prefixed to each case number in Table 1 to symbolize a way of designating the numerical and dynamical aspects of iterative map (42) .

4. Results and Discussion on Numerical and Dynamical Aspects

We first investigate numerical aspects of the local convergence of (1) with schemes W3G7 and W2A1–W2F4 for various test functions; then we explore the dynamical aspects underlying extraneous fixed points based on iterative map (45) applied to $f(z) = (z^2-1)^m$, whose attractor basins give useful information on the global convergence.

Results of numerical experiments are tabulated for all selected methods in Tables 2–4. Computational experiments on dynamical aspects have been illustrated through attractor basins in Figures 1–7. Both numerical and dynamical aspects have strongly confirmed the desired convergence.

Throughout the computational experiments with the aid of Mathematica, $\$MinPrecision = 400$ has been assigned to maintain 400 digits of minimum number of precision. If α is not exact, then it is given by an approximate value with 416 digits of precision higher than $\$MinPrecision$.

Limited paper space allows us to list x_n and α with up to 15 significant digits. We set error bound $\epsilon = \frac{1}{2} \times 10^{-360}$ to meet $|x_n - \alpha| < \epsilon$. Due to the high-order of convergence and root multiplicity, close initial guesses have been selected to achieve a moderate number of accurate digits of the asymptotic error constants.

Methods W3G7, W2A1, W2C2 and W2F2 successfully located desired zeros of test functions $F_1 - F_4$:

$$\begin{cases} \mathbf{W3G7}: F_1(x) = [\cos(\frac{\pi x}{2}) + 2x^2 - 3\pi]^4, \ \alpha \approx 2.27312045629419, m = 4, \\ \mathbf{W2A1}: F_2(x) = [\cos(x^2+1) - x\log(x^2 - \pi + 2) + 1]^4 \cdot (x^2 + 1 - \pi), \ \alpha = \sqrt{\pi - 1}, m = 5, \\ \mathbf{W2C2}: F_3(x) = [\sin^{-1}(x^2 - 1) + 3e^x - 2x - 3]^2, \ \alpha \approx 0.477696831914490, m = 2, \\ \mathbf{W2F2}: F_4(x) = (x^2+1)^4 + \log[1 + (x^2+1)^3], \ \alpha = i, m = 3, \\ \text{where } \log z (z \in \mathbb{C}) \text{ is a principal analytic branch with } -\pi < Im(\log z) \le \pi. \end{cases} \tag{46}$$

We find that Table 2 ensures sixteenth-order convergence. The computational asymptotic error constant $|e_n|/|e_{n-1}|^{16}$ is in agreement with the theoretical one $\eta = \lim_{n\to\infty} |e_n|/|e_{n-1}|^{16}$ up to 4 significant digits. The computational convergence order $p_n = \log|e_n/\eta|/\log|e_{n-1}|$ well approaches 16.

Additional test functions in Table 3 confirm the convergence of Scheme (1). The errors $|x_n - \alpha|$ are listed in Table 4 for comparison among the listed methods W3G7 and W2A1–W2F4. In the current experiments, W3G7 has slightly better convergence for f_5 and slightly poor convergence for all other test functions than the rest of the listed methods. No specific method performs better than the other among methods W2A1–W2F4 of Case 2.

According to the definition of the asymptotic error constant $\eta(c_i, Q_f, K_f, J_f) = \lim_{n\to\infty} |R_f(x_n) - \alpha|/|x_n - \alpha|^{16}$, the convergence is dependent on iterative map $R_f(x_n)$, $f(x)$, x_0, α and the weight functions Q_f, K_f and J_f. It is clear that no particular method always achieves better convergence than the others for any test functions.

Table 2. Convergence of methods W3G7, W2A1, W2C2, W2F2 for test functions $F_1(x) - F_4(x)$.

Method	F	n	x_n	$\lvert F(x_n)\rvert$	$\lvert x_n - \alpha\rvert$	$\lvert e_n/e_{n-1}^{16}\rvert$	η	p_n
		0	2.2735	1.873×10^{-10}	0.000379544			
W3G7	F_1	1	2.27312045629419	1.927×10^{-233}	6.798×10^{-60}	0.00003666355445	0.00003666729357	
		2	2.27312045629419	0.0×10^{-400}	1.004×10^{-237}			16.00000
		0	1.4634	1.93×10^{-21}	0.0000181404			
W2A1	F_2	1	1.46341814037882	3.487×10^{-366}	2.040×10^{-74}	148.4148965	148.4575003	
		2	1.46341814037882	0.0×10^{-400}	0.0×10^{-399}			16.00000
		0	0.4777	1.890×10^{-10}	3.168×10^{-6}			
W2C2	F_3	1	0.477696831914490	6.116×10^{-183}	1.802×10^{-92}	0.0001750002063	0.0001749999826	
		2	0.477696831914490	0.0×10^{-400}	8.522×10^{-367}			16.00000
		0	$0.99995i$	1.000×10^{-12}	0.00005			
W2F2	F_4	1	$1.00000000000000\,i$	4.820×10^{-215}	1.391×10^{-72}	0.001037838436	0.001041219259	
		2	$1.00000000000000\,i$	0.0×10^{-400}	0.0×10^{-400}			16.00030

$i = \sqrt{-1}$, $\eta = \lim_{n\to\infty} \frac{|e_n|}{|e_{n-1}|^{16}}$, $p_n = \frac{\log|e_n/\eta|}{\log|e_{n-1}|}$.

Table 3. Additional test functions $f_i(x)$ with zeros α and initial values x_0 and multiplicities.

i	$f_i(x)$	α	x_0	m
1	$[4 + 3\sin x - 2x^2]^4$	≈ 1.85471014256339	1.86	4
2	$[2x - Pi + x\cos x]^5$	$\frac{\pi}{2}$	1.5707	5
3	$[2x^3 + 3e^{-x} + 4\sin(x^2) - 5]^2$	$\approx -0.402282449584416$	-0.403	2
4	$[\sqrt{3}x^2 \cdot \cos\frac{\pi x}{6} + \frac{1}{x^3+1} - \frac{1}{28}] \cdot (x-3)^3$	3	3.0005	4
5	$(x-1)^2 + \frac{1}{12} - \log[\frac{25}{12} - 2x + x^2]$	$1 - i\frac{\sqrt{3}}{6}$	$0.99995 - 0.28i$	2
6	$[x\log x - \sqrt{x} + x^3]^3$	1	1.0001	3

Here, $\log z$ ($z \in \mathbb{C}$) represents a principal analytic branch with $-\pi \le Im(\log z) < \pi$.

The proposed family of methods (1) has efficiency index EI [38], which is $16^{1/5} \approx 1.741101$ and larger than that of Newton's method. In general, the local convergence of iterative methods (45) is guaranteed with good initial values x_0 that are close to α. Selection of good initial values is a difficult task, depending on precision digits, error bound, and the given function $f(x)$.

Table 4. Comparison of $|x_n - \alpha|$ among selected methods applied to various test functions.

Method	$\lvert x_n - \alpha \rvert$	$f(x)$					
		f_1	f_2	f_3	f_4	f_5	f_6
W3G7	$\lvert x_1 - \alpha \rvert$	1.77×10^{-40} *	1.62×10^{-57}	4.89×10^{-51}	1.50×10^{-61}	5.76×10^{-7}	1.19×10^{-62}
	$\lvert x_2 - \alpha \rvert$	1.02×10^{-159}	1.13×10^{-225}	1.24×10^{-201}	3.27×10^{-245}	1.08×10^{-95}	2.40×10^{-247}
W2A1	$\lvert x_1 - \alpha \rvert$	2.83×10^{-42}	1.05×10^{-58}	1.92×10^{-52}	1.23×10^{-62}	1.29×10^{-6}	1.11×10^{-63}
	$\lvert x_2 - \alpha \rvert$	0.0×10^{-399}	4.24×10^{-230}	0.0×10^{-400}	0.0×10^{-399}	6.62×10^{-90}	3.61×10^{-251}
W2A2	$\lvert x_1 - \alpha \rvert$	1.63×10^{-41}	2.33×10^{-58}	1.45×10^{-51}	1.05×10^{-61}	1.32×10^{-6}	2.34×10^{-63}
	$\lvert x_2 - \alpha \rvert$	0.0×10^{-399}	1.11×10^{-228}	0.0×10^{-400}	0.0×10^{-399}	2.53×10^{-89}	7.39×10^{-250}
W2A3	$\lvert x_1 - \alpha \rvert$	2.53×10^{-43}	1.82×10^{-60}	4.56×10^{-54}	1.20×10^{-63}	4.43×10^{-6}	3.85×10^{-65}
	$\lvert x_2 - \alpha \rvert$	0.0×10^{-399}	1.40×10^{-236}	8.40×10^{-213}	0.0×10^{-399}	3.03×10^{-83}	1.53×10^{-256}
W2A4	$\lvert x_1 - \alpha \rvert$	1.24×10^{-42}	1.35×10^{-59}	1.53×10^{-52}	8.38×10^{-63}	1.58×10^{-4}	1.12×10^{-64}
	$\lvert x_2 - \alpha \rvert$	1.70×10^{-125}	5.96×10^{-424}	5.22×10^{-155}	8.16×10^{-187}	3.34×10^{-57}	2.79×10^{-424}
W2B1	$\lvert x_1 - \alpha \rvert$	2.39×10^{-42}	2.28×10^{-59}	1.30×10^{-52}	7.80×10^{-63}	2.14×10^{-6}	2.81×10^{-64}
	$\lvert x_2 - \alpha \rvert$	0.0×10^{-399}	1.57×10^{-232}	0.0×10^{-400}	0.0×10^{-399}	6.04×10^{-87}	2.23×10^{-253}
W2B2	$\lvert x_1 - \alpha \rvert$	4.44×10^{-42}	2.73×10^{-59}	3.03×10^{-52}	1.79×10^{-62}	4.30×10^{-6}	3.29×10^{-64}
	$\lvert x_2 - \alpha \rvert$	0.0×10^{-399}	6.69×10^{-232}	0.0×10^{-400}	0.0×10^{-399}	6.01×10^{-82}	$7.78 \times 10^{e-253}$
W2B3	$\lvert x_1 - \alpha \rvert$	9.85×10^{-43}	3.11×10^{-61}	4.26×10^{-53}	3.01×10^{-63}	4.46×10^{-6}	3.26×10^{-65}
	$\lvert x_2 - \alpha \rvert$	0.0×10^{-399}	1.17×10^{-239}	0.0×10^{-400}	0.0×10^{-399}	3.06×10^{-83}	7.91×10^{-257}
W2B4	$\lvert x_1 - \alpha \rvert$	1.04×10^{-42}	1.92×10^{-59}	1.77×10^{-52}	1.12×10^{-62}	1.77×10^{-4}	1.53×10^{-64}
	$\lvert x_2 - \alpha \rvert$	1.12×10^{-125}	4.68×10^{-405}	9.06×10^{-155}	2.23×10^{-186}	2.32×10^{-56}	0.0×10^{-400}
W2C1	$\lvert x_1 - \alpha \rvert$	4.87×10^{-42}	4.27×10^{-59}	2.95×10^{-52}	1.50×10^{-62}	1.08×10^{-6}	5.14×10^{-64}
	$\lvert x_2 - \alpha \rvert$	0.0×10^{-399}	0.0×10^{-399}	0.0×10^{-400}	0.0×10^{-399}	2.41×10^{-91}	2.22×10^{-191}
W2C2	$\lvert x_1 - \alpha \rvert$	9.31×10^{-42}	1.01×10^{-58}	5.94×10^{-52}	2.61×10^{-62}	1.47×10^{-6}	1.18×10^{-63}
	$\lvert x_2 - \alpha \rvert$	0.0×10^{-399}	4.23×10^{-230}	0.0×10^{-400}	0.0×10^{-399}	7.11×10^{-89}	4.95×10^{-251}
W2C3	$\lvert x_1 - \alpha \rvert$	9.17×10^{-42}	8.88×10^{-59}	6.85×10^{-52}	4.30×10^{-62}	4.36×10^{-6}	9.66×10^{-64}
	$\lvert x_2 - \alpha \rvert$	0.0×10^{-399}	7.84×10^{-230}	0.0×10^{-400}	0.0×10^{-399}	2.11×10^{-81}	6.05×10^{-251}
W2C4	$\lvert x_1 - \alpha \rvert$	5.36×10^{-42}	6.72×10^{-59}	6.55×10^{-52}	4.38×10^{-62}	1.02×10^{-4}	6.13×10^{-64}
	$\lvert x_2 - \alpha \rvert$	9.92×10^{-124}	5.96×10^{-424}	2.97×10^{-153}	8.55×10^{-185}	1.30×10^{-59}	0.0×10^{-400}
W2F1	$\lvert x_1 - \alpha \rvert$	4.36×10^{-42}	8.67×10^{-60}	2.55×10^{-52}	1.29×10^{-62}	4.34×10^{-6}	1.60×10^{-64}
	$\lvert x_2 - \alpha \rvert$	0.0×10^{-399}	7.12×10^{-234}	0.0×10^{-400}	0.0×10^{-399}	4.57×10^{-82}	4.61×10^{-254}
W2F2	$\lvert x_1 - \alpha \rvert$	1.20×10^{-42}	1.74×10^{-60}	5.52×10^{-53}	3.67×10^{-63}	4.08×10^{-6}	5.06×10^{-65}
	$\lvert x_2 - \alpha \rvert$	0.0×10^{-399}	1.16×10^{-236}	0.0×10^{-400}	0.0×10^{-399}	2.33×10^{-83}	4.56×10^{-256}
W2F3	$\lvert x_1 - \alpha \rvert$	1.08×10^{-41}	1.54×10^{-58}	1.55×10^{-51}	1.14×10^{-61}	8.67×10^{-5}	1.396×10^{-63}
	$\lvert x_2 - \alpha \rvert$	1.41×10^{-424}	5.27×10^{-172}	8.47×10^{-424}	0.0×10^{-399}	1.66×10^{-60}	5.22×10^{-188}
W2F4	$\lvert x_1 - \alpha \rvert$	3.80×10^{-42}	5.01×10^{-60}	2.15×10^{-52}	1.07×10^{-62}	4.35×10^{-6}	1.18×10^{-64}
	$\lvert x_2 - \alpha \rvert$	0.0×10^{-399}	1.16×10^{-235}	0.0×10^{-400}	0.0×10^{-399}	3.65×10^{-82}	1.38×10^{-254}

The global convergence with appropriate initial values x_0 is effectively described by means of a basin of attraction that is the set of initial values leading to long-time behavior approaching the attractors under the iterative action of R_f. Basins of attraction contain information about the region of convergence. A method occupying a larger region of convergence is likely to be a more robust method. A quantitative analysis will play the important role for measuring the region of convergence.

The basins of attraction, as well as the relevant statistical data, are constructed in a similar manner shown in the work of Geum–Kim–Neta [19]. Because of the high order, we take a smaller square $[-1.5, 1.5]^2$ and use 601×601 initial points uniformly distributed in the domain. Maple software has been used to perform the desired dynamics with convergence stopping criteria satisfying $|x_{n+1} - x_n| < 10^{-6}$ within the maximum number of 40 iterations. An initial point is painted with a color whose intensity measures the number of iterations converging to a root. The brighter color implies the faster convergence. The black point means that its orbit did not converge within 40 iterations.

Despite the limited space, we will explore the dynamics of all listed maps W3G7 and W2A1–W2F4, with applications to $p_k(z)$, $(1 \leq k \leq 7)$ through the following seven examples. In each example, we have shown dynamical planes for the convergence behavior of iterative map $x_{n-1} = R_f(x_n)$ (42) with $f(z) = p_k(z)$ by illustrating the relevant basins of attraction through Figures 1–7 and displaying relevant statistical data in Tables 5–7 with colored fonts indicating best results.

Example 1. *As a first example, we have taken a quadratic polynomial raised to the power of two with all real roots:*

$$p_1(z) = (z^2 - 1)^2. \tag{47}$$

Clearly the roots are ± 1. *Basins of attraction for W3G7, W2A1–W2F4 are given in Figure 1. Consulting Tables 5–7, we find that the methods* **W2B2** *and* **W2F4** *use the least number (2.71) of iterations per point on average (ANIP) followed by W2F1 with 2.72 ANIP, W2C3 with 2.73 and W2B1 with 2.74. The fastest method is W2A2 with 969.374 s followed closely by W2A3 with 990.341 s. The slowest is W2A4 with 4446.528 s. Method W2C4 has the lowest number of black points (601) and W2A4 has the highest number (78843). We will not include* **W2A4** *in the coming examples.*

Table 5. Average number of iterations per point for each example (1–7).

Map	Example							
	1: m = 2	2: m = 3	3: m = 3	4: m = 4	5: m = 3	6: m = 3	7: m = 3	Average
W3G7	2.94	3.48	3.83	3.93	3.95	3.97	6.77	4.12
W2A1	2.84	3.50	3.70	4.04	6.84	3.74	5.49	4.31
W2A2	2.76	3.15	3.52	3.84	3.62	3.66	4.84	3.63
W2A3	2.78	3.21	3.61	3.89	3.70	3.74	4.98	3.70
W2A4	11.40	-	-	-	-	-	-	-
W2B1	2.74	3.25	3.70	3.88	3.67	3.72	5.01	3.71
W2B2	2.95	3.42	3.66	4.01	3.75	3.77	5.15	3.82
W2B3	2.78	3.28	3.64	3.89	3.69	4.65	5.13	3.86
W2B4	3.29	3.91	4.99	-	-	-	-	-
W2C1	2.88	3.66	3.87	4.08	3.89	5.45	6.25	4.30
W2C2	2.93	3.68	3.95	4.15	6.70	4.67	5.75	4.55
W2C3	2.73	3.22	3.53	3.98	3.60	3.62	4.94	3.66
W2C4	3.14	3.81	4.96	-	-	-	-	-
W2F1	2.72	3.24	3.55	3.84	3.49	3.57	5.41	3.69
W2F2	2.81	3.28	3.80	4.06	5.02	4.50	5.29	4.10
W2F3	2.91	3.54	4.36	4.41	-	-	-	-
W2F4	2.71	3.19	3.50	3.86	3.42	3.53	5.52	3.68

Example 2. *As a second example, we have taken the same quadratic polynomial now raised to the power of three:*

$$p_2(z) = (z^2 - 1)^3. \tag{48}$$

The basins for the best methods are plotted in Figure 2. This is an example to demonstrate the effect of raising the multiplicity from two to three. In one case, namely W3G7, we also have $m = 5$ *with CPU time of 4128.379 s. Based on the figure we see that W2B4, W2C4 and W2F3 were chaotic. The worst are W2B4, W2C4 and W2F3. In terms of ANIP, the best was W2A2 (3.15) followed by W2F4 (3.19) and the worst was W2B4 (3.91). The fastest was W2B3 using (2397.111 s) followed by W2F1 using 2407.158 s and the slowest was W2C4 (4690.295 s) preceded by W3G7 (2983.035 s). Four methods have the highest number of black points (617). Those were W2A1, W2B4, W2C1 and W2F2. The lowest number was 601 for W2A2, W2C2, W2C4 and W2F1.*

Comparing the CPU time for the cases $m = 2$ *and* $m = 3$ *of W3G7, we find it is about doubled. But when increasing from three to five, we only needed about 50% more.*

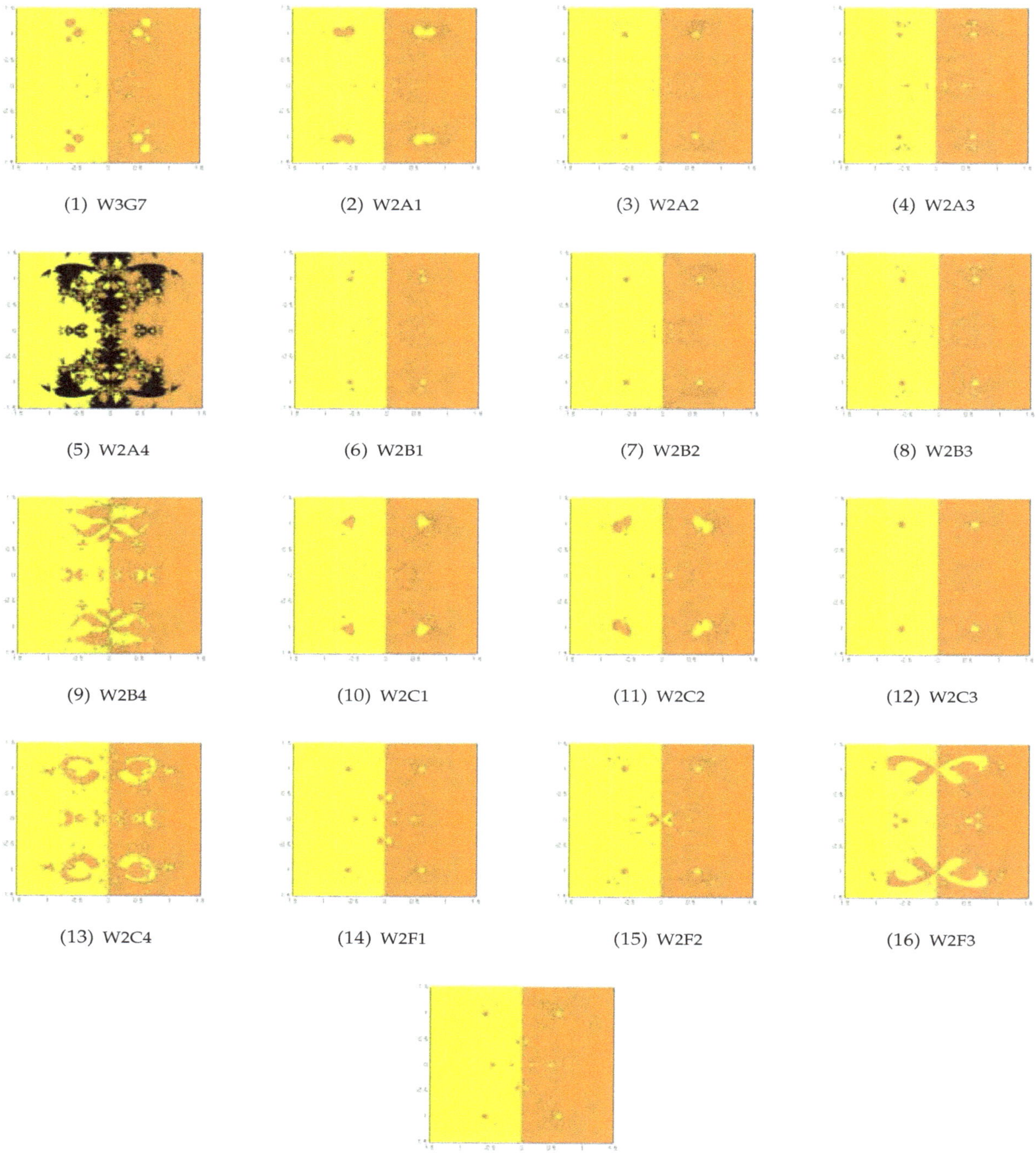

(1) W3G7 (2) W2A1 (3) W2A2 (4) W2A3

(5) W2A4 (6) W2B1 (7) W2B2 (8) W2B3

(9) W2B4 (10) W2C1 (11) W2C2 (12) W2C3

(13) W2C4 (14) W2F1 (15) W2F2 (16) W2F3

(17) W2F4

Figure 1. The top row for W3G7 (**left**), W2A1 (**center left**), W2A2 (**center right**) and W2A3 (**right**). The second row for W2A4 (**left**), W2B1 (**center left**), W2B2 (**center right**) and W2B3 (**right**). The third row for W2B4 (**left**), W2C1 (**center left**), W2C2 (**center right**) and W2C3 (**right**). The third row for W2C4 (**left**), W2F1 (**center left**), W2F2 (**center right**), and W2F3 (**right**). The bottom row for W2F4 (**center**), for the roots of the polynomial equation $(z^2-1)^2$.

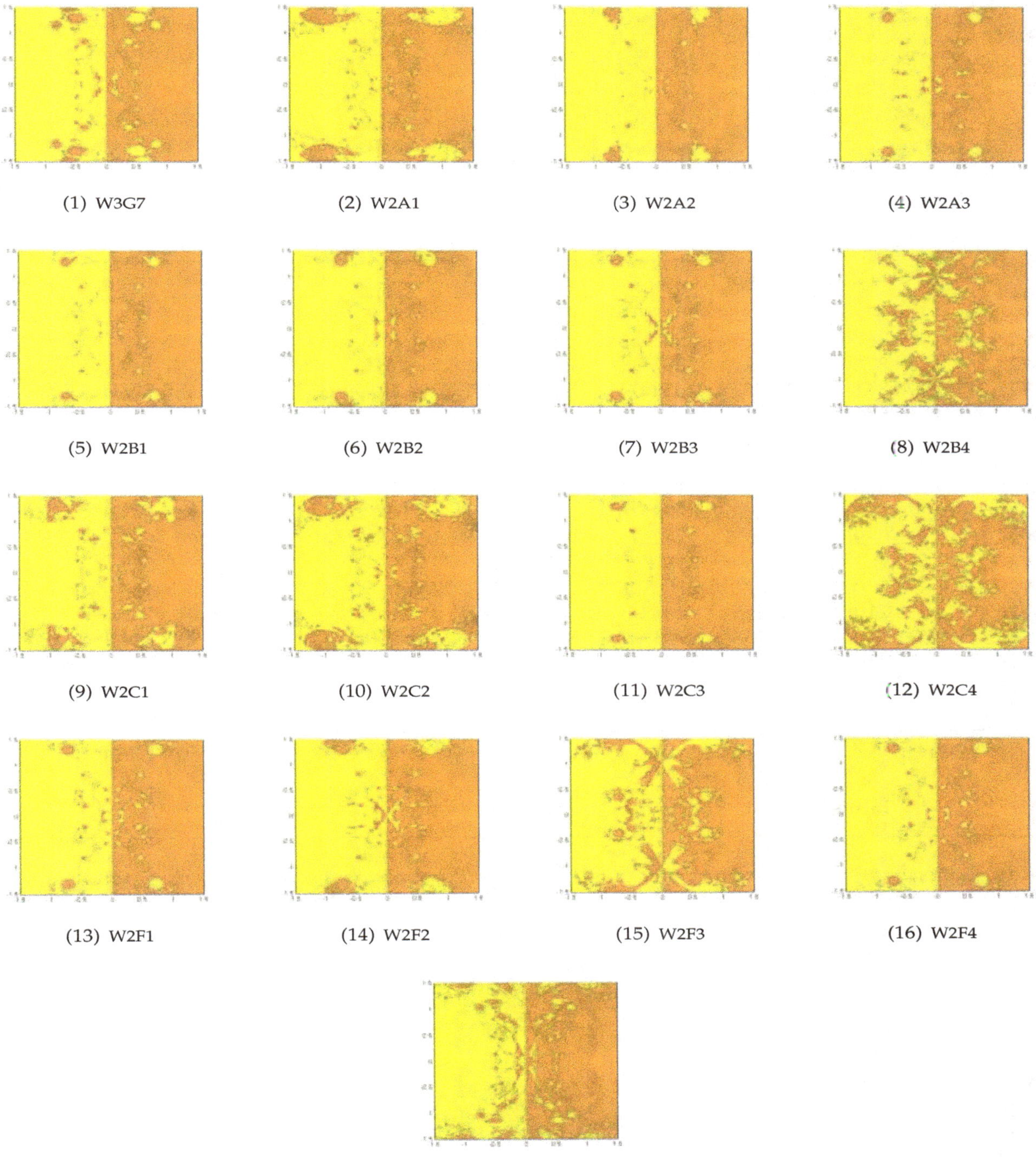

(1) W3G7 (2) W2A1 (3) W2A2 (4) W2A3

(5) W2B1 (6) W2B2 (7) W2B3 (8) W2B4

(9) W2C1 (10) W2C2 (11) W2C3 (12) W2C4

(13) W2F1 (14) W2F2 (15) W2F3 (16) W2F4

(17) W3G7m5

Figure 2. The top row for W3G7 (**left**), W2A1 (**center left**), W2A2 (**center right**) and W2A3 (**right**). The second row for W2B1 (**left**), W2B2 (**center left**), W2B3 (**center right**) and W2B4 (**right**). The third row for W2C1 (**left**), W2C2 (**center left**), W2C3 (**center right**) and W2C4 (**right**). The fourth row for W2F1 (**left**), W2F2 (**center left**), W2F3 (**center right**), and W2F4 (**right**). The bottom row for W3G7m5 (**center**), for the roots of the polynomial equation $(z^2-1)^k$, $k \in \{3,5\}$.

Example 3. *In our third example, we have taken a cubic polynomial raised to the power of three:*

$$p_3(z) = (3z^3 + 4z^2 - 10)^3. \tag{49}$$

Basins of attraction are given in Figure 3. It is clear that W2B4, W2C4 and W2F3 were too chaotic and they should be eliminated from further consideration. In terms of ANIP, the best was W2F4 (3.50) followed by W2A2 (3.52), W2C3 (3.53) and W2F1 (3.55) and the worst were W2B4 and W2C4 with 4.99 and 4.96 ANIP, respectively. The fastest was W2C3 using 2768.362 s and the slowest was W2B3 (7193.034 s). There were 13 methods with only one black point and one with two points. The highest number of black points was 101 for W2F2.

Table 6. CPU time (in seconds) required for each example(1–7) using a Dell Multiplex-990.

Map	Example							
	1: m = 2	2: m = 3	3: m = 3	4: m = 4	5: m = 3	6: m = 3	7: m = 3	Average
W3G7	1254.077	2983.035	3677.848	3720.670	3944.937	3901.679	4087.102	3367.050
W2A1	1079.667	2694.537	3528.149	3119.911	5896.635	2938.747	3526.840	3254.927
W2A2	969.374	2471.727	3287.081	2956.702	3218.223	2891.478	2981.179	2682.252
W2A3	990.341	2843.789	2859.093	2999.712	3002.146	3074.811	3155.307	2703.600
W2A4	4446.528	-	-	-	-	-	-	-
W2B1	1084.752	2634.826	3295.162	3051.941	2835.755	3238.363	3272.667	2773.352
W2B2	1075.393	2429.996	3130.223	3051.192	2929.106	3581.456	3155.619	2764.712
W2B3	1180.366	2397.111	7193.034	3000.383	2970.711	3739.766	3139.084	3374.351
W2B4	1274.653	2932.008	4872.972	-	-	-	-	-
W2C1	1132.069	2685.355	3242.637	3287.066	3147.663	4080.019	4802.662	3196.782
W2C2	1112.162	2881.697	3189.706	3873.037	5211.619	3665.773	3950.896	3412.127
W2C3	1052.570	2421.026	2768.362	3014.033	2778.518	2914.941	3953.346	2700.399
W2C4	2051.710	4690.295	7193.034	-	-	-	-	-
W2F1	1071.540	2407.158	2909.965	3472.317	2832.230	3490.896	3246.584	2775.813
W2F2	1015.051	2438.483	3031.802	3061.270	3703.152	3737.394	3324.537	2901.670
W2F3	1272.188	2596.200	3603.655	4130.158	-	-	-	-
W2F4	1216.839	2620.052	3589.177	3233.168	3534.312	3521.660	3934.845	3092.865

Table 7. Number of points requiring 40 iterations for each example (1–7).

Map	Example							
	1: m = 2	2: m = 3	3: m = 3	4: m = 4	5: m = 3	6: m = 3	7: m = 3	Average
W3G7	677	605	1	250	40	1265	33,072	5130.000
W2A1	657	617	1	166	34,396	1201	18,939	7996.714
W2A2	697	601	1	162	1	1201	15,385	2578.286
W2A3	675	605	55	152	9	1221	14,711	2489.714
W2A4	78,843	-	-	-	-	-	-	-
W2B1	679	613	1	204	9	1201	13,946	2379.000
W2B2	635	609	1	116	1	1217	15,995	2653.429
W2B3	679	613	1	146	3	10,645	16,342	4061.286
W2B4	659	617	2	-	-	-	-	-
W2C1	689	617	1	400	20	18,157	24,239	6303.286
W2C2	669	601	1	174	17,843	1265	18,382	5562.143
W2C3	659	609	1	184	1	1201	14,863	2502.571
W2C4	601	601	1	-	-	-	-	-
W2F1	681	601	1	126	10	1225	18,772	3059.429
W2F2	679	617	101	614	3515	1593	17,469	3512.571
W2F3	663	605	1	78	-	-	-	-
W2F4	645	605	1	130	12	1217	20,020	3232.857

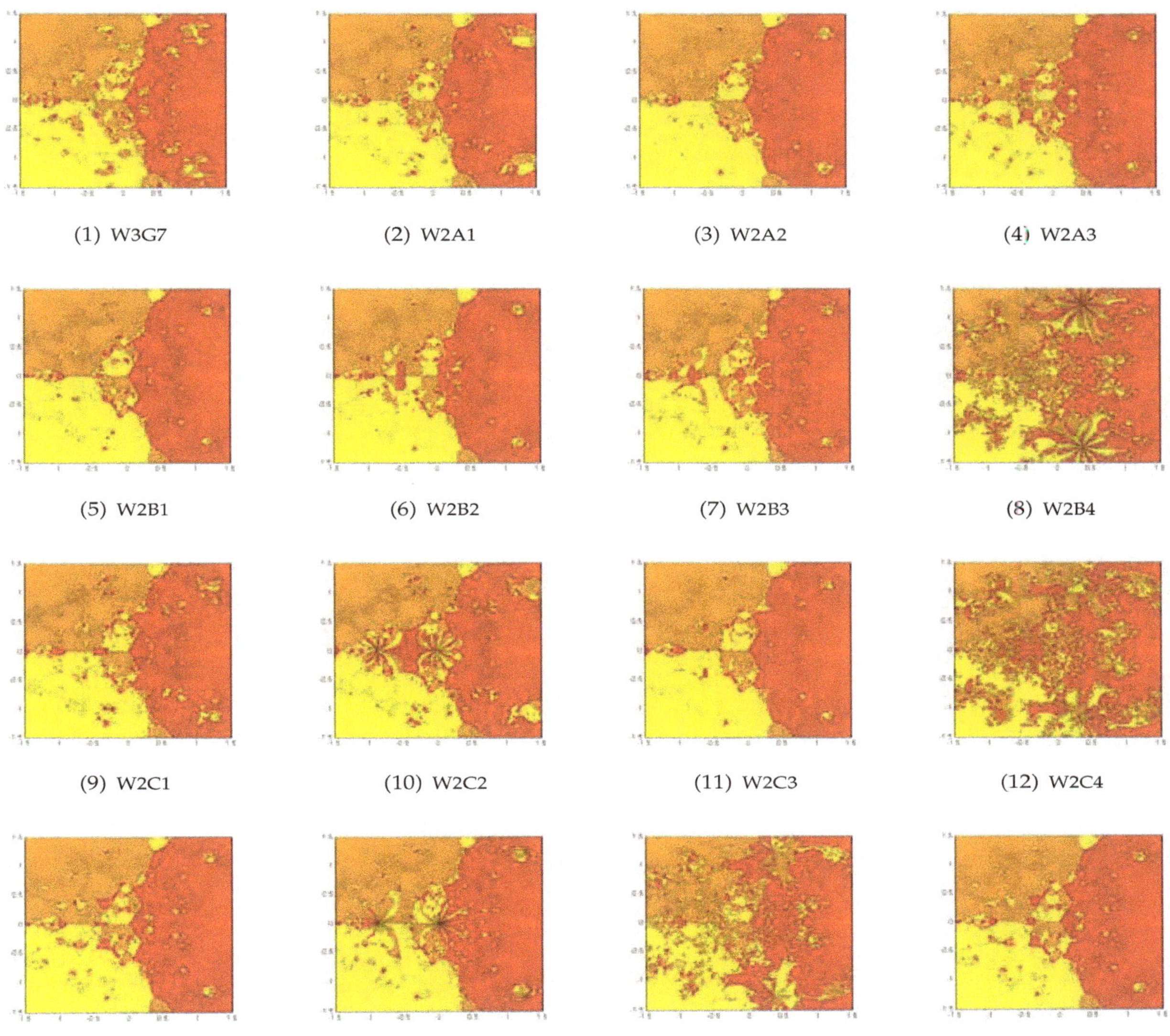

(1) W3G7 (2) W2A1 (3) W2A2 (4) W2A3

(5) W2B1 (6) W2B2 (7) W2B3 (8) W2B4

(9) W2C1 (10) W2C2 (11) W2C3 (12) W2C4

(13) W2F1 (14) W2F2 (15) W2F3 (16) W2F4

Figure 3. The top row for W3G7 (**left**), W2A1 (**center left**), W2A2 (**center right**) and W2A3 (**right**). The second row for W2B1 (**left**), W2B2 (**center left**), W2B3 (**center right**) and W2B4 (**right**). The third row for W2C1 (**left**), W2C2 (**center left**), W2C3 (**center right**) and W2C4 (**right**). The bottom row for W2F1 (**left**), W2F2 (**center left**), W2F3 (**center right**), and W2F4 (**right**), for the roots of the polynomial equation $(3z^3 + 4z^2 - 10)^3$.

Example 4. *As a fourth example, we have taken a different cubic polynomial raised to the power of four:*

$$p_4(z) = (z^3 - z)^4. \tag{50}$$

The basins are given in Figure 4. We now see that W2F3 is the worst. In terms of ANIP, W2A2 and W2F1 were the best (3.84 each) and the worst was W2F3 (4.41). The fastest was W2A2 (2956.702 s) and the slowest was W2F3 (4130.158 s). The lowest number of black points (78) was for method W2F3 and the highest number (614) for W2F2. We did not include W2F3 in the rest of the experiments.

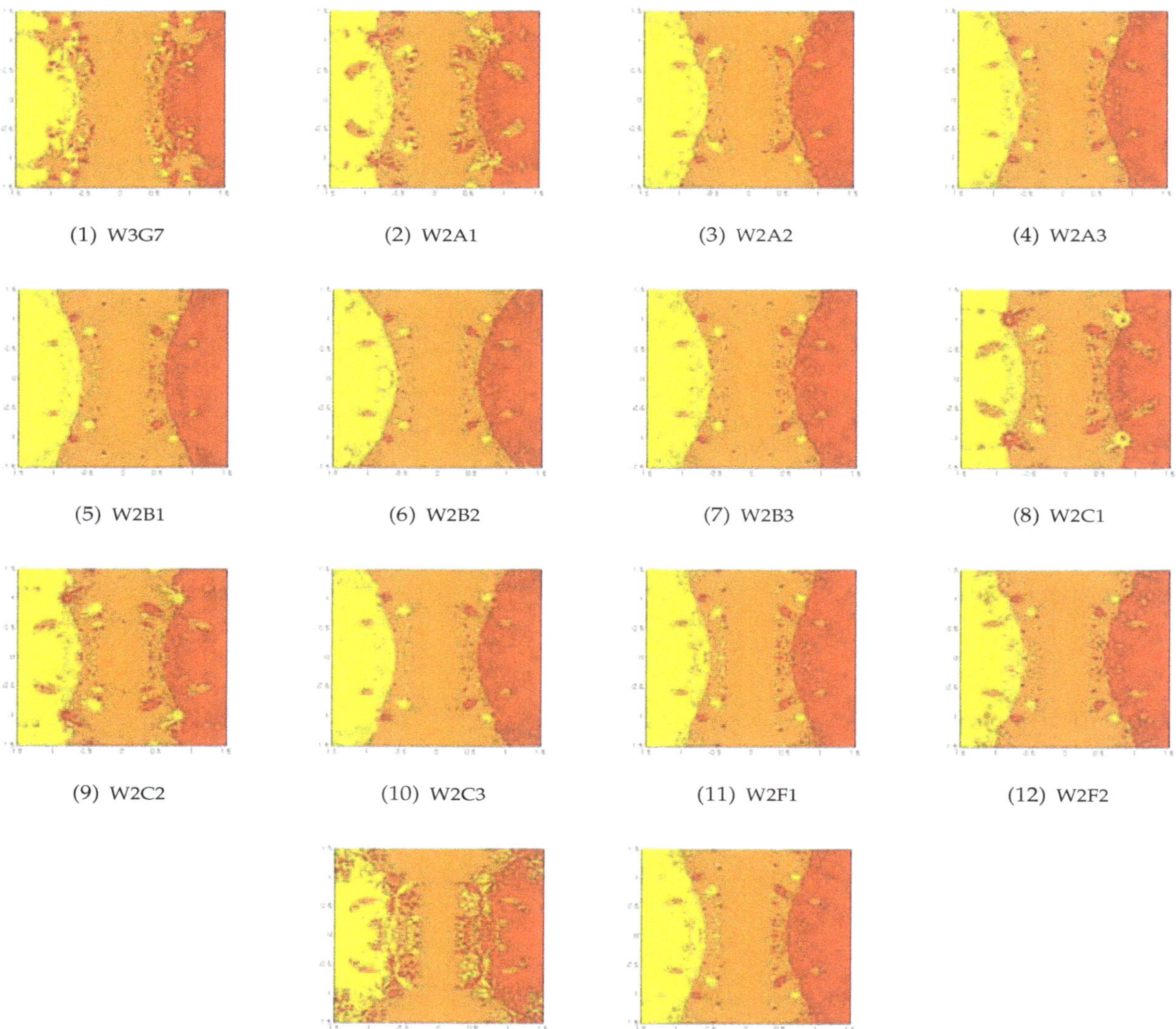

(1) W3G7 (2) W2A1 (3) W2A2 (4) W2A3

(5) W2B1 (6) W2B2 (7) W2B3 (8) W2C1

(9) W2C2 (10) W2C3 (11) W2F1 (12) W2F2

(13) W2F3 (14) W2F4

Figure 4. The top row for W3G7 (**left**), W2A1 (**center left**), W2A2 (**center right**) and W2A3 (**right**). The second row for W2B1 (**left**), W2B2 (**center left**), W2B3 (**center right**) and W2C1 (**right**). The third row for W2C2 (**left**), W2C3 (**center left**), W2F1 (**center right**) and W2F2 (**right**). The bottom row for W2F3 (**left**) and W2F4 (**right**), for the roots of the polynomial equation $(z^3 - z)^4$.

Example 5. *As a fifth example, we have taken a quintic polynomial raised to the power of three:*

$$p_3(z) = (z^5 - 1)^3. \tag{51}$$

The basins for the best methods left are plotted in Figure 5. The worst were W2A1 and W2C2. In terms of ANIP, the best was W2F4 (3.42) followed by W2F1 (3.49) and the worst were W2A1 (6.84) and W2C2 (6.70). The fastest was W2C3 using 2778.518 s followed by W2F1 using 2832.23 s and W2B1 using 2835.755 s. The slowest was W2A1 using 5896.635 s. There were three methods with one black point (W2A2, W2B2 and W2C3) and four others with 10 or less such points, namely W2B3 (3), W2A3 and W2B1 (9) and W2F1 (10). The highest number was for W2A1 (34,396) preceded by W2C2 with 17,843 black points.

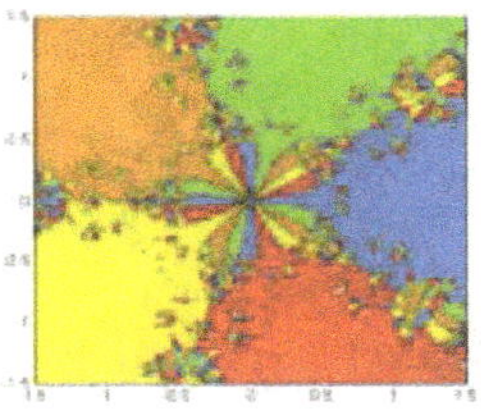
(1) W3G7

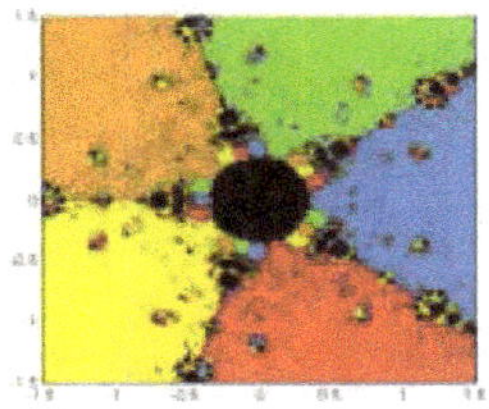
(2) W2A1

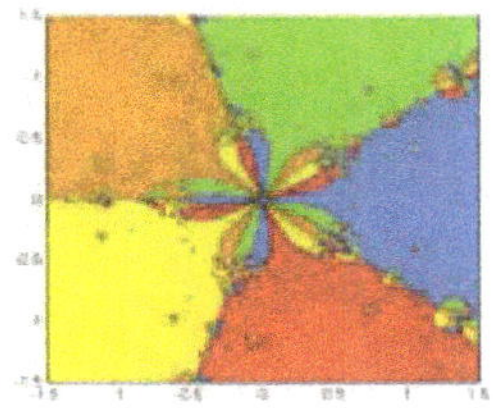
(3) W2A2

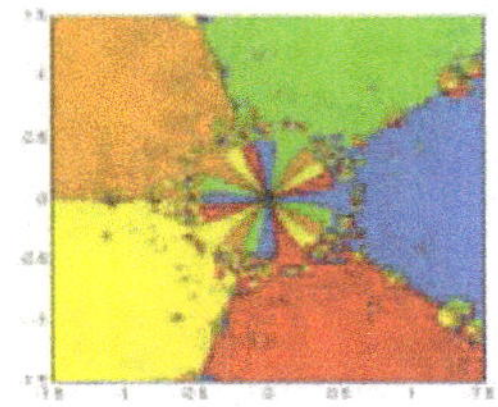
(4) W2A3

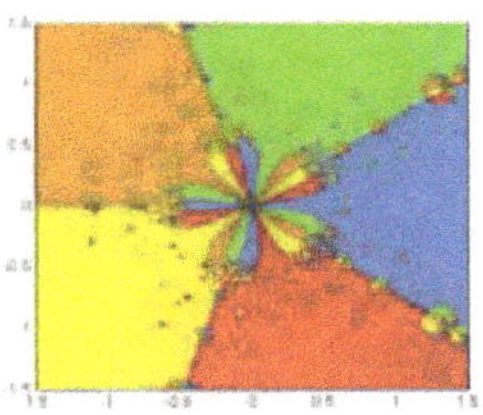
(5) W2B1

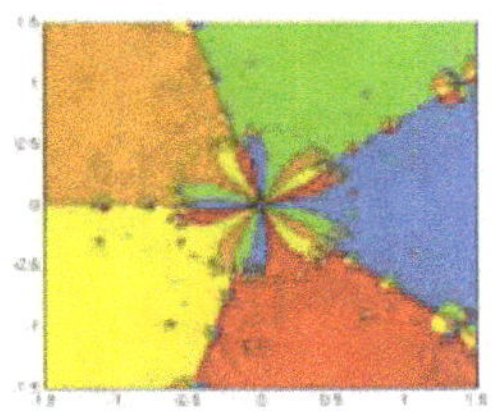
(6) W2B2

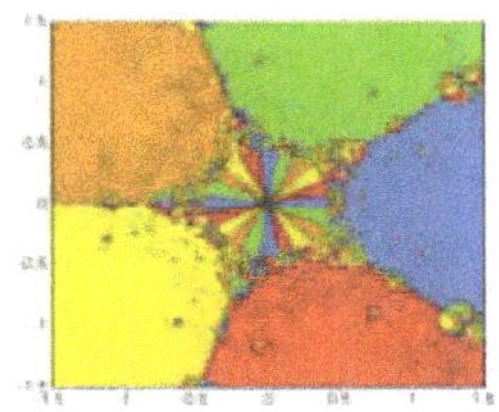
(7) W2B3

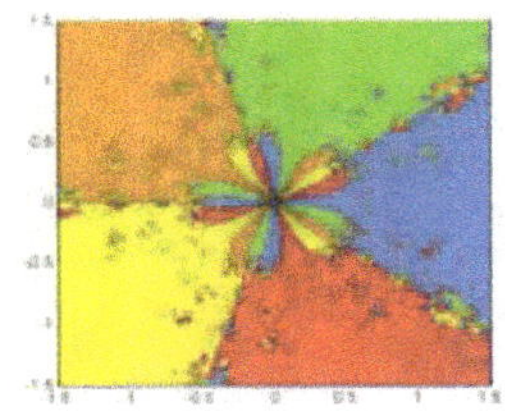
(8) W2C1

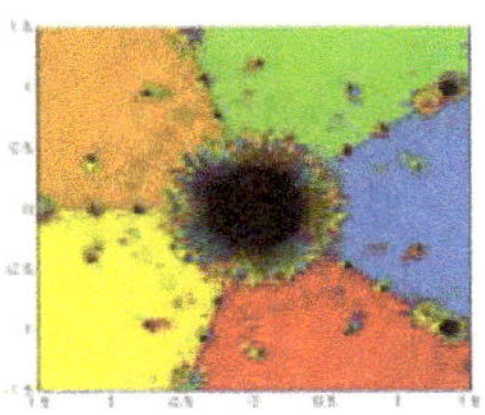
(9) W2C2

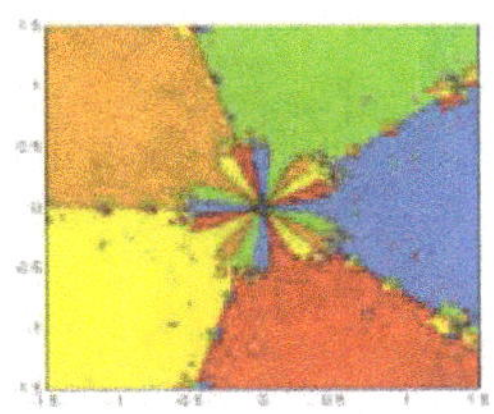
(10) W2C3

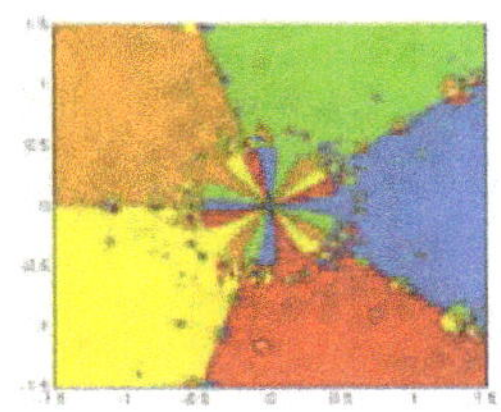
(11) W2F1

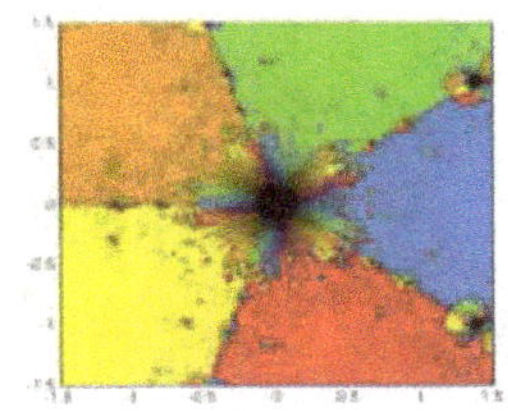
(12) W2F2

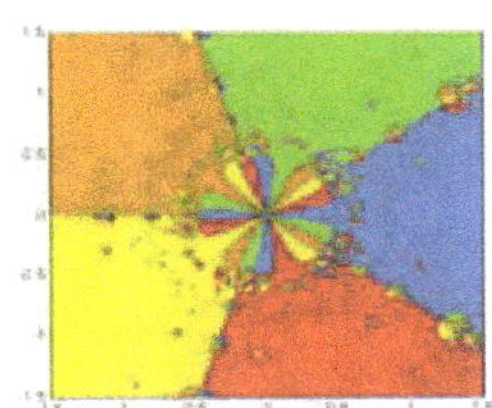
(13) W2F4

Figure 5. The top row for W3G7 (**left**), W2A1 (**center left**), W2A2 (**center right**) and W2A3 (**right**). The second row for W2B1 (**left**), W2B2 (**center left**), W2B3 (**center right**) and W2C1 (**right**). The third row for W2C2 (**left**), W2C3 (**center left**), W2F1 (**center right**) and W2F2 (**right**). The bottom row for W2F4 (**center**), for the roots of the polynomial equation $(z^5 - 1)^3$.

Example 6. *As a sixth example, we have taken a quartic polynomial raised to the power of three:*

$$p_6(z) = (z^4 - 1)^3. \tag{52}$$

The basins for the best methods left are plotted in Figure 6. It seems that most of the methods left were good except W2B3 and W2C1. Based on Table 5 we find that W2F4 has the lowest ANIP (3.53) followed by W2F1 (3.57). The fastest method was W2A2 (2891.478 s) followed by W2C3 (2914.941 s). The slowest was W2C1 (4080.019 s) preceded by W3G7 using 3901.679 s. The lowest number of black points was for W2A1, W2A2, W2B1 and W2C3 (1201) and the highest number was for W2C1 with 18,157 black points.

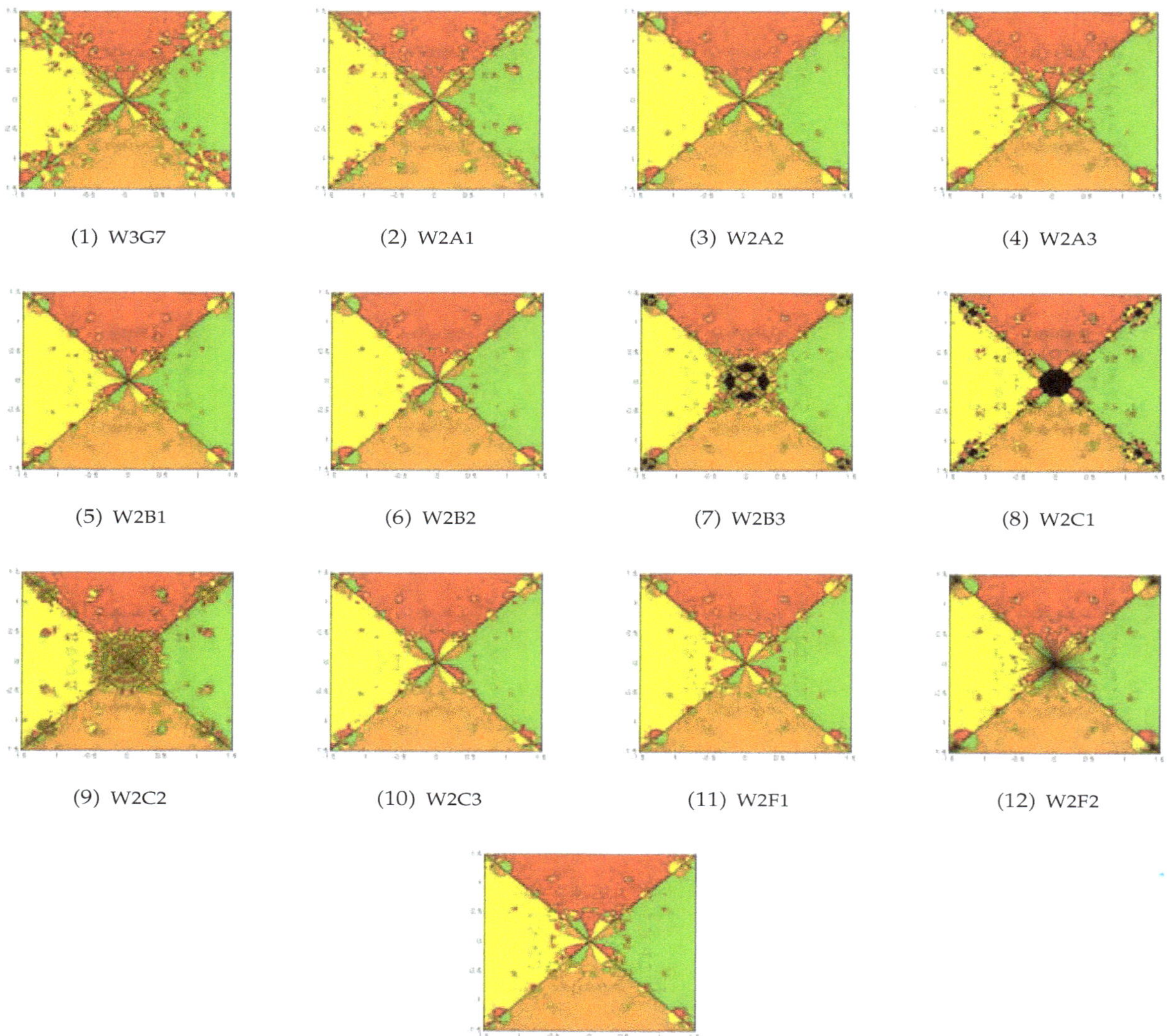

(1) W3G7 (2) W2A1 (3) W2A2 (4) W2A3

(5) W2B1 (6) W2B2 (7) W2B3 (8) W2C1

(9) W2C2 (10) W2C3 (11) W2F1 (12) W2F2

(13) W2F4

Figure 6. The top row for W3G7 (**left**), W2A1 (**center left**), W2A2 (**center right**) and W2A3 (**right**). The second row for W2B1 (**left**), W2B2 (**center left**), W2B3 (**center right**) and W2C1 (**right**). The third row for W2C2 (**left**), W2C3 (**center left**), W2F1 (**center right**) and W2F2 (**right**). The bottom row for W2F4 (**center**), for the roots of the polynomial equation $(z^4-1)^3$.

Example 7. *As a seventh example, we have taken a non-polynomial equation having* $\pm i$ *as its triple roots:*

$$p_6(z) = (z+i)^3(e^{z-i}-1)^3, \text{ with } i = \sqrt{-1}. \tag{53}$$

The basins for the best methods left are plotted in Figure 7. It seems that most of the methods left have a larger basin for the root $-i$*, i.e., the boundary does not match the real line exactly. Based on Table 5 we find that W2A2 has the lowest ANIP (4.84) followed by W2C3 (4.94) and W2A3 (4.98). The fastest method was W2A2 (2981.179 seconds) followed by W2B3 (3139.084 s), W2A3 (3155.307 s) and W2B2 (3155.619 s). The slowest was W2C1 (4802.662 s). The lowest number of black points was for W2B1 (13,946) and the highest number was for W3G7 with 33,072 black points. In general all methods had higher number of black points compared to the polynomial examples.*

We now average all these results across the seven examples to try and pick the best method. W2A2 had the lowest ANIP (3.63), followed by W2C3 with 3.66, W2F4 with 3.68 and W2F1 with 3.69. The fastest method was W2A2 (2682.252 seconds), followed by W2C3 (2700.399 s) and W2A3 using 2703.600 s of CPU. W2B1 has the lowest number of black points on average (2379), followed by W2A3 (2490 black points). The highest number of black points was for W2A1.

Based on these seven examples we see that **W2F4** has four examples with the lowest ANIP, W2A2 had three examples and W2F1 has one example. On average, though, W2A2 had the lowest ANIP. W2A2 was the fastest in four examples and on average. W2C3 was the fastest in two examples and W2B3 in one example. In terms of black points, W2A2, W2B1 and W2B3 had the lowest number in three examples and W2F1 in two examples. On average W2B1 has the lowest number. Thus, we recommend W2A2, since it is in the top in all categories.

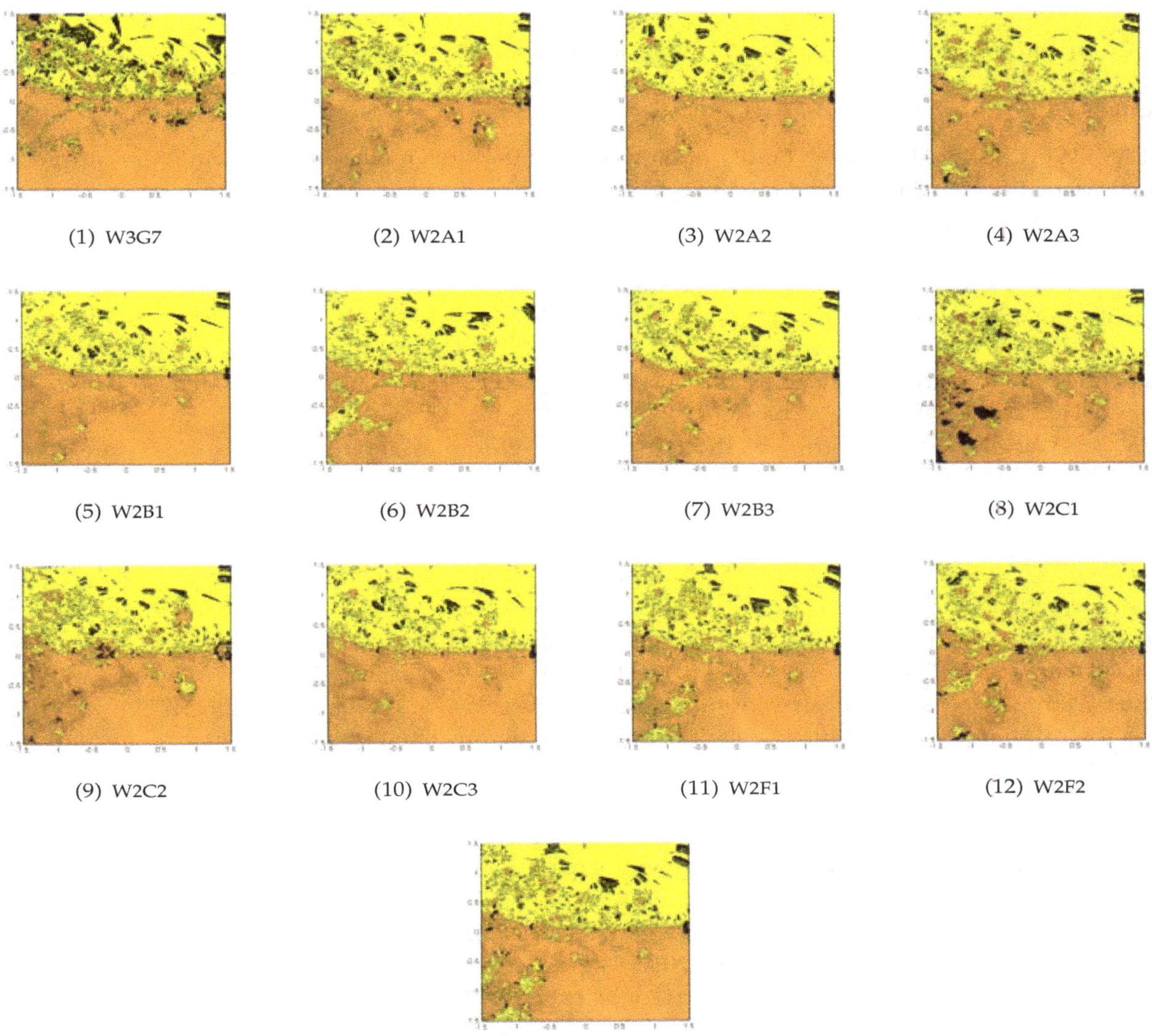

(1) W3G7 (2) W2A1 (3) W2A2 (4) W2A3

(5) W2B1 (6) W2B2 (7) W2B3 (8) W2C1

(9) W2C2 (10) W2C3 (11) W2F1 (12) W2F2

(13) W2F4

Figure 7. The top row for W3G7 (**left**), W2A1 (**center left**), W2A2 (**center right**) and W2A3 (**right**). The second row for W2B1 (**left**), W2B2 (**center left**), W2B3 (**center right**) and W2C1 (**right**). The third row for W2C2 (**left**), W2C3 (**center left**), W2F1 (**center right**) and W2F2 (**right**). The bottom row for W2F4 (**center**), for the roots of the non-polynomial equation $(z+i)^3(e^{z-i}-1)^3$.

5. Conclusions

Both numerical and dynamical aspects of iterative map (1) support the main theorem well through a number of test equations and examples. The W2C2 and W2B3 methods were observed to occupy relatively slower CPU time. Such dynamical aspects would be greatly strengthened if we could include a study of parameter planes with reference to appropriate parameters in Table 1.

The proposed family of methods (1) employing generic weight functions favorably cover most of optimal sixteenth-order multiple-root finders with a number of feasible weight functions. The dynamics behind the purely imaginary extraneous fixed points will choose best members of the family with improved convergence behavior. However, due to the high order of convergence, the algebraic difficulty might arise resolving its increased complexity. The current work is limited to univariate nonlinear equations; its extension to multivariate ones becomes another task.

Author Contributions: investigation, M.–Y.L.; formal analysis, Y.I.K.; supervision, B.N.

References

1. Bi, W.; Wu, Q.; Ren, H. A new family of eighth-order iterative methods for solving nonlinear equations. *Appl. Math. Comput.* **2009**, *214*, 236–245. [CrossRef]
2. Cordero, A.; Torregrosa, J.R.; Vassileva, M.P. Three-step iterative methods with optimal eighth-order convergence. *J. Comput. Appl. Math.* **2011**, *235*, 3189–3194. [CrossRef]
3. Geum, Y.H.; Kim, Y.I. A uniparametric family of three-step eighth-order multipoint iterative methods for simple roots. *Appl. Math. Lett.* **2011**, *24*, 929–935. [CrossRef]
4. Lee, S.D.; Kim, Y.I.; Neta, B. An optimal family of eighth-order simple-root finders with weight functions dependent on function-to-function ratios and their dynamics underlying extraneous fixed points. *J. Comput. Appl. Math.* **2017**, *317*, 31–54. [CrossRef]
5. Liu, L.; Wang, X. Eighth-order methods with high efficiency index for solving nonlinear equations. *Appl. Math. Comput.* **2010**, *215*, 3449–3454. [CrossRef]
6. Petković, M.S.; Neta, B.; Petković, L.D.; Džunić, J. *Multipoint Methods for Solving Nonlinear Equations*; Elsevier: New York, NY, USA, 2012.
7. Petković, M.S.; Neta, B.; Petković, L.D.; Džunić, J.; Multipoint methods for solving nonlinear equations: A survey. *Appl. Math. Comput.* **2014**, *226*, 635–660. [CrossRef]
8. Sharma, J.R.; Arora, H. A new family of optimal eighth order methods with dynamics for nonlinear equations. *Appl. Math. Comput.* **2016**, *273*, 924–933. [CrossRef]
9. Rhee, M.S.; Kim, Y.I.; Neta, B. An optimal eighth-order class of three-step weighted Newton's methods and their dynamics behind the purely imaginary extraneous fixed points. *Int. J. Comput. Math.* **2017**, *95*, 2174–2211. [CrossRef]
10. Kung, H.T.; Traub, J.F. Optimal order of one-point and multipoint iteration. *J. Assoc. Comput. Mach.* **1974**, *21*, 643–651. [CrossRef]
11. Maroju, P.; Behl, R.; Motsa, S.S. Some novel and optimal families of King's method with eighth and sixteenth-order of convergence. *J. Comput. Appl. Math.* **2017**, *318*, 136–148. [CrossRef]
12. Sharma, J.R.; Argyros, I.K.; Kumar, D. On a general class of optimal order multipoint methods for solving nonlinear equations. *J. Math. Anal. Appl.* **2017**, *449*, 994–1014. [CrossRef]
13. Neta, B. On a family of Multipoint Methods for Non-linear Equations. *Int. J. Comput. Math.* **1981**, *9*, 353–361. [CrossRef]
14. Geum, Y.H.; Kim, Y.I.; Neta, B. Constructing a family of optimal eighth-order modified Newton-type multiple-zero finders along with the dynamics behind their purely imaginary extraneous fixed points. *J. Comput. Appl. Math.* **2018**, *333*, 131–156. [CrossRef]
15. Ahlfors, L.V. *Complex Analysis*; McGraw-Hill Book, Inc.: New York, NY, USA, 1979.
16. Hörmander, L. *An Introduction to Complex Analysis in Several Variables*; North-Holland Publishing Company: Amsterdam, The Netherlands, 1973.

17. Shabat, B.V. *Introduction to Complex Analysis PART II, Functions of several Variables*; American Mathematical Society: Providence, RI, USA, 1992.
18. Wolfram, S. *The Mathematica Book*, 5th ed.; Wolfram Media: Champaign, IL, USA, 2003.
19. Geum, Y.H.; Kim, Y.I.; Neta, B. Developing an Optimal Class of Generic Sixteenth-Order Simple-Root Finders and Investigating Their Dynamics. *Mathematics* **2019**, *7*, 8. [CrossRef]
20. Stewart, B.D. Attractor Basins of Various Root-Finding Methods. Master's Thesis, Naval Postgraduate School, Department of Applied Mathematics, Monterey, CA, USA, June 2001.
21. Vrscay, E.R.; Gilbert, W.J. Extraneous Fixed Points, Basin Boundaries and Chaotic Dynamics for shröder and König rational iteration Functions. *Numer. Math.* **1988**, *52*, 1–16. [CrossRef]
22. Amat, S.; Busquier, S.; Plaza, S. Review of some iterative root-finding methods from a dynamical point of view. *Scientia* **2004**, *10*, 3–35.
23. Argyros, I.K.; Magreñán, A.Á. On the convergence of an optimal fourth-order family of methods and its dynamics. *Appl. Math. Comput.* **2015**, *252*, 336–346. [CrossRef]
24. Chun, C.; Lee, M.Y.; Neta, B.; Džunić, J. On optimal fourth-order iterative methods free from second derivative and their dynamics. *Appl. Math. Comput.* **2012**, *218*, 6427–6438. [CrossRef]
25. Chicharro, F.; Cordero, A.; Gutiérrez, J.M.; Torregrosa, J.R. Complex dynamics of derivative-free methods for nonlinear equations. *Appl. Math. Comput.* **2013**, *219*, 7023–7035. [CrossRef]
26. Chun, C.; Neta, B. Comparison of several families of optimal eighth order methods. *Appl. Math. Comput.* **2016**, *274*, 762–773. [CrossRef]
27. Cordero, A.; García-Maimó, J.; Torregrosa, J.R.; Vassileva, M.P.; Vindel, P. Chaos in King's iterative family. *Appl. Math. Lett.* **2013**, *26*, 842–848. [CrossRef]
28. Geum, Y.H.; Kim, Y.I.; Magreñán, Á.A. A biparametric extension of King's fourth-order methods and their dynamics. *Appl. Math. Comput.* **2016**, *282*, 254–275. [CrossRef]
29. Geum, Y.H.; Kim, Y.I.; Neta, B. A class of two-point sixth-order multiple-zero finders of modified double-Newton type and their dynamics. *Appl. Math. Comput.* **2015**, *270*, 387–400. [CrossRef]
30. Geum, Y.H.; Kim, Y.I.; Neta, B. A sixth-order family of three-point modified Newton-like multiple-root finders and the dynamics behind their extraneous fixed points. *Appl. Math. Comput.* **2016**, *283*, 120–140. [CrossRef]
31. Magreñan, Á.A. Different anomalies in a Jarratt family of iterative root-finding methods. *Appl. Math. Comput.* **2014**, *233*, 29–38.
32. Neta, B.; Scott, M.; Chun, C. Basin attractors for various methods for multiple roots. *Appl. Math. Comput.* **2012**, *218*, 5043–5066. [CrossRef]
33. Neta, B.; Chun, C.; Scott, M. Basins of attraction for optimal eighth order methods to find simple roots of nonlinear equations. *Appl. Math. Comput.* **2014**, *227*, 567–592. [CrossRef]
34. Scott, M.; Neta, B.; Chun, C. Basin attractors for various methods. *Appl. Math. Comput.* **2011**, *218*, 2584–2599. [CrossRef]
35. Chun, C.; Neta, B.; Basins of attraction for Zhou-Chen-Song fourth order family of methods for multiple roots. *Math. Comput. Simul.* **2015**, *109*, 74–91. [CrossRef]
36. Neta, B.; Chun, C. Basins of attraction for several optimal fourth order methods for multiple roots. *Math. Comput. Simul.* **2014**, *103*, 39–59. [CrossRef]
37. Beardon, A.F. *Iteration of Rational Functions*; Springer: New York, NY, USA, 1991.
38. Traub, J.F. *Iterative Methods for the Solution of Equations*; Chelsea Publishing Company: Chelsea, VT, USA, 1982.

14

A Modified Fletcher–Reeves Conjugate Gradient Method for Monotone Nonlinear Equations with Some Applications

Auwal Bala Abubakar [1,2], Poom Kumam [1,3,4,*], Hassan Mohammad [2], Aliyu Muhammed Awwal [1,5] and Kanokwan Sitthithakerngkiet [6]

1 KMUTTFixed Point Research Laboratory, SCL 802 Fixed Point Laboratory, Science Laboratory Building, Department of Mathematics, Faculty of Science, King Mongkut's University of Technology Thonburi (KMUTT), 126 Pracha-Uthit Road, Bang Mod, Thrung Khru, Bangkok 10140, Thailand
2 Department of Mathematical Sciences, Faculty of Physical Sciences, Bayero University, Kano 700241, Nigeria
3 Center of Excellence in Theoretical and Computational Science (TaCS-CoE), Science Laboratory Building, King Mongkut's University of Technology Thonburi (KMUTT), 126 Pracha-Uthit Road, Bang Mod, Thrung Khru, Bangkok 10140, Thailand
4 Department of Medical Research, China Medical University Hospital, China Medical University, Taichung 40402, Taiwan
5 Department of Mathematics, Faculty of Science, Gombe State University, Gombe 760214, Nigeria
6 Department of Mathematics, Faculty of Applied Science, King Mongkut's University of Technology North Bangkok, 1518 Pracharat 1 Road, Wongsawang, Bangsue, Bangkok 10800, Thailand
* Correspondence: poom.kum@kmutt.ac.th

Abstract: One of the fastest growing and efficient methods for solving the unconstrained minimization problem is the conjugate gradient method (CG). Recently, considerable efforts have been made to extend the CG method for solving monotone nonlinear equations. In this research article, we present a modification of the Fletcher–Reeves (FR) conjugate gradient projection method for constrained monotone nonlinear equations. The method possesses sufficient descent property and its global convergence was proved using some appropriate assumptions. Two sets of numerical experiments were carried out to show the good performance of the proposed method compared with some existing ones. The first experiment was for solving monotone constrained nonlinear equations using some benchmark test problem while the second experiment was applying the method in signal and image recovery problems arising from compressive sensing.

Keywords: nonlinear equations; conjugate gradient method; projection method; convex constraints; signal and image processing

MSC: 65K05; 90C52; 90C56; 94A08

---

1. Introduction

In this paper, we are considering a system of nonlinear monotone equations of the form

$$F(x) = 0, \quad \text{subject to} \quad x \in E, \tag{1}$$

where $E \subseteq R^n$ is closed and convex, $F : R^n \to R^m$, $(m \geq n)$ is continuous and monotone, which means

$$\langle F(x) - F(y), (x - y) \rangle \geq 0, \quad \forall x, y \in R^n.$$

A well-known fact is that under the above assumption, the solution set of (1) is convex unless is empty. It is important to mention that nonlinear monotone equations arise in many practical applications. These and other reasons motivate researchers to develop a large number of class of Iterative methods for solving such systems, for example, see [1–7] among others. In addition, convex constrained equations have application in many scientific fields, some of which are the economic equilibrium problems [8], the chemical equilibrium systems [9], etc. Several algorithms were developed to solve (1), among them, are the trust-region [10] and the Levenberg-Marquardt method [11]. Moreover, the requirement to compute and store the matrix in every iteration makes them ineffective for large-scale nonlinear equations.

Conjugate gradient (CG) methods are efficient for solving large-scale optimization and nonlinear systems because of their low memory requirements. This forms part of the reason several Iterative methods with CG-like directions are proposed in recent years [12,13]. Initially, CG methods and their modified versions are proposed for unconstrained optimization problems [14–19]. Inspired by them, in the last decade, many authors used the CG direction to solve nonlinear monotone equations for both constrained and unconstrained cases. Since in this article, we are interested in solving nonlinear monotone equations with convex constraints, we will only discuss existing methods with such properties.

Many methods for solving nonlinear monotone equations with convex constraints have been presented in the last decade. For examples, Xiao and Zhu [20] presented a CG method, which combines the well-known CG-DESCENT method in [17] and the projection method by Solodov and Svaiter [21]. Liu et al. [22] proposed two CG methods with projection strategy for solving (1). In [23], a modification of the method in [20] was presented by Liu and Li. One of the reasons for the modification was to improve the numerical performance of the method in [20]. Also, Sun and Liu [24] presented derivative-free projection methods for solving nonlinear equations with convex constraints. These methods are the combination of some existing CG methods and the well-known projection method. In addition, a hybrid CG projection method for convex constrained equations was developed in [25]. Ou and Li [26] proposed a combination of a scaled CG method and the projection strategy to solve (1). Furthermore, Ding et al. [27] extended the Dai and Kou (DK) CG method to solve (1) by also combining it with the projection method. Just recently, to popularize the Dai-Yuan (DY) method, Liu and Feng [28] proposed a modified DY method for solving convex constraints monotone equation. The global convergence was also obtained under certain assumptions and finally, some numerical results were reported to show its efficiency.

Inspired by some the above proposals, we present a simple modification of the Fletcher–Reeves (FR) conjugate gradient method [19] considered in [12] to solve nonlinear monotone equations with convex constraints. The modification ensures that the direction is automatically descent, improves its numerical performance and still inherits the nice convergence properties of the method. Under suitable assumptions, we establish the global convergence of the proposed algorithm. Numerical experiments presented show the good performance and competitiveness of the method. In addition, the proposed method has the advantages of the direct methods [29] such as boundary control method by Belishev and Kuryiev [30], the globally convergent method proposed by Beilina and Klibanov [31] and method based on the multidimensional analogs of Gelfand–Levitan–Krein equations [32,33]. The proposed method can be seen as a local method that looks for the closest root. However, there are several global nonlinear solvers that guarantee finding all roots inside a domain and within a very fine double-float accuracy. In some cases a combination of subdivision-based polynomial solver with a decomposition algorithm are employed in order to handle large and complex systems (see for examples [34–36] and references therein).

The remaining part of this article is organized as follows. In Section 2, we mention some preliminaries and present the proposed method. The global convergence of the method is established in Section 3. Finally, Section 4 reports some numerical results to show the performance of the method in solving monotone nonlinear equations with convex constraints, and also apply it to recover a noisy signal and a blurred image.

2. Algorithm

In this section, we define the projection map together with its well-known properties, give some useful assumptions and finally present the proposed algorithm. Throughout this article, $\|\cdot\|$ denotes the Euclidean norm.

Definition 1. *Let $E \subset R^n$ be nonempty closed and convex set. Then for any $x \in R^n$, its projection onto E is defined as*

$$P_E(x) = \arg\min\{\|x-y\| \ : y \in E.\}$$

The following lemma gives some properties of the projection map.

Lemma 1 ([37]). *Suppose $E \subset R^n$ is nonempty, closed and convex set. Then the following statements are true:*

1. $\langle x - P_E(x), P_E(x) - z\rangle \geq 0, \quad \forall x, z \in R^n.$
2. $\|P_E(x) - P_E(y)\| \leq \|x-y\|, \quad \forall x, y \in R^n.$
3. $\|P_E(x) - z\|^2 \leq \|x-z\|^2 - \|x - P_E(x)\|^2, \quad \forall x, z \in R^n.$

Throughout, we suppose the followings

(C_1) The solution set of (1), denoted by E', is nonempty.
(C_2) The mapping F is monotone.
(C_3) The mapping F is Lipschitz continuous, that is there exists a positive constant L such that $\|F(x) - F(y)\| \leq L\|x-y\|, \quad \forall x, y \in R^n.$

Our algorithm is motivated by the work of Papp and Rapajić in [12]. In the paper, they modified the well known Fletcher–Reeves conjugate gradient method to solve unconstrained nonlinear monotone equation. The modification was adding the term $-\theta_k F(x_k)$ to the direction of Fletcher–Reeves. The parameter θ_k was then determined in three different ways and three different directions were proposed, namely, M3TFR1, M3TFR2 and M3TFR3. The direction we are interested in is M3TFR1 and is defined as:

$$d_k = \begin{cases} -F(x_k), & \text{if } k = 0, \\ -F(x_k) + \beta_k^{FR} w_{k-1} + \theta_k F(x_k), & \text{if } k \geq 1, \end{cases} \tag{2}$$

where,

$$\beta_k^{FR} = \frac{\|F(x_k)\|^2}{\|F(x_{k-1})\|^2}, \quad \theta_k = -\frac{F(x_k)^T w_{k-1}}{\|F(x_{k-1})\|^2}, \quad w_{k-1} = z_{k-1} - x_{k-1}, \ z_{k-1} = x_{k-1} + \alpha_{k-1} d_{k-1}.$$

It follows that

$$F(x_k)^T d_k = -\|F(x_k)\|^2.$$

Using same modification proposed in [3], we modify the direction (2) as follows

$$d_k = \begin{cases} -F(x_k), & \text{if } k = 0, \\ -F(x_k) + \frac{\|F(x_k)\|^2 w_{k-1} - F(x_k)^T w_{k-1} F(x_k)}{\max\{\mu\|w_{k-1}\|\|F(x_k)\|, \|F(x_{k-1})\|^2\}}, & \text{if } k \geq 1, \end{cases} \tag{3}$$

where $\mu > 0$ is a positive constant. The difference between the M3TFR1 direction and the direction proposed in this paper is the scaling term appearing in the denominator of Equation (3) i.e., $\max\{\mu\|w_{k-1}\|\|F(x_k)\|, \|F(x_{k-1})\|^2\}$. This modification was shown to have a very good numerical performance in [3] and also helps in obtaining the boundedness of the direction easily.

Remark 1. *Note the the parameter μ is chosen to be strictly positive because if $\mu \leq 0$ then*

$$\max\{\mu\|w_{k-1}\|\|F(x_k)\|, \|F(x_{k-1})\|^2\} = \|F(x_{k-1})\|^2.$$

This means that the direction d_k will always be M3TFR1 given by (2).

3. Convergence Analysis

To prove the global convergence of Algorithm 1, the following results are needed.

Algorithm 1: A modified descent Fletcher–Reeves CG method (MFRM).

Step 0. Select the initial point $x_0 \in R^n$, parameters $\mu > 0, \sigma > 0, 0 < \rho < 1, Tol > 0$, and set $k := 0$.
Step 1. If $\|F(x_k)\| \leq Tol$, stop, otherwise go to **Step 2**.
Step 2. Find d_k using (3).
Step 3. Find the step length $\alpha_k = \gamma\rho^{m_k}$ where m_k is the smallest non-negative integer m such that

$$-\langle F(x_k + \alpha_k d_k), d_k\rangle \geq \sigma\alpha_k\|F(x_k + \alpha_k d_k)\|\|d_k\|^2. \tag{4}$$

Step 4. Set $z_k = x_k + \alpha_k d_k$. If $z_k \in E$ and $\|F(z_k)\| \leq Tol$, stop. Else compute

$$x_{k+1} = P_E[x_k - \zeta_k F(z_k)]$$

where

$$\zeta_k = \frac{F(z_k)^T(x_k - z_k)}{\|F(z_k)\|^2}.$$

Step 5. Let $k = k + 1$ and go to Step 1.

Lemma 2. *Let d_k be defined by Equation (3), then*

$$d_k^T F(x_k) = -\|F(x_k)\|^2 \tag{5}$$

and

$$\|F(x_k)\| \leq \|d_k\| \leq \left(1 + \frac{2}{\mu}\right)\|F(x_k)\|. \tag{6}$$

Proof. By Equation (3), suppose $k = 0$,

$$d_k^T F(x_k) = -F(x_k)^T F(x_k) = -\|F(x_k)\|^2.$$

Now suppose $k > 0$,

$$\begin{aligned} d_k^T F(x_k) &= -F(x_k)^T F(x_k) + \frac{(\|F(x_k)\|^2 w_{k-1})^T F(x_k) - (F(x_k)^T w_{k-1} F(x_k))^T F(x_k)}{\max\{\mu\|w_{k-1}\|\|F(x_k)\|, \|F(x_{k-1})\|^2\}} \\ &= -\|F(x_k)\|^2 + \frac{\|F(x_k)\|^2 w_{k-1}^T F(x_k) - F(x_k)^T (w_{k-1}^T F(x_k)) F(x_k)}{\max\{\mu\|w_{k-1}\|\|F(x_k)\|, \|F(x_{k-1})\|^2\}} \\ &= -\|F(x_k)\|^2 + \frac{\|F(x_k)\|^2 w_{k-1}^T F(x_k) - \|F(x_k)\|^2 w_{k-1}^T F(x_k)}{\max\{\mu\|w_{k-1}\|\|F(x_k)\|, \|F(x_{k-1})\|^2\}} \\ &= -\|F(x_k)\|^2. \end{aligned} \tag{7}$$

Using Cauchy–Schwartz inequality, we get

$$\|F(x_k)\| \leq \|d_k\|. \tag{8}$$

Furthermore, since $\max\{\mu\|w_{k-1}\|\|F(x_k)\|, \|F(x_{k-1})\|^2\} \geq \mu\|w_{k-1}\|\|F(x_k)\|$, then,

$$\begin{aligned}
\|d_k\| &= \left\| -F(x_k) + \frac{\|F(x_k)\|^2 w_{k-1} - (F(x_k)^T w_{k-1})F(x_k)}{\max\{\mu\|w_{k-1}\|\|F(x_k)\|, \|F(x_{k-1})\|^2\}} \right\| \\
&\leq \| - F(x_k)\| + \frac{\| \|F(x_k)\|^2 w_{k-1} - (F(x_k)^T w_{k-1})F(x_k)\|}{\max\{\mu\|w_{k-1}\|\|F(x_k)\|, \|F(x_{k-1})\|^2\}} \\
&\leq \|F(x_k)\| + \frac{\|F(x_k)\|^2\|w_{k-1}\|}{\mu\|w_{k-1}\|\|F(x_k)\|} + \frac{\|F(x_k)^T w_{k-1} F(x_k)\|}{\mu\|w_{k-1}\|\|F(x_k)\|} \\
&\leq \|F(x_k)\| + \frac{\|F(x_k)\|^2\|w_{k-1}\|}{\mu\|w_{k-1}\|\|F(x_k)\|} + \frac{\|F(x_k)\|^2\|w_{k-1}\|}{\mu\|w_{k-1}\|\|F(x_k)\|} \\
&= \|F(x_k)\| + \frac{2\|F(x_k)\|}{\mu} \\
&= \left(1 + \frac{2}{\mu}\right) \|F(x_k)\|.
\end{aligned} \tag{9}$$

Combining (8) and (9), we get the desired result. □

Lemma 3. *Suppose that assumptions (C_1)–(C_3) hold and the sequences $\{x_k\}$ and $\{z_k\}$ are generated by Algorithm 1. Then we have*

$$\alpha_k \geq \rho \min \left\{ 1, \frac{\|F(x_k)\|^2}{(L+\sigma)\|F(x_k + \frac{\alpha_k}{\rho} d_k)\|\|d_k\|^2} \right\}$$

Proof. Suppose $\alpha_k \neq \rho$, then $\frac{\alpha_k}{\rho}$ does not satisfy Equation (4), that is

$$-F\left(x_k + \frac{\alpha_k}{\rho} d_k\right) < \sigma\frac{\alpha_k}{\rho}\|F(x_k + \frac{\alpha_k}{\rho} d_k)\|\|d_k\|^2.$$

This combined with (7) and the fact that F is Lipschitz continuous yields

$$\begin{aligned}
\|F(x_k)\|^2 &= -F(x_k)^T d_k \\
&= \left(F(x_k + \frac{\alpha_k}{\rho} d_k) - F(x_k)\right)^T d_k - F^T\left(x_k + \frac{\alpha_k}{\rho} d_k\right) d_k \\
&\leq L\frac{\alpha_k}{\rho}\|F(x_k + \frac{\alpha_k}{\rho} d_k)\|\|d_k\|^2 + \sigma\frac{\alpha_k}{\rho}\|F(x_k + \frac{\alpha_k}{\rho} d_k)\|\|d_k\|^2 \\
&= \frac{L+\sigma}{\rho}\alpha_k\|F(x_k + \frac{\alpha_k}{\rho} d_k)\|\|d_k\|^2.
\end{aligned} \tag{10}$$

The above equation implies

$$\alpha_k \geq \rho \min \frac{\|F(x_k)\|^2}{(L+\sigma)\|F(x_k + \frac{\alpha_k}{\rho} d_k)\|\|d_k\|^2},$$

which completes the proof. □

Lemma 4. *Suppose that assumptions (C_1)–(C_3) holds, then the sequences $\{x_k\}$ and $\{z_k\}$ generated by Algorithm 1 are bounded. Moreover, we have*

$$\lim_{k\to\infty} \|x_k - z_k\| = 0 \tag{11}$$

and

$$\lim_{k\to\infty} \|x_{k+1} - x_k\| = 0. \tag{12}$$

Proof. We will start by showing that the sequences $\{x_k\}$ and $\{z_k\}$ are bounded. Suppose $\bar{x} \in E'$, then by monotonicity of F, we get

$$\langle F(z_k), x_k - \bar{x}\rangle \geq \langle F(z_k), x_k - z_k\rangle. \tag{13}$$

Also by definition of z_k and the line search (4), we have

$$\langle F(z_k), x_k - z_k\rangle \geq \sigma\alpha_k^2\|F(z_k)\|\|d_k\|^2 \geq 0. \tag{14}$$

So, we have

$$\begin{aligned}
\|x_{k+1} - \bar{x}\|^2 &= \|P_E[x_k - \zeta_k F(z_k)] - \bar{x}\|^2 \leq \|x_k - \zeta_k F(z_k) - \bar{x}\|^2 \\
&= \|x_k - \bar{x}\|^2 - 2\zeta\langle F(z_k), x_k - \bar{x}\rangle + \|\zeta F(z_k)\|^2 \\
&\leq \|x_k - \bar{x}\|^2 - 2\zeta_k\langle F(z_k), x_k - z_k\rangle + \|\zeta F(z_k)\|^2 \\
&= \|x_k - \bar{x}\|^2 - \left(\frac{\langle F(z_k), x_k - z_k\rangle}{\|F(z_k)\|}\right)^2 \\
&\leq \|x_k - \bar{x}\|^2.
\end{aligned} \tag{15}$$

Thus the sequence $\{\|x_k - \bar{x}\|\}$ is non increasing and convergent, and hence $\{x_k\}$ is bounded. Furthermore, from Equation (15), we have

$$\|x_{k+1} - \bar{x}\|^2 \leq \|x_k - \bar{x}\|^2, \tag{16}$$

and we can deduce recursively that

$$\|x_k - \bar{x}\|^2 \leq \|x_0 - \bar{x}\|^2, \quad \forall k \geq 0.$$

Then from assumption (C_3), we obtain

$$\|F(x_k)\| = \|F(x_k) - F(\bar{x})\| \leq L\|x_k - \bar{x}\| \leq L\|x_0 - \bar{x}\|.$$

If we let $L\|x_0 - \bar{x}\| = \kappa$, then the sequence $\{F(x_k)\}$ is bounded, that is,

$$\|F(x_k)\| \leq \kappa, \quad \forall k \geq 0. \tag{17}$$

By the definition of z_k, Equation (14), monotonicity of F and the Cauchy–Schwatz inequality, we get

$$\sigma\|x_k - z_k\| = \frac{\sigma\|\alpha_k d_k\|^2}{\|x_k - z_k\|} \leq \frac{\langle F(z_k), x_k - z_k\rangle}{\|x_k - z_k\|} \leq \frac{\langle F(x_k), x_k - z_k\rangle}{\|x_k - z_k\|} \leq \|F(x_k)\|. \tag{18}$$

The boundedness of the sequence $\{x_k\}$ together with Equations (17) and (18), implies the sequence $\{z_k\}$ is bounded.

Now, as $\{z_k\}$ is bounded, then for any $\bar{x} \in E'$, the sequence $\{z_k - \bar{x}\}$ is also bounded, that is, there exists a positive constant $\nu > 0$ such that

$$\|z_k - \bar{x}\| \leq \nu.$$

This together with assumption (C_3), this yields

$$\|F(z_k)\| = \|F(z_k) - F(\bar{x})\| \leq L\|z_k - \bar{x}\| \leq L\nu.$$

Therefore, using Equation (15), we have

$$\frac{\sigma^2}{(L\nu)^2}\|x_k - z_k\|^4 \leq \|x_k - \bar{x}\|^2 - \|x_{k+1} - \bar{x}\|^2,$$

which implies

$$\frac{\sigma^2}{(L\nu)^2}\sum_{k=0}^{\infty}\|x_k - z_k\|^4 \leq \sum_{k=0}^{\infty}(\|x_k - \bar{x}\|^2 - \|x_{k+1} - \bar{x}\|^2) \leq \|x_0 - \bar{x}\| < \infty. \tag{19}$$

Equation (19) implies

$$\lim_{k\to\infty}\|x_k - z_k\| = 0.$$

However, using statement 2 of Lemma 1, the definition of ζ_k and the Cauchy-Schwartz inequality, we have

$$\begin{aligned}
\|x_{k+1} - x_k\| &= \|P_E[x_k - \zeta_k F(z_k)] - x_k\| \\
&\leq \|x_k - \zeta_k F(z_k) - x_k\| \\
&= \|\zeta_k F(z_k)\| \\
&= \|x_k - z_k\|,
\end{aligned} \tag{20}$$

which yields

$$\lim_{k\to\infty}\|x_{k+1} - x_k\| = 0.$$

□

Remark 2. *By Equation (11) and definition of z_k, then*

$$\lim_{k\to\infty}\alpha_k\|d_k\| = 0. \tag{21}$$

Theorem 1. *Suppose that assumption (C_1)–(C_3) holds and let the sequence $\{x_k\}$ be generated by Algorithm 1, then*

$$\liminf_{k\to\infty}\|F(x_k)\| = 0. \tag{22}$$

Proof. Assume that Equation (22) is not true, then there exists a constant $\epsilon > 0$ such that

$$\|F(x_k)\| \geq \epsilon, \quad \forall k \geq 0. \tag{23}$$

Combining (8) and (23), we have

$$\|d_k\| \geq \|F(x_k)\| \geq \epsilon, \quad \forall k \geq 0.$$

As $w_k = x_k + \alpha_k d_k$ and $\lim_{k\to\infty} \|x_k - z_k\| = 0$, we get $\lim_{k\to\infty} \alpha_k \|d_k\| = 0$ and

$$\lim_{k\to\infty} \alpha_k = 0. \tag{24}$$

On the other side, if $M = \left(1 + \frac{2}{\mu}\right)\kappa$, Lemma 3 and Equation (9) implies $\alpha_k \|d_k\| \geq \rho \frac{\epsilon^2}{(L+\sigma)ML\nu}$, which contradicts with (24). Therefore, (22) must hold. □

4. Numerical Experiments

To test the performance of the proposed method, we compare it with accelerated conjugate gradient descent (ACGD) and projected Dai-Yuan (PDY) methods in [27,28], respectively. In addition, MFRM method is applied to solve signal and image recovery problems arising in compressive sensing. All codes were written in MATLAB R2018b and run on a PC with intel COREi5 processor with 4GB of RAM and CPU 2.3GHZ. All runs were stopped whenever $\|F(x_k)\| < 10^{-5}$. The parameters chosen for each method are as follows:

MFRM method: $\gamma = 1, \rho = 0.9, \mu = 0.01, \sigma = 0.0001$.
ACGD method: all parameters are chosen as in [27].
PDY method: all parameters are chosen as in [28].

We tested eight problems with dimensions of $n = 1000, 5000, 10{,}000, 50{,}000, 100{,}000$ and 6 initial points: $x_1 = (0.1, 0.1, \cdots, 1)^T$, $x_2 = (0.2, 0.2, \cdots, 0.2)^T$, $x_3 = (0.5, 0.5, \cdots, 0.5)^T$, $x_4 = (1.2, 1.2, \cdots, 1.2)^T$, $x_5 = (1.5, 1.5, \cdots, 1.5)^T$, $x_6 = (2, 2, \cdots, 2)^T$. In Tables 1–8, the number of Iterations (Iter), number of function evaluations (Fval), CPU time in seconds (time) and the norm at the approximate solution (NORM) were reported. The symbol '−' is used when the number of Iterations exceeds 1000 and/or the number of function evaluations exceeds 2000.

The test problems are listed below, where the function F is taken as $F(x) = (f_1(x), f_2(x), \ldots, f_n(x))^T$.

Problem 1 [38] Exponential Function.

$$\begin{aligned} &f_1(x) = e^{x_1} - 1, \\ &f_i(x) = e^{x_i} + x_i - 1, \text{ for } i = 2, 3, ..., n, \\ &\text{and } E = \mathbb{R}^n_+. \end{aligned}$$

Problem 2 [38] Modified Logarithmic Function.

$$\begin{aligned} &f_i(x) = \ln(x_i + 1) - \frac{x_i}{n}, \text{ for } i = 2, 3, ..., n, \\ &\text{and } E = \{x \in \mathbb{R}^n : \sum_{i=1}^n x_i \leq n, x_i > -1, i = 1, 2, \ldots, n\}. \end{aligned}$$

Problem 3 [6] Nonsmooth Function.

$$\begin{aligned} &f_i(x) = 2x_i - \sin|x_i|, \ i = 1, 2, 3, ..., n, \\ &\text{and } E = \{x \in \mathbb{R}^n : \sum_{i=1}^n x_i \leq n, x_i \geq 0, i = 1, 2, \ldots, n\}. \end{aligned}$$

It is clear that problem 3 is nonsmooth at $x = 0$.

Problem 4 [38] Strictly Convex Function I.

$$\begin{aligned} &f_i(x) = e^{x_i} - 1, \text{ for } i = 1, 2, ..., n, \\ &\text{and } E = \mathbb{R}^n_+. \end{aligned}$$

Problem 5 [38] Strictly Convex Function II.

$$\begin{aligned} &f_i(x) = \frac{i}{n} e^{x_i} - 1, \text{ for } i = 1, 2, ..., n, \\ &\text{and } E = \mathbb{R}^n_+. \end{aligned}$$

Problem 6 [39] Tridiagonal Exponential Function

$$\begin{aligned} f_1(x) &= x_1 - e^{\cos(h(x_1+x_2))}, \\ f_i(x) &= x_i - e^{\cos(h(x_{i-1}+x_i+x_{i+1}))}, \text{ for } i = 2, ..., n-1, \\ f_n(x) &= x_n - e^{\cos(h(x_{n-1}+x_n))}, \\ h &= \frac{1}{n+1} \text{ and } E = \mathbb{R}^n_+. \end{aligned}$$

Problem 7 [40] Nonsmooth Function

$$\begin{aligned} &f_i(x) = x_i - \sin|x_i - 1|, \ i = 1, 2, 3, ..., n. \\ &\text{and } E = \{x \in \mathbb{R}^n : \sum_{i=1}^{n} x_i \leq n, x_i \geq -1, i = 1, 2, \ldots, n\}. \end{aligned}$$

Problem 8 [27] Penalty 1

$$\begin{aligned} t_i &= \sum_{i=1}^{n} x_i^2, \quad c = 10^{-5} \\ f_i(x) &= 2c(x_i - 1) + 4(t_i - 0.25)x_i, \ i = 1, 2, 3, ..., n. \\ \text{and } E &= \mathbb{R}^n_+. \end{aligned}$$

To show in detail the efficiency and robustness of all methods, we employ the performance profile developed in [41], which is a helpful process of standardizing the comparison of methods. Suppose that we have n_s solvers and n_l problems and we are interested in using either number of Iterations, CPU time or number of function evaluations as our measure of performance; so we let $k_{l,s}$ to be the number of iterations, CPU time or number of function evaluations required to solve problem by solver s. To compare the performance on problem l by a solver s with the best performance by any other solver on this problem, we use the performance ratio $r_{l,s}$ defined as

$$r_{l,s} = \frac{k_{l,s}}{\min\{k_{l,s} : s \in S\}},$$

where S is the set of solvers.

The overall performance of the solver is obtained using the (cumulative) distribution function for the performance ratio P. So if we let

$$P(t) = \frac{1}{n_l} size\{l \in L : r_{l,s} \leq t\},$$

then $P(t)$ is the probability for solver $s \in S$ that a performance ratio $r_{l,s}$ is within a factor $t \in R$ of the best possible ratio. If the set of problems L is large enough, then the solvers with the large probability $P(t)$ are considered as the best.

Table 1. Numerical results for modified Fletcher–Reeves (MFRM), accelerated conjugate gradient descent (ACGD) and projected Dai-Yuan (PDY) for problem 1 with given initial points and dimensions.

		MFRM				ACGD				PDY			
Dimension	Initial Point	Iter	Fval	Time	Norm	Iter	Fval	Time	Norm	Iter	Fval	Time	Norm
1000	x_1	23	98	0.42639	9.01×10^{-6}	8	34	0.21556	9.26×10^{-6}	12	49	0.19349	9.18×10^{-6}
	x_2	7	35	0.019885	8.82×10^{-6}	9	39	0.086582	3.01×10^{-6}	13	53	0.07318	6.35×10^{-6}
	x_3	8	40	0.011238	9.74×10^{-6}	9	38	0.034359	4.02×10^{-6}	14	57	0.01405	5.59×10^{-6}
	x_4	15	70	0.066659	6.01×10^{-6}	16	67	0.017188	9.22×10^{-6}	15	61	0.01421	4.07×10^{-6}
	x_5	5	31	0.16103	0	18	75	0.11646	4.46×10^{-6}	14	57	0.08690	9.91×10^{-6}
	x_6	31	134	0.03232	7.65×10^{-6}	25	104	0.042967	6.74×10^{-6}	40	162	0.04060	9.70×10^{-6}
5000	x_1	8	38	0.053865	5.63×10^{-6}	9	38	0.023729	3.89×10^{-6}	13	53	0.02775	6.87×10^{-6}
	x_2	8	40	0.036653	2.59×10^{-6}	9	38	0.021951	6.65×10^{-6}	14	57	0.02974	4.62×10^{-6}
	x_3	8	40	0.030089	6.41×10^{-6}	9	39	0.019317	8.01×10^{-6}	15	61	0.04353	4.18×10^{-6}
	x_4	16	74	0.081741	4.71×10^{-6}	17	71	0.05235	8.12×10^{-6}	15	61	0.03238	9.08×10^{-6}
	x_5	5	31	0.030748	0	18	75	0.038894	8.14×10^{-6}	15	61	0.03556	7.30×10^{-6}
	x_6	31	134	0.087531	8.1×10^{-6}	26	108	0.053473	7.96×10^{-6}	39	158	0.10419	9.86×10^{-6}
10,000	x_1	5	26	0.03829	3.7×10^{-6}	9	39	0.044961	5.5×10^{-6}	13	53	0.05544	9.70×10^{-6}
	x_2	8	40	0.055099	3.64×10^{-6}	9	39	0.0358	9.39×10^{-6}	14	57	0.06201	6.53×10^{-6}
	x_3	8	40	0.049974	5.44×10^{-6}	10	43	0.04176	2.12×10^{-6}	15	61	0.08704	5.90×10^{-6}
	x_4	16	74	0.125	6.61×10^{-6}	18	75	0.066316	4.58×10^{-6}	16	65	0.07797	4.28×10^{-6}
	x_5	5	31	0.048751	0	18	75	0.11807	7.86×10^{-6}	39	158	0.20751	7.97×10^{-6}
	x_6	28	122	0.13649	7.18×10^{-6}	27	112	0.10593	6.22×10^{-6}	87	351	0.36678	9.93×10^{-6}
50,000	x_1	5	26	0.1584	3.58×10^{-6}	10	43	0.15918	2.33×10^{-6}	14	57	0.23129	7.12×10^{-6}
	x_2	8	40	0.18044	8.1×10^{-6}	10	43	0.16252	3.97×10^{-6}	15	61	0.23975	4.91×10^{-6}
	x_3	8	40	0.186	4.54×10^{-6}	10	43	0.15707	4.67×10^{-6}	16	65	0.24735	4.37×10^{-6}
	x_4	17	78	0.31567	5.47×10^{-6}	19	79	0.27474	4.1×10^{-6}	38	154	0.55277	7.54×10^{-6}
	x_5	5	31	0.18586	0	18	75	0.27118	5.06×10^{-6}	177	712	2.29950	9.44×10^{-6}
	x_6	20	90	0.39237	6.44×10^{-6}	28	116	0.35197	7.69×10^{-6}	361	1449	4.63780	9.74×10^{-6}
100,000	x_1	5	26	0.26116	4.59×10^{-6}	10	42	0.28038	3.29×10^{-6}	15	61	0.50090	3.39×10^{-6}
	x_2	9	43	0.35288	1.59×10^{-6}	10	42	0.28999	5.62×10^{-6}	15	61	0.45876	6.94×10^{-6}
	x_3	8	40	0.35809	4.96×10^{-6}	10	42	0.29255	6.59×10^{-6}	16	65	0.51380	6.18×10^{-6}
	x_4	17	78	0.59347	7.73×10^{-6}	19	79	0.51261	5.79×10^{-6}	175	704	4.48920	9.47×10^{-6}
	x_5	32	138	0.98463	7.09×10^{-6}	18	75	0.46086	4.05×10^{-6}	176	708	4.49410	9.91×10^{-6}
	x_6	17	78	0.57701	9.31×10^{-6}	29	120	0.71678	6.05×10^{-6}	360	1445	9.10170	9.99×10^{-6}

Table 2. Numerical results for MFRM, ACGD and PDY for problem 2 with given initial points and dimensions.

		MFRM				ACGD				PDY			
Dimension	Initial Point	Iter	Fval	Time	Norm	Iter	Fval	Time	Norm	Iter	Fval	Time	Norm
1000	x_1	3	8	0.007092	5.17×10^{-7}	3	8	0.036061	5.17×10^{-7}	10	39	0.01053	6.96×10^{-6}
	x_2	3	8	0.012401	6.04×10^{-6}	3	8	0.006143	6.04×10^{-6}	11	43	0.00937	9.23×10^{-6}
	x_3	4	11	0.003993	4.37×10^{-7}	4	11	0.006476	4.37×10^{-7}	13	51	0.01111	6.26×10^{-6}
	x_4	5	14	0.010363	1.52×10^{-7}	5	14	0.005968	1.52×10^{-7}	14	55	0.02154	9.46×10^{-6}
	x_5	5	14	0.007234	1.1×10^{-6}	5	14	0.02349	1.1×10^{-6}	15	59	0.01850	4.60×10^{-6}
	x_6	6	17	0.006496	1.74×10^{-8}	6	17	0.00677	1.74×10^{-8}	15	59	0.01938	7.71×10^{-6}
5000	x_1	3	8	0.011561	1.75×10^{-7}	3	8	0.009794	1.75×10^{-7}	11	43	0.03528	4.86×10^{-6}
	x_2	3	8	0.010452	3.13×10^{-6}	3	8	0.009591	3.13×10^{-6}	12	47	0.04032	6.89×10^{-6}
	x_3	4	11	0.01516	1.42×10^{-7}	4	11	0.013767	1.42×10^{-7}	14	55	0.04889	4.61×10^{-6}
	x_4	5	14	0.019733	3.94×10^{-8}	5	14	0.014274	3.94×10^{-8}	15	59	0.04826	6.96×10^{-6}
	x_5	5	14	0.018462	4.05×10^{-7}	5	14	0.011728	4.05×10^{-7}	16	63	0.05969	3.37×10^{-6}
	x_6	6	17	0.028536	2.36×10^{-9}	6	17	0.016345	2.36×10^{-9}	16	63	0.06253	5.64×10^{-6}
10,000	x_1	3	8	0.019053	1.21×10^{-7}	3	8	0.0135	1.21×10^{-7}	11	43	0.06732	6.85×10^{-6}
	x_2	3	8	0.01791	2.79×10^{-6}	3	8	0.015807	2.79×10^{-6}	12	47	0.12232	9.72×10^{-6}
	x_3	4	11	0.033042	9.73×10^{-8}	4	11	0.020752	9.73×10^{-8}	14	55	0.08288	6.51×10^{-6}
	x_4	5	14	0.031576	2.56×10^{-8}	5	14	0.04483	2.56×10^{-8}	15	59	0.08413	9.82×10^{-6}
	x_5	5	14	0.032747	2.93×10^{-7}	5	14	0.026975	2.93×10^{-7}	16	63	0.09589	4.75×10^{-6}
	x_6	6	17	0.036002	1.24×10^{-9}	6	17	0.032445	1.24×10^{-9}	16	64	0.11499	8.55×10^{-6}
50,000	x_1	3	8	0.0737	6.32×10^{-8}	7	26	0.16925	2.94×10^{-6}	12	47	0.27826	5.23×10^{-6}
	x_2	3	8	0.06964	3.37×10^{-6}	9	34	0.18801	2.78×10^{-6}	13	51	0.29642	7.11×10^{-6}
	x_3	4	11	0.093027	4.87×10^{-8}	7	25	0.15375	9.11×10^{-6}	15	59	0.35602	4.82×10^{-6}
	x_4	5	14	0.11219	1.11×10^{-8}	7	24	0.15382	9.18×10^{-6}	35	141	0.69470	6.69×10^{-6}
	x_5	5	14	0.1173	1.84×10^{-7}	9	32	0.18164	6.71×10^{-6}	35	141	0.68488	9.12×10^{-6}
	x_6	6	17	0.13794	4.01×10^{-10}	6	19	0.11216	5.2×10^{-6}	35	141	0.70973	9.91×10^{-6}
100,000	x_1	3	8	0.13021	5.4×10^{-8}	7	26	0.2609	4.14×10^{-6}	12	47	0.44541	7.39×10^{-6}
	x_2	3	8	0.13267	4.27×10^{-6}	9	34	0.32666	3.93×10^{-6}	14	55	0.53299	3.39×10^{-6}
	x_3	4	11	0.17338	4.05×10^{-8}	8	29	0.3113	3.33×10^{-6}	15	60	0.58603	8.71×10^{-6}
	x_4	5	14	0.20036	8.15×10^{-9}	8	28	0.2997	3.34×10^{-6}	72	290	2.70630	8.31×10^{-6}
	x_5	5	14	0.25274	1.8×10^{-7}	9	32	0.32098	9.46×10^{-6}	72	290	2.72220	8.68×10^{-6}
	x_6	6	17	0.24952	2.71×10^{-10}	6	19	0.21972	7.01×10^{-6}	72	290	2.75850	8.96×10^{-6}

Table 3. Numerical results for MFRM, ACGD and PDY for problem 3 with given initial points and dimensions.

		MFRM				ACGD				PDY			
Dimension	Initial Point	Iter	Fval	Time	Norm	Iter	Fval	Time	Norm	Iter	Fval	Time	Norm
1000	x_1	6	24	0.024062	3.11×10^{-6}	6	40	0.02951	4.44×10^{-6}	12	48	0.01255	4.45×10^{-6}
	x_2	6	24	0.005345	5.94×10^{-6}	6	40	0.0077681	8.75×10^{-6}	12	48	0.01311	9.02×10^{-6}
	x_3	6	24	0.006109	9.94×10^{-6}	6	44	0.0067049	5.09×10^{-6}	13	52	0.01486	8.34×10^{-6}
	x_4	8	33	0.006127	3.1×10^{-6}	8	44	0.007142	5.04×10^{-6}	14	56	0.01698	8.04×10^{-6}
	x_5	11	46	0.010427	2.71×10^{-6}	11	40	0.010411	3.12×10^{-6}	14	56	0.01551	9.72×10^{-6}
	x_6	16	68	0.010682	8.38×10^{-6}	16	77	0.014759	5.98×10^{-6}	14	56	0.01534	9.42×10^{-6}
5000	x_1	6	24	0.020455	6.96×10^{-6}	6	40	0.020368	9.93×10^{-6}	12	48	0.03660	9.94×10^{-6}
	x_2	7	28	0.021552	1.33×10^{-6}	7	44	0.029622	5.09×10^{-6}	13	52	0.03616	6.85×10^{-6}
	x_3	7	28	0.023056	2.22×10^{-6}	7	48	0.030044	2.96×10^{-6}	14	56	0.04594	6.14×10^{-6}
	x_4	8	33	0.022984	6.92×10^{-6}	8	48	0.022777	2.93×10^{-6}	15	60	0.04342	6.01×10^{-6}
	x_5	11	46	0.031466	6.06×10^{-6}	11	40	0.019226	6.97×10^{-6}	15	60	0.04296	7.25×10^{-6}
	x_6	17	72	0.049308	7.67×10^{-6}	17	81	0.036095	6.05×10^{-6}	32	129	0.10081	8.85×10^{-6}
10,000	x_1	6	24	0.03064	9.85×10^{-6}	6	44	0.03997	3.65×10^{-6}	13	52	0.06192	4.77×10^{-6}
	x_2	7	28	0.035806	1.88×10^{-6}	7	44	0.037221	7.19×10^{-6}	13	52	0.06442	9.68×10^{-6}
	x_3	7	28	0.035795	3.14×10^{-6}	7	48	0.053226	4.18×10^{-6}	14	56	0.09499	8.69×10^{-6}
	x_4	8	33	0.041017	9.79×10^{-6}	8	48	0.057984	4.15×10^{-6}	15	60	0.07696	8.5×10^{-6}
	x_5	11	46	0.06448	8.58×10^{-6}	11	40	0.047413	9.85×10^{-6}	33	133	0.18625	6.45×10^{-6}
	x_6	18	76	0.09651	4.44×10^{-6}	18	81	0.085238	8.56×10^{-6}	33	133	0.15548	7.51×10^{-6}
50,000	x_1	7	28	0.14323	2.2×10^{-6}	7	44	0.17175	8.17×10^{-6}	14	56	0.23642	3.51×10^{-6}
	x_2	7	28	0.13625	4.2×10^{-6}	7	48	0.18484	4.18×10^{-6}	14	56	0.24813	7.12×10^{-6}
	x_3	7	28	0.13246	7.03×10^{-6}	7	48	0.1827	9.36×10^{-6}	15	60	0.27049	6.53×10^{-6}
	x_4	9	37	0.18261	4.16×10^{-6}	9	48	0.18993	9.27×10^{-6}	34	137	0.54545	7.13×10^{-6}
	x_5	12	50	0.21743	5.2×10^{-6}	12	44	0.17043	5.73×10^{-6}	68	274	1.02330	9.99×10^{-6}
	x_6	18	76	0.34645	9.93×10^{-6}	18	85	0.32938	8.66×10^{-6}	69	278	1.03810	8.05×10^{-6}
100,000	x_1	7	28	0.27078	3.11×10^{-6}	7	48	0.36144	3×10^{-6}	14	56	0.45475	4.96×10^{-6}
	x_2	7	28	0.26974	5.94×10^{-6}	7	48	0.37515	5.91×10^{-6}	15	60	0.49018	3.39×10^{-6}
	x_3	7	28	0.25475	9.94×10^{-6}	7	52	0.39071	3.44×10^{-6}	15	60	0.49016	9.24×10^{-6}
	x_4	9	37	0.3089	5.88×10^{-6}	9	52	0.35961	3.41×10^{-6}	139	559	4.03110	9.01×10^{-6}
	x_5	12	50	0.41839	7.35×10^{-6}	12	44	0.33105	8.1×10^{-6}	70	282	2.07100	8.54×10^{-6}
	x_6	19	80	0.64773	5.75×10^{-6}	19	89	0.61329	5.54×10^{-6}	139	559	4.02440	9.38×10^{-6}

Table 4. Numerical results for MFRM, ACGD and PDY for problem 4 with given initial points and dimensions.

		MFRM				ACGD				PDY			
Dimension	Initial Point	Iter	Fval	Time	Norm	Iter	Fval	Time	Norm	Iter	Fval	Time	Norm
1000	x_1	6	24	0.00855	1.65×10^{-6}	10	40	0.014662	3.65×10^{-6}	12	48	0.00989	4.60×10^{-6}
	x_2	5	20	0.004234	2.32×10^{-6}	10	40	0.0064115	5.79×10^{-6}	12	48	0.00966	9.57×10^{-6}
	x_3	10	42	0.007426	6.42×10^{-6}	10	40	0.0054818	3.29×10^{-6}	13	52	0.00887	8.49×10^{-6}
	x_4	21	90	0.011603	5.84×10^{-6}	27	110	0.012854	8.97×10^{-6}	12	48	0.01207	5.83×10^{-6}
	x_5	16	71	0.010735	8.48×10^{-6}	26	106	0.015603	5.97×10^{-6}	29	117	0.05371	9.43×10^{-6}
	x_6	1	15	0.005932	0	36	147	0.025039	9.56×10^{-6}	29	117	0.02396	6.65×10^{-6}
5000	x_1	6	24	0.019995	3.68×10^{-6}	10	40	0.018283	8.15×10^{-6}	13	52	0.02503	3.49×10^{-6}
	x_2	5	20	0.00934	5.2×10^{-6}	11	44	0.016733	3.36×10^{-6}	13	52	0.02626	7.24×10^{-6}
	x_3	11	46	0.02156	3.89×10^{-6}	10	40	0.017073	7.37×10^{-6}	14	56	0.03349	6.29×10^{-6}
	x_4	22	94	0.043325	6.81×10^{-6}	29	118	0.047436	7.09×10^{-6}	13	52	0.02258	4.25×10^{-6}
	x_5	18	79	0.096692	6.15×10^{-6}	27	110	0.058405	7.95×10^{-6}	31	125	0.05471	7.59×10^{-6}
	x_6	1	15	0.012199	0	39	159	0.059448	7.33×10^{-6}	63	254	0.10064	8.54×10^{-6}
10,000	x_1	6	24	0.019264	5.2×10^{-6}	11	44	0.026877	3×10^{-6}	13	52	0.03761	4.93×10^{-6}
	x_2	5	20	0.017891	7.35×10^{-6}	11	44	0.03118	4.76×10^{-6}	14	56	0.04100	3.37×10^{-6}
	x_3	11	46	0.036079	5.5×10^{-6}	11	44	0.034673	2.71×10^{-6}	14	56	0.03919	8.90×10^{-6}
	x_4	22	94	0.069778	9.63×10^{-6}	30	122	0.069971	5.97×10^{-6}	32	129	0.09613	6.02×10^{-6}
	x_5	18	79	0.062821	8.69×10^{-6}	28	114	0.066866	6.68×10^{-6}	32	129	0.09177	6.44×10^{-6}
	x_6	1	15	0.017237	0	40	163	0.093749	7.26×10^{-6}	64	258	0.20791	9.39×10^{-6}
50,000	x_1	7	28	0.093473	1.16×10^{-6}	11	44	0.16749	6.7×10^{-6}	14	56	0.17193	3.63×10^{-6}
	x_2	6	24	0.072206	1.64×10^{-6}	12	48	0.11391	2.77×10^{-6}	14	56	0.15237	7.54×10^{-6}
	x_3	12	50	0.14285	3.33×10^{-6}	11	44	0.11036	6.06×10^{-6}	15	60	0.16549	6.66×10^{-6}
	x_4	24	102	0.30313	5.86×10^{-6}	31	126	0.30903	7.94×10^{-6}	67	270	0.76283	7.81×10^{-6}
	x_5	20	87	0.28955	6.31×10^{-6}	29	118	0.30266	8.89×10^{-6}	67	270	0.76157	8.80×10^{-6}
	x_6	1	15	0.061327	0	42	171	0.41158	7.96×10^{-6}	269	1080	2.92510	9.41×10^{-6}
100,000	x_1	7	28	0.15038	1.65×10^{-6}	11	44	0.2434	9.48×10^{-6}	14	56	0.30229	5.13×10^{-6}
	x_2	6	24	0.13126	2.32×10^{-6}	12	48	0.2614	3.91×10^{-6}	15	60	0.31648	3.59×10^{-6}
	x_3	12	50	0.31585	4.71×10^{-6}	11	44	0.2161	8.57×10^{-6}	32	129	0.72838	9.99×10^{-6}
	x_4	24	102	0.58023	8.29×10^{-6}	32	130	0.65289	6.68×10^{-6}	135	543	2.86780	9.73×10^{-6}
	x_5	20	87	0.5122	8.92×10^{-6}	30	122	0.61637	7.48×10^{-6}	272	1092	5.74140	9.91×10^{-6}
	x_6	1	15	0.11696	0	43	175	0.82759	7.88×10^{-6}	548	2197	11.44130	9.87×10^{-6}

Table 5. Numerical results for MFRM, ACGD and PDY for problem 5 with given initial points and dimensions.

		MFRM				ACGD				PDY			
Dimension	Initial Point	Iter	Fval	Time	Norm	Iter	Fval	Time	Norm	Iter	Fval	Time	Norm
1000	x_1	26	98	0.023555	3.51×10^{-6}	39	154	0.022285	9.7×10^{-6}	16	63	0.07575	6.03×10^{-6}
	x_2	40	154	0.024539	5.9×10^{-6}	22	85	0.015671	5.03×10^{-6}	16	63	0.01470	5.42×10^{-6}
	x_3	37	144	0.021659	7.11×10^{-6}	43	173	0.029569	7.96×10^{-6}	33	132	0.02208	6.75×10^{-6}
	x_4	49	206	0.030696	9.52×10^{-6}	30	122	0.014942	6.05×10^{-6}	30	121	0.01835	8.39×10^{-6}
	x_5	46	194	0.11589	7.06×10^{-6}	29	118	0.040406	6.5×10^{-6}	32	129	0.02700	8.47×10^{-6}
	x_6	43	182	0.027471	8.7×10^{-6}	40	163	0.0311	9.83×10^{-6}	30	121	0.01712	6.95×10^{-6}
5000	x_1	38	147	0.073315	4.96×10^{-6}	30	117	0.060877	9.56×10^{-6}	17	67	0.04394	5.64×10^{-6}
	x_2	20	77	0.056225	4.98×10^{-6}	16	60	0.027911	5.91×10^{-6}	17	67	0.04635	5.07×10^{-6}
	x_3	41	157	0.082151	8.92×10^{-6}	78	315	0.12774	9.7×10^{-6}	35	140	0.08311	9.74×10^{-6}
	x_4	48	202	0.10166	9.19×10^{-6}	31	126	0.067911	8.39×10^{-6}	33	133	0.08075	6.02×10^{-6}
	x_6	147	562	3.308158	8.44×10^{-7}	31	126	0.067856	7.81×10^{-6}	35	141	0.10091	7.51×10^{-6}
	x_7	45	190	0.090276	7.14×10^{-6}	44	179	0.09371	7.37×10^{-6}	32	129	0.08054	8.55×10^{-6}
10,000	x_1	37	143	0.12665	9.28×10^{-6}	77	308	0.28678	9.85×10^{-6}	17	67	0.06816	8.81×10^{-6}
	x_2	22	84	0.077288	9.78×10^{-6}	16	60	0.071657	7.52×10^{-6}	17	67	0.08833	7.80×10^{-6}
	x_3	39	149	0.1297	6.74×10^{-6}	105	424	0.34212	9.08×10^{-6}	37	148	0.14732	6.36×10^{-6}
	x_4	60	250	0.2175	7.56×10^{-6}	32	130	0.11937	7.17×10^{-6}	37	149	0.14293	8.25×10^{-6}
	x_5	44	186	0.1727	7.68×10^{-6}	32	130	0.11921	8.26×10^{-6}	36	145	0.14719	8.23×10^{-6}
	x_6	46	194	0.1728	8.62×10^{-6}	45	183	0.15634	9.01×10^{-6}	74	298	0.26456	7.79×10^{-6}
50,000	x_1	44	170	0.62202	1×10^{-5}	90	539	31.75299	2.56×10^{-7}	42	169	0.58113	7.78×10^{-6}
	x_2	69	280	0.9662	6.87×10^{-6}	31	122	0.33817	7.09×10^{-6}	42	169	0.58456	7.13×10^{-6}
	x_3	119	464	25.87657	9.34×10^{-7}	260	1047	2.8824	9.67×10^{-6}	41	165	0.58717	8.87×10^{-6}
	x_4	50	210	0.71599	8.38×10^{-6}	33	134	0.39039	9.98×10^{-6}	40	161	0.56431	7.17×10^{-6}
	x_5	46	194	0.65538	8.47×10^{-6}	35	142	0.40807	7.19×10^{-6}	82	330	1.08920	8.44×10^{-6}
	x_6	50	210	0.69117	8.12×10^{-6}	49	199	0.57702	8.97×10^{-6}	80	322	1.06570	7.82×10^{-6}
100,000	x_1	31	121	0.84183	4.48×10^{-6}	88	530	61.97806	5.53×10^{-7}	43	173	1.09520	8.47×10^{-6}
	x_2	135	518	59.19294	8.37×10^{-7}	110	442	2.2661	9.55×10^{-6}	43	173	1.10040	7.77×10^{-6}
	x_3	46	178	1.1322	6.99×10^{-6}	345	1388	7.1938	9.76×10^{-6}	42	169	1.08330	9.66×10^{-6}
	x_4	50	210	1.3737	8.85×10^{-6}	34	138	0.74362	8.65×10^{-6}	85	342	2.11880	9.22×10^{-6}
	x_5	47	198	1.3879	8.31×10^{-6}	36	146	0.79012	8.09×10^{-6}	84	338	2.10640	9.78×10^{-6}
	x_6	52	218	1.4318	7.37×10^{-6}	51	207	1.1601	8.42×10^{-6}	167	671	4.06200	9.90×10^{-6}

Table 6. Numerical Results for MFRM, ACGD and PDY for problem 6 with given initial points and dimensions.

		MFRM				ACGD				PDY			
Dimension	Initial Point	Iter	Fval	Time	Norm	Iter	Fval	Time	Norm	Iter	Fval	Time	Norm
1000	x_1	11	44	0.011156	8.32×10^{-6}	12	48	0.02786	7.88×10^{-6}	15	60	0.01671	4.35×10^{-6}
	x_2	11	44	0.016092	7.32×10^{-6}	12	48	0.01042	7.58×10^{-6}	15	60	0.01346	4.18×10^{-6}
	x_3	11	44	0.010446	8.83×10^{-6}	12	48	0.0092	6.68×10^{-6}	15	60	0.01630	3.68×10^{-6}
	x_4	10	40	0.011233	7.38×10^{-6}	12	48	0.013617	4.57×10^{-6}	14	56	0.01339	7.48×10^{-6}
	x_5	9	36	0.011325	8.29×10^{-6}	12	48	0.011492	3.67×10^{-6}	14	56	0.01267	6.01×10^{-6}
	x_6	7	28	0.009452	8.25×10^{-6}	11	44	0.016351	8.32×10^{-6}	14	56	0.01685	3.54×10^{-6}
5000	x_1	8	32	0.026924	1.87×10^{-6}	13	52	0.036025	4.59×10^{-6}	15	60	0.05038	9.73×10^{-6}
	x_2	8	32	0.043488	1.8×10^{-6}	13	52	0.040897	4.42×10^{-6}	15	60	0.04775	9.36×10^{-6}
	x_3	8	32	0.02709	1.59×10^{-6}	13	52	0.039937	3.89×10^{-6}	15	60	0.04923	8.25×10^{-6}
	x_4	8	32	0.026351	1.1×10^{-6}	13	52	0.033013	2.66×10^{-6}	15	60	0.05793	5.64×10^{-6}
	x_5	7	28	0.023442	8.62×10^{-6}	12	48	0.030462	8.22×10^{-6}	15	60	0.04597	4.53×10^{-6}
	x_6	7	28	0.022952	5.08×10^{-6}	12	48	0.028786	4.85×10^{-6}	14	56	0.05070	7.93×10^{-6}
10,000	x_1	8	32	0.061374	2.62×10^{-6}	13	52	0.092372	6.5×10^{-6}	68	274	0.40724	9.06×10^{-6}
	x_2	8	32	0.06285	2.52×10^{-6}	13	52	0.059778	6.25×10^{-6}	68	274	0.41818	8.72×10^{-6}
	x_3	8	32	0.059913	2.22×10^{-6}	13	52	0.077326	5.5×10^{-6}	34	137	0.21905	6.22×10^{-6}
	x_4	8	32	0.057003	1.52×10^{-6}	13	52	0.087745	3.77×10^{-6}	15	60	0.10076	7.98×10^{-6}
	x_5	8	32	0.070377	1.22×10^{-6}	13	52	0.077217	3.02×10^{-6}	15	60	0.12680	6.40×10^{-6}
	x_6	7	28	0.052718	7.18×10^{-6}	12	48	0.067375	6.85×10^{-6}	15	60	0.11984	3.78×10^{-6}
50,000	x_1	8	32	0.21258	5.85×10^{-6}	14	56	0.32965	3.78×10^{-6}	143	575	3.09120	9.42×10^{-6}
	x_2	8	32	0.21203	5.63×10^{-6}	14	56	0.31297	3.63×10^{-6}	143	575	3.06200	9.06×10^{-6}
	x_3	8	32	0.20885	4.96×10^{-6}	14	56	0.30089	3.2×10^{-6}	142	571	3.04950	9.04×10^{-6}
	x_4	8	32	0.20483	3.4×10^{-6}	13	52	0.26855	8.42×10^{-6}	69	278	1.53920	9.14×10^{-6}
	x_5	8	32	0.21467	2.72×10^{-6}	13	52	0.26304	6.76×10^{-6}	68	274	1.49490	9.43×10^{-6}
	x_6	8	32	0.20933	1.61×10^{-6}	13	52	0.26143	3.99×10^{-6}	15	60	0.38177	8.44×10^{-6}
100,000	x_1	8	32	0.41701	8.28×10^{-6}	14	56	0.58853	5.34×10^{-6}	292	1172	13.59530	9.53×10^{-6}
	x_2	8	32	0.41511	7.96×10^{-6}	14	56	0.58897	5.14×10^{-6}	290	1164	13.30930	9.75×10^{-6}
	x_3	8	32	0.44061	7.01×10^{-6}	14	56	0.57318	4.53×10^{-6}	144	579	6.68150	9.96×10^{-6}
	x_4	8	32	0.43805	4.8×10^{-6}	14	56	0.58712	3.1×10^{-6}	141	567	6.50800	9.92×10^{-6}
	x_5	8	32	0.41147	3.85×10^{-6}	13	52	0.56384	9.56×10^{-6}	70	282	3.30510	8.07×10^{-6}
	x_6	8	32	0.43925	2.27×10^{-6}	13	52	0.53343	5.64×10^{-6}	34	137	1.64510	6.37×10^{-6}

Table 7. Numerical Results for MFRM, ACGD and PDY for problem 7 with given initial points and dimensions.

		MFRM				ACGD				PDY			
Dimension	Initial Point	Iter	Fval	Time	Norm	Iter	Fval	Time	Norm	Iter	Fval	Time	Norm
1000	x_1	4	21	0.011834	3.24×10^{-7}	10	42	0.008528	2.46×10^{-6}	14	57	0.00953	5.28×10^{-6}
	x_2	4	21	0.006228	1.43×10^{-7}	9	38	0.008289	3.91×10^{-6}	13	53	0.00896	9.05×10^{-6}
	x_3	3	17	0.004096	5.81×10^{-8}	8	34	0.006702	7.43×10^{-6}	3	12	0.00426	8.47×10^{-6}
	x_4	7	34	0.00585	3.89×10^{-6}	11	46	0.009579	5.94×10^{-6}	15	61	0.01169	6.73×10^{-6}
	x_5	7	34	0.006133	6.36×10^{-6}	11	46	0.015328	8.97×10^{-6}	31	126	0.03646	9.03×10^{-6}
	x_6	8	37	0.006106	1.9×10^{-6}	12	49	0.01426	2.87×10^{-6}	15	60	0.01082	3.99×10^{-6}
5000	x_1	4	21	0.015836	7.25×10^{-7}	10	42	0.023953	5.49×10^{-6}	15	61	0.03215	4.25×10^{-6}
	x_2	4	21	0.014521	3.2×10^{-7}	9	38	0.021065	8.74×10^{-6}	14	57	0.02942	7.40×10^{-6}
	x_3	3	17	0.014517	1.3×10^{-7}	9	38	0.025437	4.01×10^{-6}	4	16	0.01107	1.01×10^{-7}
	x_4	7	34	0.028388	8.71×10^{-6}	12	50	0.028607	3.21×10^{-6}	16	65	0.04331	5.43×10^{-6}
	x_5	8	38	0.02787	1.49×10^{-6}	12	50	0.037806	4.84×10^{-6}	33	134	0.09379	7.78×10^{-6}
	x_6	8	37	0.027898	4.26×10^{-6}	12	49	0.029226	6.43×10^{-6}	15	60	0.04077	8.92×10^{-6}
10,000	x_1	4	21	0.028528	1.02×10^{-6}	10	42	0.045585	7.77×10^{-6}	15	61	0.06484	6.01×10^{-6}
	x_2	4	21	0.033782	4.52×10^{-7}	10	42	0.041715	2.98×10^{-6}	15	61	0.07734	3.77×10^{-6}
	x_3	3	17	0.029265	1.84×10^{-7}	9	38	0.036422	5.67×10^{-6}	4	16	0.02707	1.42×10^{-7}
	x_4	8	38	0.043301	1.29×10^{-6}	12	50	0.063527	4.53×10^{-6}	16	65	0.07941	7.69×10^{-6}
	x_5	8	38	0.043741	2.1×10^{-6}	12	50	0.049604	6.85×10^{-6}	34	138	0.14942	6.83×10^{-6}
	x_6	8	37	0.053666	6.02×10^{-6}	12	49	0.050153	9.09×10^{-6}	34	138	0.15224	8.81×10^{-6}
50,000	x_1	4	21	0.10816	2.29×10^{-6}	11	46	0.20624	4.19×10^{-6}	16	65	0.25995	4.89×10^{-6}
	x_2	4	21	0.11969	1.01×10^{-6}	10	42	0.16364	6.67×10^{-6}	15	61	0.24674	8.42×10^{-6}
	x_3	3	17	0.068644	4.11×10^{-7}	10	42	0.1539	3.06×10^{-6}	4	16	0.09405	3.18×10^{-7}
	x_4	8	38	0.16067	2.88×10^{-6}	13	54	0.20728	2.45×10^{-6}	36	146	0.55207	6.39×10^{-6}
	x_5	8	38	0.14484	4.7×10^{-6}	13	54	0.19421	3.69×10^{-6}	35	142	0.54679	9.05×10^{-6}
	x_6	9	41	0.161	1.41×10^{-6}	13	53	0.19386	4.9×10^{-6}	36	146	0.55764	7.59×10^{-6}
100,000	x_1	4	21	0.21825	3.24×10^{-6}	11	46	0.32512	5.93×10^{-6}	17	69	0.52595	5.68×10^{-6}
	x_2	4	21	0.16435	1.43×10^{-6}	10	42	0.30949	9.43×10^{-6}	16	65	0.52102	4.34×10^{-6}
	x_3	3	17	0.13072	5.81×10^{-7}	10	42	0.31031	4.32×10^{-6}	4	16	0.14864	4.50×10^{-7}
	x_4	8	38	0.29012	4.07×10^{-6}	13	54	0.38833	3.46×10^{-6}	36	146	1.05360	9.04×10^{-6}
	x_5	8	38	0.32821	6.65×10^{-6}	13	54	0.3522	5.22×10^{-6}	74	299	2.10730	8.55×10^{-6}
	x_6	9	41	0.43649	1.99×10^{-6}	13	53	0.3561	6.94×10^{-6}	37	150	1.08240	6.66×10^{-6}

Table 8. Numerical results for MFRM, ACGD and PDY for problem 8 with given initial points and dimensions.

		MFRM				ACGD				PDY			
Dimension	Initial Point	Iter	Fval	Time	Norm	Iter	Fval	Time	Norm	Iter	Fval	Time	Norm
1000	x_1	8	27	0.1502	1.52×10^{-6}	8	26	0.049826	6.09×10^{-6}	69	279	0.05538	8.95×10^{-6}
	x_2	8	27	0.042248	1.52×10^{-6}	8	26	0.017594	6.09×10^{-6}	270	1085	0.18798	9.72×10^{-6}
	x_3	26	114	0.03877	7.85×10^{-6}	8	26	0.010888	6.09×10^{-6}	24	52	0.02439	6.57×10^{-6}
	x_4	26	114	0.017542	7.85×10^{-6}	8	26	0.007873	6.09×10^{-6}	27	58	0.01520	7.59×10^{-6}
	x_5	26	114	0.067692	7.85×10^{-6}	8	26	0.060733	6.09×10^{-6}	28	61	0.04330	9.21×10^{-6}
	x_6	26	114	0.045173	7.85×10^{-6}	8	26	0.006889	6.09×10^{-6}	40	85	0.02116	8.45×10^{-6}
5000	x_1	6	28	0.023925	8.77×10^{-6}	4	13	0.011005	5.76×10^{-6}	658	2639	1.13030	9.98×10^{-6}
	x_2	15	70	0.043512	7.94×10^{-6}	4	13	0.009131	5.76×10^{-6}	27	58	0.05101	7.59×10^{-6}
	x_3	15	70	0.046458	7.94×10^{-6}	4	13	0.011311	5.76×10^{-6}	49	104	0.08035	8.11×10^{-6}
	x_4	15	70	0.044788	7.94×10^{-6}	4	13	0.010475	5.75×10^{-6}	40	85	0.07979	8.45×10^{-6}
	x_5	15	70	0.044639	7.94×10^{-6}	4	13	0.011034	5.77×10^{-6}	18	40	0.09128	9.14×10^{-6}
	x_6	15	70	0.043974	7.94×10^{-6}	4	13	0.00785	5.76×10^{-6}	17	38	0.18528	8.98×10^{-6}
10,000	x_1	11	54	0.06595	6.15×10^{-6}	5	20	0.024232	2.19×10^{-6}	49	104	0.20443	7.62×10^{-6}
	x_2	11	54	0.068125	6.15×10^{-6}	5	20	0.023511	2.19×10^{-6}	40	85	0.15801	8.45×10^{-6}
	x_3	11	54	0.065486	6.15×10^{-6}	5	20	0.023004	2.19×10^{-6}	19	42	0.37880	7.66×10^{-6}
	x_4	11	54	0.064515	6.15×10^{-6}	5	20	0.030435	2.19×10^{-6}	90	187	1.25802	9.7×10^{-6}
	x_5	11	54	0.056261	6.15×10^{-6}	5	20	0.021963	2.19×10^{-6}	988	1988	12.68259	9.93×10^{-6}
	x_6	11	54	0.067785	6.15×10^{-6}	5	20	0.021889	2.21×10^{-6}	27	58	0.32859	7.59×10^{-6}
50,000	x_1	7	38	0.17856	4.5×10^{-6}	5	23	0.087544	2.45×10^{-6}	19	42	0.52291	6.42×10^{-6}
	x_2	7	38	0.17862	4.5×10^{-6}	5	23	0.093227	2.45×10^{-6}	148	304	3.93063	9.92×10^{-6}
	x_3	7	38	0.17746	4.5×10^{-6}	5	23	0.087484	2.45×10^{-6}	937	1886	22.97097	9.87×10^{-6}
	x_4	7	38	0.17392	4.5×10^{-6}	5	23	0.086329	2.4×10^{-6}	27	58	0.68467	7.59×10^{-6}
	x_5	7	38	0.18035	4.5×10^{-6}	5	23	0.08954	2.4×10^{-6}	346	702	8.45043	9.79×10^{-6}
	x_6	7	38	0.17504	4.5×10^{-6}	5	23	0.093203	2.5×10^{-6}	40	85	0.99230	8.45×10^{-6}
100,000	x_1	28	122	0.91448	8.61×10^{-6}	4	20	0.14743	2.71×10^{-6}	-	-	-	-
	x_2	28	122	0.93662	8.61×10^{-6}	4	20	0.14823	2.7×10^{-6}	-	-	-	-
	x_3	28	122	0.90604	8.61×10^{-6}	4	20	0.1497	2.79×10^{-6}	-	-	-	-
	x_4	28	122	0.92351	8.61×10^{-6}	4	20	0.14844	2.37×10^{-6}	-	-	-	-
	x_5	28	122	0.91896	8.61×10^{-6}	4	20	0.12346	1.66×10^{-6}	-	-	-	-
	x_6	28	122	0.91294	8.61×10^{-6}	4	20	0.12522	2.11×10^{-6}	-	-	-	-

Figure 1 reveals that MFRM performed better in terms of number of Iterations, as it solves and wins over 70 percent of the problems with less number of Iterations, while ACGD and PDY solve and win over 40 and almost 10 percent respectively. The story is a little bit different in Figure 2 as ACGD method was very competitive. However, MFRM method performed a little bit better by solving and winning over 50 percent of the problems with less CPU time as against ACGD method which solves and wins less than 50 percent of the problems considered. The PDY method had the least performance with just 10 percent success. The interpretation of Figure 3 was similar to that of Figure 1. Finally, in Table 11 we report numerical results for MFRM, ACGD and PDY for problem 2 with given initial points and dimensions with double float (10^{-16}) accuracy.

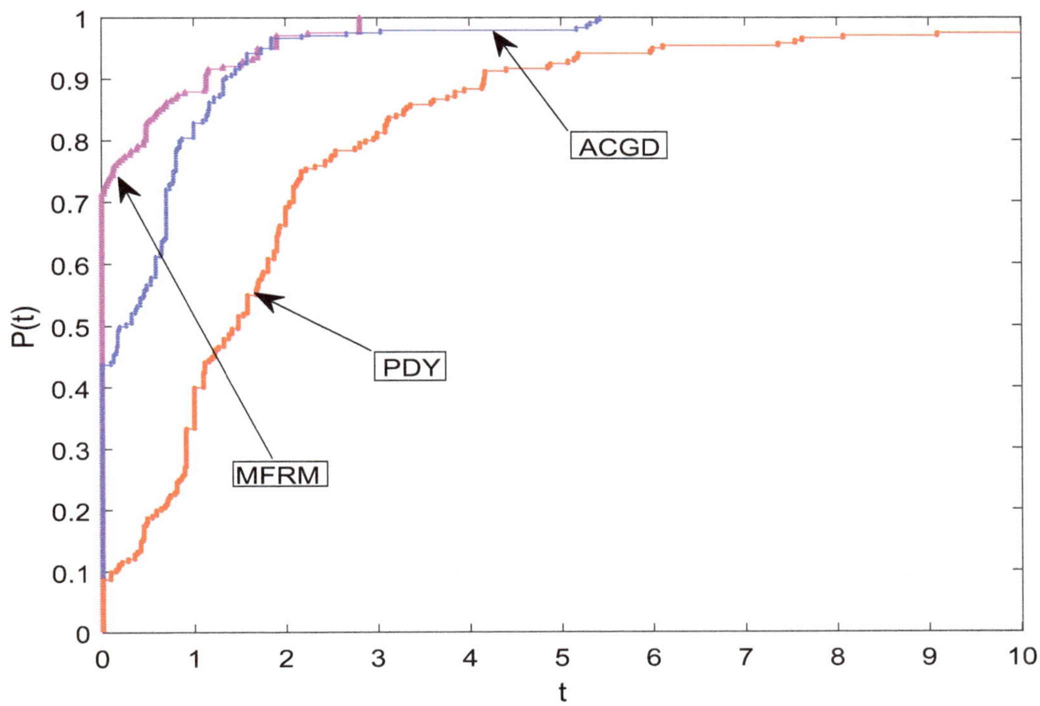

Figure 1. Performance profiles for the number of iterations.

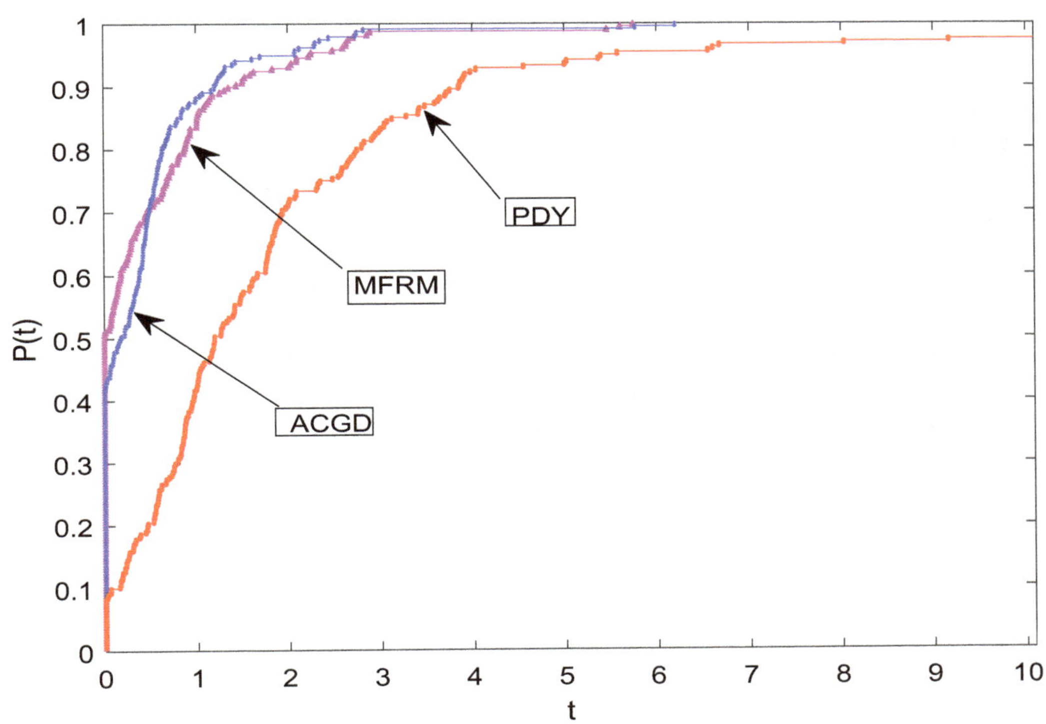

Figure 2. Performance profiles for the CPU time (in seconds).

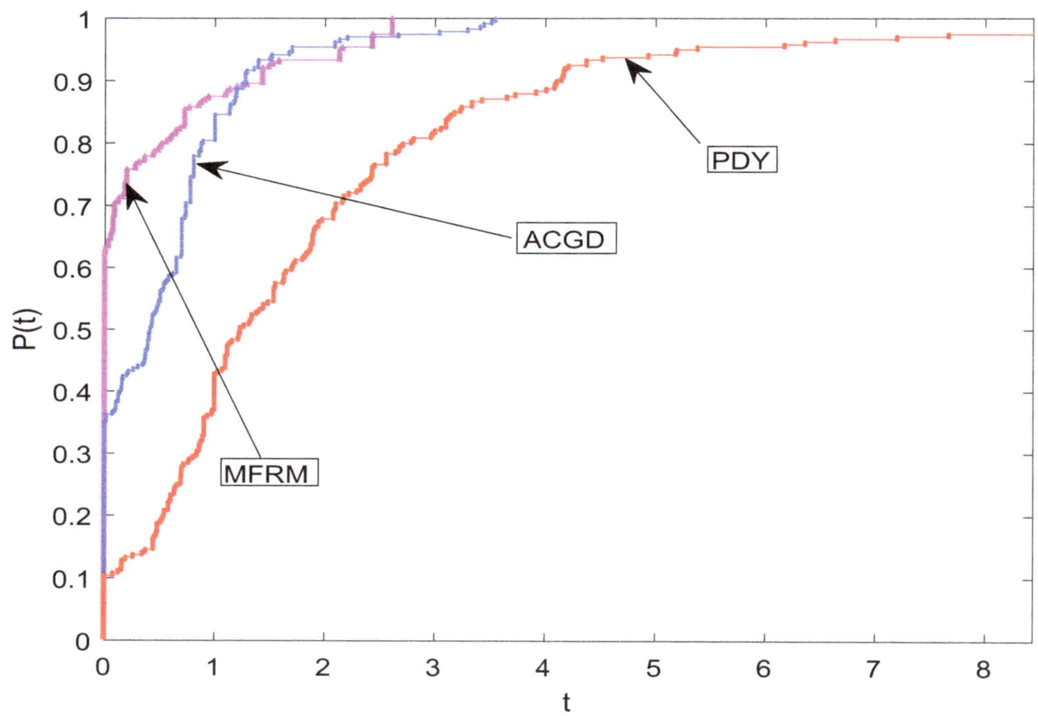

Figure 3. Performance profiles for the number of function evaluations.

4.1. Experiments on Solving Sparse Signal Problems

There were many problems in signal processing and statistical inference involving finding sparse solutions to ill-conditioned linear systems of equations. Among popular approaches was minimizing an objective function which contains quadratic (ℓ_2) error term and a sparse ℓ_1-regularization term, i.e.,

$$\min_x \frac{1}{2}\|y - Bx\|_2^2 + \eta\|x\|_1, \tag{25}$$

where $x \in R^n$, $y \in R^k$ is an observation, $B \in R^{k\times n}$ $(k << n)$ is a linear operator, η is a non-negative parameter, $\|x\|_2$ denotes the Euclidean norm of x and $\|x\|_1 = \sum_{i=1}^{n} |x_i|$ is the ℓ_1-norm of x. It is easy to see that problem (25) is a convex unconstrained minimization problem. Due to the fact that if the original signal is sparse or approximately sparse in some orthogonal basis, problem (25) frequently appears in compressive sensing, and hence an exact restoration can be produced by solving (25).

Iterative methods for solving (25) have been presented in many papers (see [42–45]). The most popular method among these methods is the gradient-based method and the earliest gradient projection method for sparse reconstruction (GPRS) was proposed by Figueiredo et al. [44]. The first step of the GPRS method is to express (25) as a quadratic problem using the following process. Consider a point $x \in \mathbb{R}^n$ such that $x = u - v$, where $u, v \geq 0$. u and v are chosen in such a way that x is splitted into its positive and negative parts as follows $u_i = (x_i)_+$, $v_i = (-x_i)_+$ for all $i = 1, 2, ..., n$, and $(.)_+ = \max\{0, .\}$. By definition of ℓ_1-norm, we have $\|x\|_1 = e_n^T u + e_n^T v$, where $e_n = (1, 1, ..., 1)^T \in R^n$. Now (25) can be written as

$$\min_{u,v} \frac{1}{2}\|y - B(u - v)\|_2^2 + \eta e_n^T u + \eta e_n^T v, \qquad u \geq 0, \quad v \geq 0, \tag{26}$$

which is a bound-constrained quadratic program. However, from [44], Equation (26) can be written in standard form as

$$\min_{z} \frac{1}{2} z^T D z + c^T z, \qquad \text{such that} \quad z \geq 0, \tag{27}$$

where $z = \begin{pmatrix} u \\ v \end{pmatrix}$, $c = \omega e_{2n} + \begin{pmatrix} -b \\ b \end{pmatrix}$, $b = B^T y$, $D = \begin{pmatrix} B^T B & -B^T B \\ -B^T B & B^T B \end{pmatrix}$. Clearly, D is a positive semi-definite matrix, which implies that Equation (27) is a convex quadratic problem.

Xiao et al. [20] translated (27) into a linear variable inequality problem which is equivalent to a linear complementarity problem. Moreover, z is a solution of the linear complementarity problem if and only if it is a solution of the following nonlinear equation:

$$F(z) = \min\{z, Dz + c\} = 0. \tag{28}$$

The function F is a vector-valued function and the "min" was interpreted as component wise minimum. Furthermore, F was proved to be continuous and monotone in [46]. Therefore problem (25) can be translated into problem (1) and thus MFRM method can be applied to solve it.

In this experiment, we consider a simple compressive sensing possible situation, where our goal is to reconstruct a sparse signal of length n from k observations. The quality of recovery is assessed by mean of squared error (MSE) to the original signal $\tilde{x}$,

$$MSE = \frac{1}{n} \|\tilde{x} - x_*\|^2,$$

where x_* is the recovered signal. The signal size is chosen as $n = 2^{11}$, $k = 2^9$ and the original signal contains 2^6 randomly nonzero elements. In addition, the measurement y is distributed with noise, that is, $y = B\tilde{x} + \varrho$, where B is a randomly generated Gaussian matrix and ϱ is the Gaussian noise distributed normally with mean 0 and variance 10^{-4}.

To demonstrate the performance of the MFRM method in signal recovery problems, we compare it with the conjugate gradient descent CGD [20] and projected conjugate gradient PCG [23] methods. The parameters in PCG and CGD methods are chosen as $\gamma = 10$, $\sigma = 10^{-4}$, $\rho = 0.5$. However, we chose $\gamma = 1$, $\sigma = 10^{-4}$, $\rho = 0.9$ and $\mu = 0.01$ in MFRM method. For fairness in comparison, each code was run from the same initial point, same continuation technique on the parameter η, and observed only the behavior of the convergence of each method to have a similar accurate solution. The experiment was initialized with $x_0 = B^T y$ and terminates when

$$\frac{\|f(x_k) - f(x_{k-1})\|}{\|f(x_{k-1})\|} < 10^{-5},$$

where $f(x_k) = \frac{1}{2}\|y - Bx_k\|_2^2 + \eta \|x_k\|_1$.

In Figures 4 and 5, MFRM, CGD and PCG methods recovered the disturbed signal almost exactly. The experiment was repeated for 20 different noise samples (see Table 9). It can be observed that the MFRM is more efficient in terms of the number of Iterations and CPU time than CGD and PCG methods in most cases. Furthermore, MFRM was able to achieve the least MSE in nine (9) out of the twenty (20) experiments. To reveal visually the performance of both methods, two figures were plotted to demonstrate their convergence behavior based on MSE, objective function values, the number of Iterations and CPU time (see Figures 6 and 7). It can also be observed that MFRM requires less computing time to achieve similar quality resolution. This can be seen graphically in Figures 6 and 7 which illustrate that the objective function values obtained by MFRM decrease faster throughout the entire Iteration process.

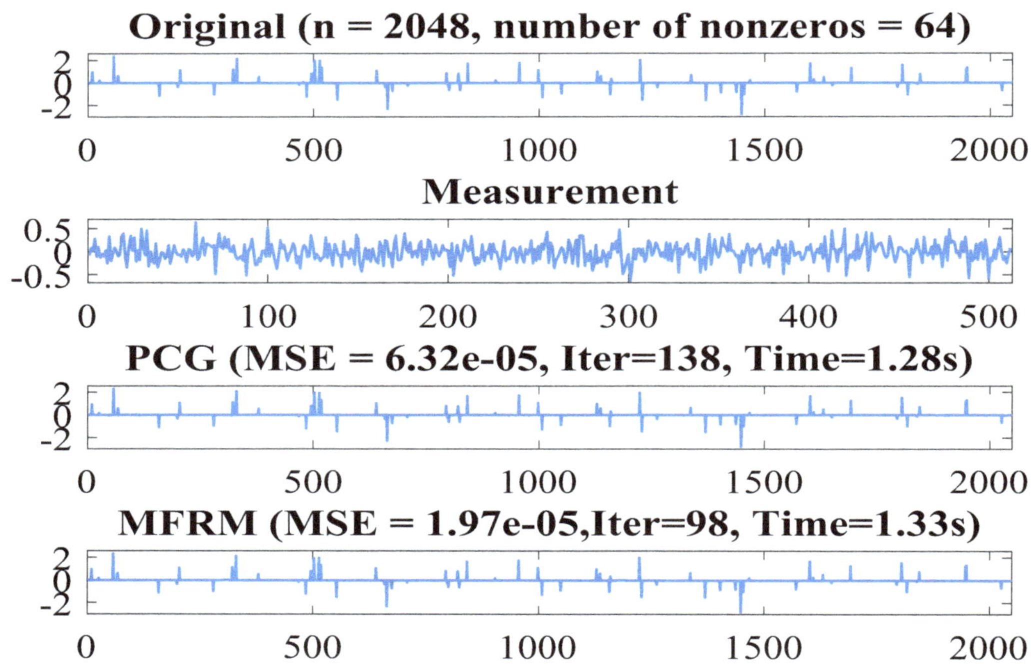

Figure 4. (**top**) to (**bottom**) The original image, the measurement, and the recovered signals by projected conjugate gradient PCG and modified descent Fletcher–Reeves CG method (MFRM) methods.

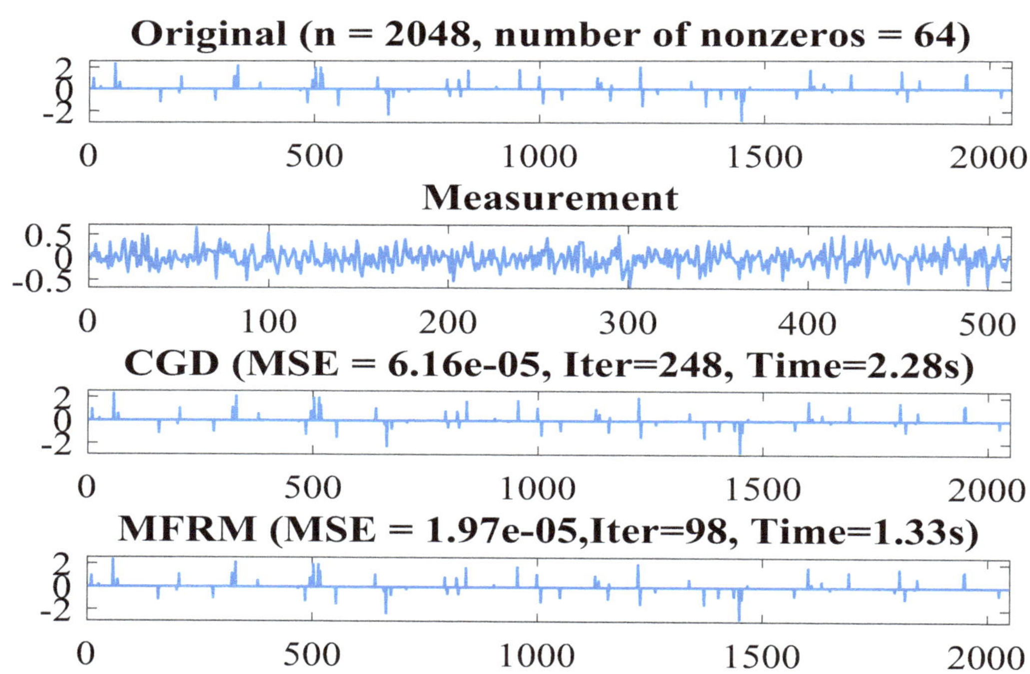

Figure 5. (**top**) to (**bottom**) The original image, the measurement, and the recovered signals by conjugate gradient descent (CGD) and MFRM methods.

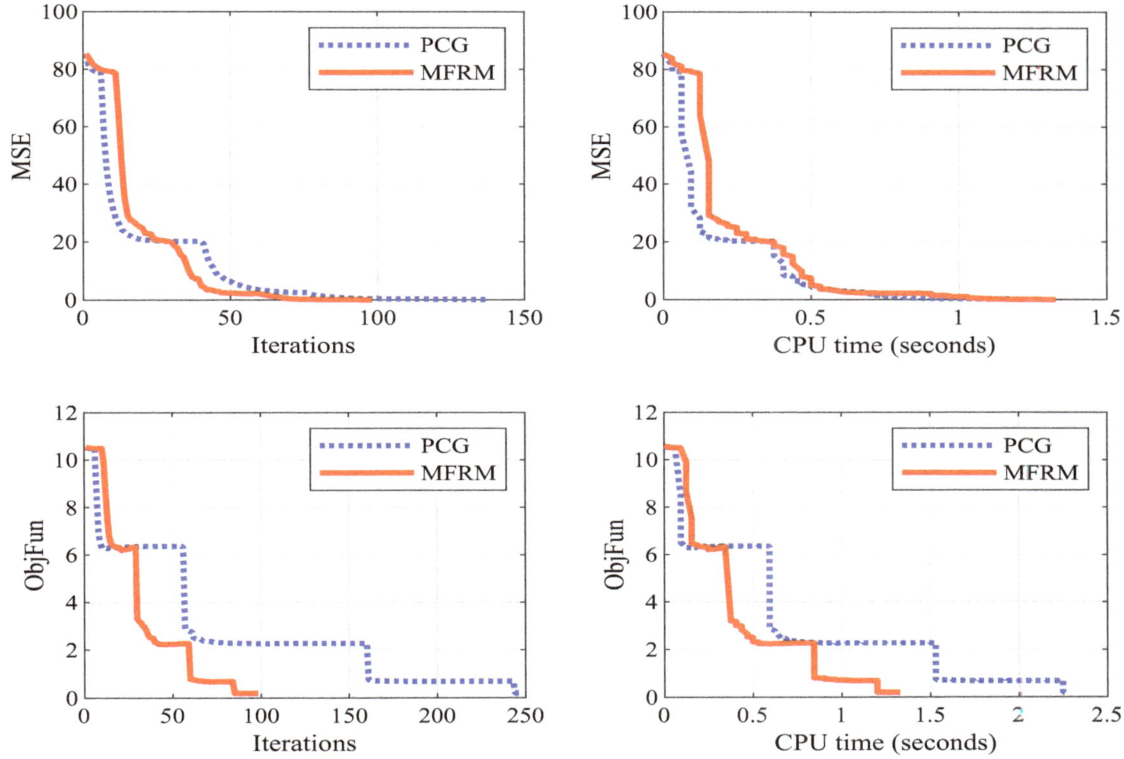

Figure 6. Comparison result of PCG and MFRM. The x-axis represent the number of Iterations ((**top left**) and (**bottom left**)) and CPU time in seconds ((**top right**) and (**bottom right**)). The y-axis represent the MSE ((**top left**) and (**top right**)) and the objective function values ((**bottom left**) and (**bottom right**)).

Table 9. Twenty experiment results of ℓ_1−norm regularization problem for CGD, PCG and MFRM methods.

S/N	Iter			Time			MSE		
	CGD	PCG	MFRM	CGD	PCG	MFRM	CGD	PCG	MFRM
1	248	138	98	2.28	1.28	1.33	6.16×10^{-5}	6.32×10^{-5}	1.97×10^{-5}
2	234	138	117	3.37	1.26	1.19	4.08×10^{-5}	3.36×10^{-5}	5.40×10^{-5}
3	224	152	104	1.90	1.29	0.97	2.78×10^{-5}	1.78×10^{-5}	1.02×10^{-5}
4	230	143	117	3.21	2.48	1.17	4.08×10^{-5}	3.36×10^{-5}	5.40×10^{-5}
5	152	119	114	1.65	1.03	1.15	1.23×10^{-5}	2.07×10^{-5}	5.49×10^{-5}
6	223	127	110	1.89	2.56	1.83	3.33×10^{-5}	6.08×10^{-5}	6.50×10^{-6}
7	156	120	125	1.37	1.01	1.20	4.25×10^{-5}	3.26×10^{-5}	1.46×10^{-5}
8	213	89	10	1.90	0.78	1.12	1.86×10^{-5}	3.77×10^{-4}	1.31×10^{-5}
9	227	152	118	2.14	1.53	1.45	2.75×10^{-5}	1.54×10^{-5}	8.11×10^{-6}
10	201	142	101	2.22	1.64	1.01	6.75×10^{-5}	1.86×10^{-5}	1.17×10^{-5}
11	200	151	90	1.70	1.42	0.90	2.36×10^{-5}	1.29×10^{-5}	3.81×10^{-5}
12	202	153	91	1.75	1.34	0.84	6.94×10^{-5}	2.99×10^{-5}	9.21×10^{-5}
13	208	128	125	1.89	1.12	1.26	1.71×10^{-5}	1.42×10^{-5}	9.20×10^{-6}
14	161	145	122	1.47	1.28	1.26	1.15×10^{-5}	8.75×10^{-6}	4.36×10^{-6}
15	227	160	100	1.97	1.42	1.00	3.41×10^{-5}	2.40×10^{-5}	1.54×10^{-5}
16	269	172	88	2.51	1.67	0.98	3.90×10^{-5}	6.59×10^{-5}	2.08×10^{-4}
17	210	129	105	1.84	1.19	1.11	2.11×10^{-5}	1.89×10^{-5}	6.22×10^{-5}
18	225	132	96	1.93	1.15	1.00	3.87×10^{-5}	7.78×10^{-5}	9.49×10^{-5}
19	152	120	92	1.37	1.09	0.87	2.12×10^{-5}	1.32×10^{-5}	4.03×10^{-5}
20	151	128	113	1.31	1.15	1.06	4.48×10^{-5}	1.85×10^{-5}	1.71×10^{-5}

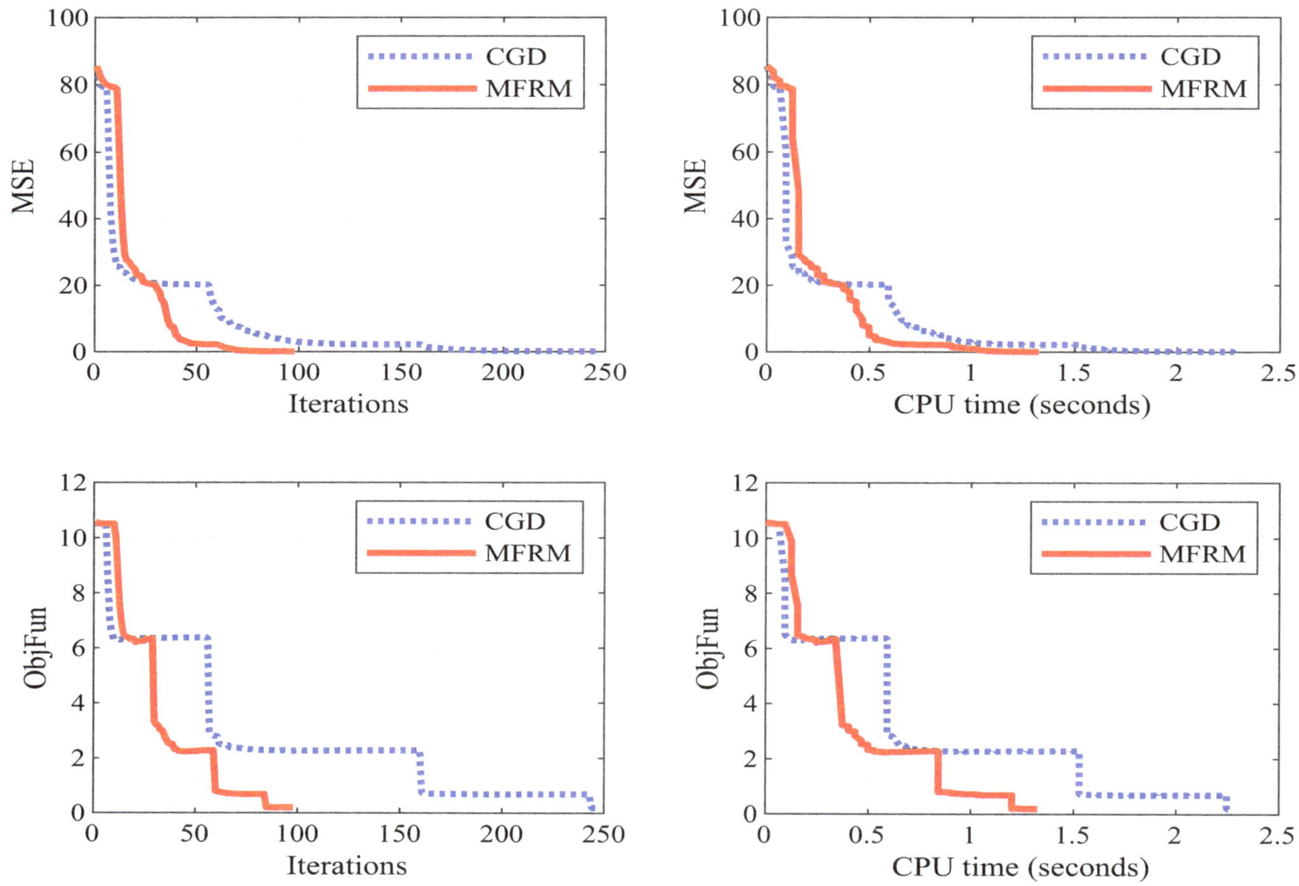

Figure 7. Comparison result of PCG and MFRM. The x-axis represent the number of Iterations ((**top left**) and (**bottom left**)) and CPU time in seconds ((**top right**) and (**bottom right**)). The y-axis represent the MSE ((**top left**) and (**top right**)) and the objective function values ((**bottom left**) and (**bottom right**)).

4.2. Experiments on Blurred Image Restoration

In this subsection, we test the performance of MFRM in restoring a blurred image. We use the following well-known gray test images; (P1) Cameraman, (P2) Lena, (P3) House and (P4) Peppers for the experiments. We use 4 different Gaussian blur kernels with a standard deviation v to compare the robustness of MFRM method with CGD method proposed in [20].

To assess the performance of each algorithm tested with respect to the metrics that indicate better quality of restoration, in Table 10 we reported the objective function (ObjFun) at the approximate solution, the MSE, the signal-to-noise-ratio (SNR) which is defined as

$$\text{SNR} = 20 \times \log_{10}\left(\frac{||\bar{x}||}{||x - \bar{x}||}\right),$$

and the structural similarity (SSIM) index that measure the similarity between the original image and the restored image [47] for each of the 16 experiments. The MATLAB implementation of the SSIM index can be obtained at http://www.cns.nyu.edu/~lcv/ssim/.

The original, blurred and restored images by each of the algorithm are given in Figures 8–11. The figures demonstrate that both the two algorithms can restore the blurred images. In contrast to the CGD, the quality of the restored image by MFRM is superior in most cases. Table 11 reported numerical results for MFRM, ACGD and PDY for problem 2.

Table 10. Efficiency comparison based on the value of the objective function (ObjFun) mean-square-error (MSE), SNR and the SSIM index under different Pi(v).

Image	ObjFun		MSE		SNR		SSIM	
	MFRM	CGD	MFRM	CGD	MFRM	CGD	MFRM	CGD
P1(1×10^{-4})	1.43×10^6	1.47×10^6	133.90	177.57	21.28	20.05	0.86	0.83
P1(1×10^{-1})	1.43×10^6	1.48×10^6	130.60	177.69	21.39	20.5	0.86	0.83
P1(0.25)	1.47×10^6	1.48×10^6	145.27	177.72	20.93	20.05	0.85	0.83
P1(6.25)	1.58×10^6	1.65×10^6	146.06	183.96	20.9	19.9	0.75	0.79
P2(1×10^{-4})	1.61×10^6	1.65×10^6	36.88	57.55	27.59	25.65	0.88	0.86
P2(1×10^{-1})	1.61×10^6	1.65×10^6	36.85	57.61	27.59	25.65	0.88	0.86
P2(0.25)	1.62×10^6	1.66×10^6	37.78	57.68	27.48	25.64	0.88	0.86
P2(6.25)	1.77×10^6	1.82×10^6	56.65	58.96	25.72	25.55	0.76	0.83
P3(1×10^{-4})	5.74×10^6	5.89×10^6	41.63	44.48	26.26	25.97	0.9	0.88
P3(1×10^{-1})	5.75×10^6	5.90×10^6	42.42	44.54	26.17	25.96	0.89	0.88
P3(0.25)	5.76×10^6	5.91×10^6	43.33	44.65	26.08	25.95	0.88	0.88
P3(6.25)	6.35×10^6	6.60×10^6	106.79	48.47	22.16	25.6	0.63	0.85
P4(1×10^{-4})	1.40×10^6	1.48×10^6	88.81	122.44	22.9	21.5	0.87	0.84
P4(1×10^{-1})	1.41×10^6	1.48×10^6	89.22	122.56	22.88	21.5	0.87	0.84
P4(0.25)	1.41×10^6	1.49×10^6	89.86	122.56	22.85	21.5	0.87	0.84
P4(6.25)	1.56×10^6	1.69×10^6	116.79	138.97	21.71	20.95	0.76	0.82

Table 11. Numerical results for modified Fletcher-Reeves method MFRM, accelerated conjugate gradient descent (ACGD) and projected Dai-Yuan (PDY) methods for problem 2 with given initial points and dimensions with double float (10^{-16}) accuracy.

		MFRM				ACGD				PDY			
Dimension	Initial Point	Iter	Fval	Time	Norm	Iter	Fval	Time	Norm	Iter	Fval	Time	Norm
1000	$x1$	8	27	0.14061	9.47×10^{-19}	12	53	0.030479	3.32×10^{-18}	30	119	0.04027	4.76×10^{-19}
	$x2$	8	36	0.010782	1.49×10^{-18}	7	20	0.013503	1.08×10^{-18}	36	153	0.034454	3.51×10^{-18}
	$x3$	7	20	0.008263	1.21×10^{-18}	13	56	0.021302	3.26×10^{-18}	38	161	0.038168	3.51×10^{-18}
	$x4$	8	23	0.015654	1.80×10^{-19}	12	51	0.02056	3.31×10^{-18}	39	165	0.057793	3.51×10^{-18}
	$x5$	11	38	0.018461	1.59×10^{-18}	14	59	0.088858	3.34×10^{-18}	41	173	0.069756	3.51×10^{-18}
	$x6$	10	34	0.016788	1.07×10^{-18}	10	32	0.012069	5.83×10^{-19}	40	169	0.03311	3.50×10^{-18}
5000	$x1$	9	33	0.028658	7.22×10^{-19}	12	54	0.041685	1.52×10^{-18}	35	149	0.10692	1.57×10^{-18}
	$x2$	7	23	0.024046	2.18×10^{-19}	9	41	0.049194	1.55×10^{-18}	37	157	0.12219	1.57×10^{-18}
	$x3$	6	17	0.03436	3.89×10^{-19}	14	61	0.094129	1.47×10^{-18}	33	131	0.10635	1.06×10^{-19}
	$x4$	8	26	0.03133	7.17×10^{-19}	14	60	0.065147	1.47×10^{-18}	39	165	0.18361	1.57×10^{-18}
	$x5$	9	31	0.036727	5.84×10^{-19}	10	43	0.1165	1.47×10^{-18}	36	144	0.2178	7.43×10^{-20}
	$x6$	10	34	0.030168	6.41×10^{-19}	12	51	0.038218	1.51×10^{-18}	38	161	0.13144	1.57×10^{-18}
10,000	$x1$	8	28	0.064617	1.89×10^{-19}	11	50	0.068567	1.03×10^{-18}	35	149	0.2253	1.11×10^{-18}
	$x2$	6	19	0.044204	1.90×10^{-19}	14	62	0.15949	1.09×10^{-18}	32	128	0.34325	8.21×10^{-20}
	$x3$	6	17	0.045192	1.45×10^{-19}	18	78	0.10766	1.04×10^{-18}	39	165	0.23899	1.11×10^{-18}
	$x4$	10	35	0.055408	4.99×10^{-19}	12	52	0.061589	1.06×10^{-18}	39	165	0.23162	1.11×10^{-18}
	$x5$	7	20	0.038439	2.06×10^{-19}	14	60	0.087394	1.05×10^{-18}	40	169	0.28998	1.11×10^{-18}
	$x6$	9	29	0.065318	5.27×10^{-19}	16	68	0.09917	1.03×10^{-18}	40	170	0.22564	1.11×10^{-18}
50,000	$x1$	7	26	0.21017	1.93×10^{-19}	23	100	0.51879	4.79×10^{-19}	34	145	0.92896	4.96×10^{-19}
	$x2$	6	21	0.24752	2.09×10^{-19}	25	108	0.64677	4.90×10^{-19}	36	153	0.9954	4.96×10^{-19}
	$x3$	6	17	0.11243	6.27×10^{-20}	23	99	0.50402	4.93×10^{-19}	38	161	0.96768	4.96×10^{-19}
	$x4$	7	20	0.13442	1.02×10^{-19}	24	102	0.63664	4.75×10^{-19}	79	326	1.7542	4.96×10^{-19}
	$x5$	9	30	0.20288	7.25×10^{-20}	25	106	0.51116	4.78×10^{-19}	78	322	1.7246	4.96×10^{-19}
	$x6$	12	52	0.36526	2.28×10^{-19}	23	97	0.56342	4.76×10^{-19}	80	330	1.6812	4.96×10^{-19}
100,000	$x1$	7	27	0.36065	6.53×10^{-20}	23	100	0.88236	3.26×10^{-19}	30	119	1.2102	9.26×10^{-21}
	$x2$	5	14	0.20041	3.91×10^{-20}	25	108	0.90777	3.27×10^{-19}	35	149	1.5699	3.51×10^{-19}
	$x3$	7	24	0.34075	1.47×10^{-19}	25	107	0.95898	3.26×10^{-19}	40	170	1.7126	3.51×10^{-19}
	$x4$	8	31	0.40444	2.09×10^{-20}	24	102	0.83332	3.38×10^{-19}	151	614	5.8306	3.51×10^{-19}
	$x5$	8	26	0.52598	5.03×10^{-20}	25	106	1.0223	3.47×10^{-19}	151	614	5.6777	3.50×10^{-19}
	$x6$	7	20	0.33434	1.45×10^{-19}	23	97	0.87438	3.33×10^{-19}	153	622	5.7906	3.51×10^{-19}

Figure 8. The original image (**top left**), the blurred image (**top right**), the restored image by CGD (**bottom left**) with time = 3.70, signal-to-noise-ratio (SNR) = 20.05 and structural similarity (SSIM) = 0.83, and by MFRM (**bottom right**) with time = 1.97, SNR = 21.28 and SSIM = 0.86.

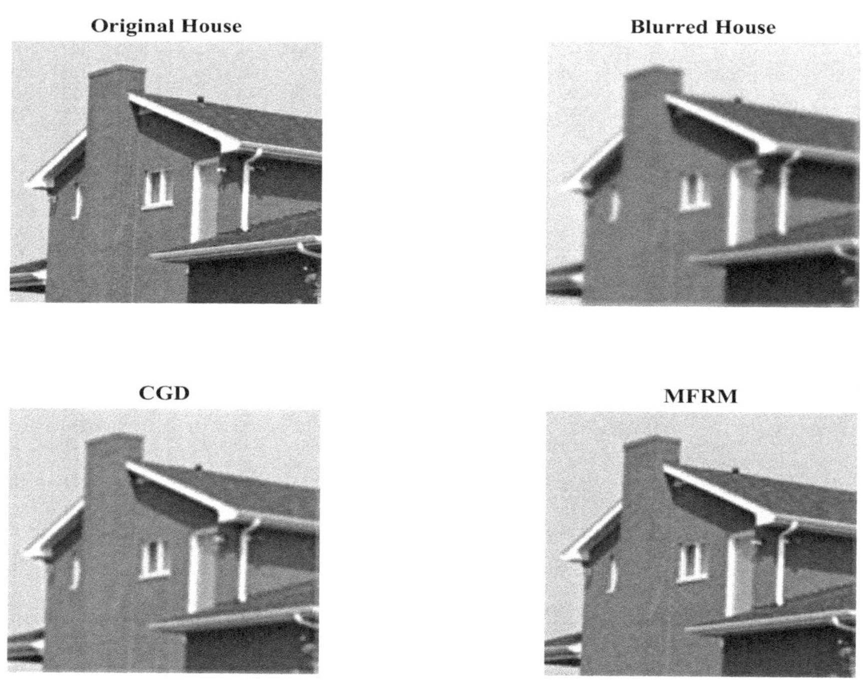

Figure 9. The original image (**top left**), the blurred image (**top right**), the restored image by CGD (**bottom left**) with Time = 1.95, SNR = 25.65 and SSIM = 0.86, and by MFRM (**bottom right**) with Time = 3.59, SNR = 27.59 and SSIM = 0.88.

Figure 10. The original image (**top left**), the blurred image (**top right**), the restored image by CGD (**bottom left**) with time = 5.38, SNR = 25.97 and SSIM = 0.88, and by MFRM (**bottom right**) with time = 38.77, SNR = 26.26 and SSIM = 0.90.

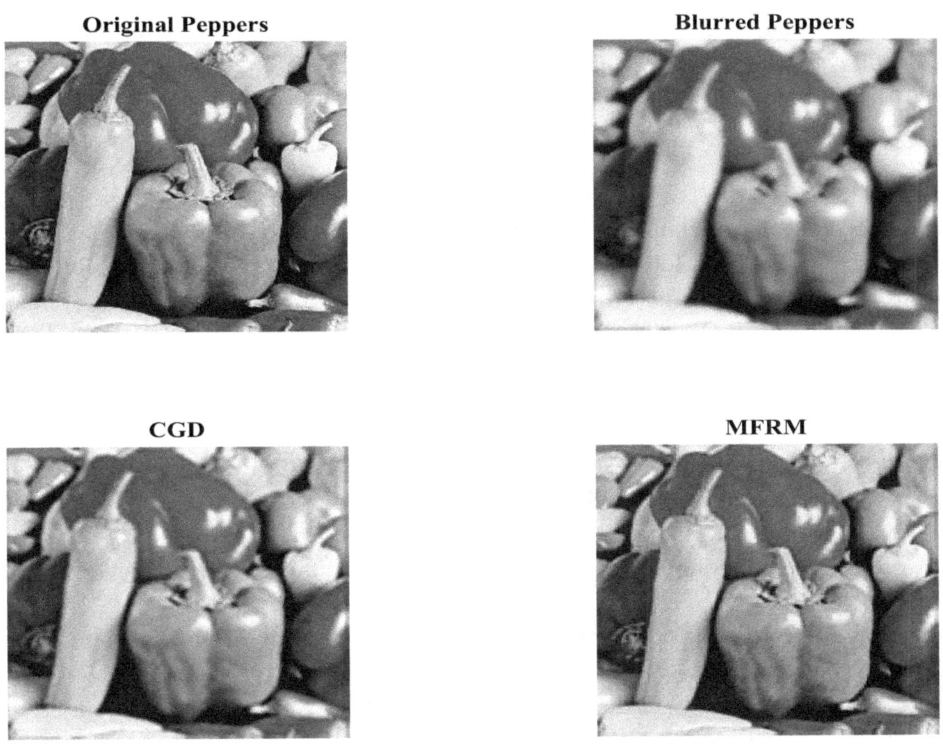

Figure 11. The original image (**top left**), the blurred image (**top right**), the restored image by CGD (**bottom left**) with Time = 2.48, SNR = 21.50 and SSIM = 0.84, and by MFRM (**bottom right**) with Time = 4.93, SNR = 22.90 and SSIM = 0.87.

5. Conclusions

In this paper, a modified conjugate gradient method for solving monotone nonlinear equations with convex constraints was presented which is similar to that in [3]. The proposed method is suitable for non-smooth equations. Under some suitable assumptions, the global convergence of the proposed method was demonstrated. Numerical results were presented to show the effectiveness of the MFRM method compared to the ACGD and PDY methods for the given constrained monotone equation problems. Finally, the MFRM was also shown to be effective in decoding sparse signals and restoration of blurred images.

Author Contributions: conceptualization, A.B.A.; methodology, A.B.A.; software, H.M.; validation, P.K., A.M.A. and K.S.; formal analysis, P.K. and K.S.; investigation, P.K. and H.M.; resources, P.K. and K.S.; data curation, H.M. and A.M.A.; writing–original draft preparation, A.B.A.; writing–review and editing, H.M.; visualization, A.M.A. and K.S.; supervision, P.K.; project administration, P.K. and K.S.; funding acquisition, P.K. and K.S.

Acknowledgments: We thank Associate Professor Jin Kiu Liu for providing us with the access of the CGD-CS MATLAB codes. The authors acknowledge the financial support provided by King Mongkut's University of Technology Thonburi through the "KMUTT 55th Anniversary Commemorative Fund". The first author was supported by the "Petchra Pra Jom Klao Ph.D. Research Scholarship from King Mongkut's University of Technology Thonburi".

References

1. Abubakar, A.B.; Kumam, P.; Awwal, A.M. A Descent Dai-Liao Projection Method for Convex Constrained Nonlinear Monotone Equations with Applications. *Thai J. Math.* **2018**, *17*, 128–152.
2. Abubakar, A.B.; Kumam, P. A descent Dai-Liao conjugate gradient method for nonlinear equations. *Numer. Algorithms* **2019**, *81*, 197–210. [CrossRef]
3. Abubakar, A.B.; Kumam, P. An improved three-term derivative-free method for solving nonlinear equations. *Comput. Appl. Math.* **2018**, *37*, 6760–6773. [CrossRef]
4. Mohammad, H.; Abubakar, A.B. A positive spectral gradient-like method for nonlinear monotone equations. *Bull. Comput. Appl. Math.* **2017**, *5*, 99–115.
5. Muhammed, A.A.; Kumam, P.; Abubakar, A.B.; Wakili, A.; Pakkaranang, N. A New Hybrid Spectral Gradient Projection Method for Monotone System of Nonlinear Equations with Convex Constraints. *Thai J. Math.* **2018**, *16*, 125–147.
6. Zhou, W.J.; Li, D.H. A globally convergent BFGS method for nonlinear monotone equations without any merit functions. *Math. Comput.* **2008**, *77*, 2231–2240. [CrossRef]
7. Yan, Q.R.; Peng, X.Z.; Li, D.H. A globally convergent derivative-free method for solving large-scale nonlinear monotone equations. *J. Comput. Appl. Math.* **2010**, *234*, 649–657. [CrossRef]
8. DiRksEandM, S.P.; FERRis, C. A collection of nonlinear mixed complementarity problems. *Optim. Methods Softw.* **1995**, *5*, 319–345.
9. Meintjes, K.; Morgan, A.P. A methodology for solving chemical equilibrium systems. *Appl. Math. Comput.* **1987**, *22*, 333–361. [CrossRef]
10. Bellavia, S.; Macconi, M.; Morini, B. STRSCNE: A Scaled Trust-Region Solver for Constrained Nonlinear Equations. *Comput. Optim. Appl.* **2004**, *28*, 31–50. [CrossRef]
11. Kanzow, C.; Yamashita, N.; Fukushima, M. Levenberg–Marquardt methods with strong local convergence properties for solving nonlinear equations with convex constraints. *J. Comput. Appl. Math.* **2004**, *172*, 375–397. [CrossRef]
12. Papp, Z.; Rapajić, S. FR type methods for systems of large-scale nonlinear monotone equations. *Appl. Math. Comput.* **2015**, *269*, 816–823. [CrossRef]
13. Zhou, W.; Wang, F. A PRP-based residual method for large-scale monotone nonlinear equations. *Appl. Math. Comput.* **2015**, *261*, 1–7. [CrossRef]
14. Dai, Y.H.; Yuan, Y. A nonlinear conjugate gradient method with a strong global convergence property. *SIAM J. Optim.* **1999**, *10*, 177–182. [CrossRef]
15. Polak, E.; Ribiere, G. Note sur la convergence de méthodes de directions conjuguées. *Revue Française D'informatique et de Recherche Opérationnelle Série Rouge* **1969**, *3*, 35–43. [CrossRef]

16. Polyak, B.T. The conjugate gradient method in extremal problems. *USSR Comput. Math. Math. Phys.* **1969**, *9*, 94–112. [CrossRef]
17. Hager, W.W.; Zhang, H. A new conjugate gradient method with guaranteed descent and an efficient line search. *SIAM J. Optim.* **2005**, *16*, 170–192. [CrossRef]
18. Dai, Y.H.; Liao, L.Z. New conjugacy conditions and related nonlinear conjugate gradient methods. *Appl. Math. Optim.* **2001**, *43*, 87–101. [CrossRef]
19. Fletcher, R.; Reeves, C.M. Function minimization by conjugate gradients. *Comput. J.* **1964**, *7*, 149–154. [CrossRef]
20. Xiao, Y.; Zhu, H. A conjugate gradient method to solve convex constrained monotone equations with applications in compressive sensing. *J. Math. Anal. Appl.* **2013**, *405*, 310–319. [CrossRef]
21. Solodov, M.V.; Svaiter, B.F. A globally convergent inexact Newton method for systems of monotone equations. In *Reformulation: Nonsmooth, Piecewise Smooth, Semismooth and Smoothing Methods*; Springer: Boston, MA, USA, 1998; pp. 355–369.
22. Liu, S.Y.; Huang, Y.Y.; Jiao, H.W. Sufficient descent conjugate gradient methods for solving convex constrained nonlinear monotone equations. In *Abstract and Applied Analysis*; Hindawi: New York, NY, USA, 2014; Volume 2014.
23. Liu, J.; Li, S. A projection method for convex constrained monotone nonlinear equations with applications. *Comput. Math. Appl.* **2015**, *70*, 2442–2453. [CrossRef]
24. Sun, M.; Liu, J. Three derivative-free projection methods for nonlinear equations with convex constraints. *J. Appl. Math. Comput.* **2015**, *47*, 265–276. [CrossRef]
25. Sun, M.; Liu, J. New hybrid conjugate gradient projection method for the convex constrained equations. *Calcolo* **2016**, *53*, 399–411. [CrossRef]
26. Ou, Y.; Li, J. A new derivative-free SCG-type projection method for nonlinear monotone equations with convex constraints. *J. Appl. Math. Comput.* **2018**, *56*, 195–216. [CrossRef]
27. Ding, Y.; Xiao, Y.; Li, J. A class of conjugate gradient methods for convex constrained monotone equations. *Optimization* **2017**, *66*, 2309–2328. [CrossRef]
28. Liu, J.; Feng, Y. A derivative-free iterative method for nonlinear monotone equations with convex constraints. *Numer. Algorithms* **2018**, 1–18. [CrossRef]
29. Kabanikhin, S.I. Definitions and examples of inverse and ill-posed problems. *J. Inverse Ill-Posed Probl.* **2008**, *16*, 317–357. [CrossRef]
30. Belishev, M.I.; Kurylev, Y.V. Boundary control, wave field continuation and inverse problems for the wave equation. *Comput. Math. Appl.* **1991**, *22*, 27–52. [CrossRef]
31. Beilina, L.; Klibanov, M.V. A Globally Convergent Numerical Method for a Coefficient Inverse Problem. *SIAM J. Sci. Comput.* **2008**, *31*, 478–509. [CrossRef]
32. Kabanikhin, S.; Shishlenin, M. Boundary control and Gel'fand–Levitan–Krein methods in inverse acoustic problem. *J. Inverse Ill-Posed Probl.* **2004**, *12*, 125–144. [CrossRef]
33. Lukyanenko, D.; Grigorev, V.; Volkov, V.; Shishlenin, M. Solving of the coefficient inverse problem for a nonlinear singularly perturbed two dimensional reaction diffusion equation with the location of moving front data. *Comput. Math. Appl.* **2019**, *77*, 1245–1254. [CrossRef]
34. Van, S.B.; Elber, G. Solving piecewise polynomial constraint systems with decomposition and a subdivision-based solver. *Computer-Aided Design* **2017**, *90*, 37–47.
35. Aizenshtein, M.; Bartoň, M.; Elber, G. Global solutions of well-constrained transcendental systems using expression trees and a single solution test. *Computer Aided Geometric Design* **2012**, *29*, 265–279.
36. Bartoň, M. Solving polynomial systems using no-root elimination blending schemes. *Computer-Aided Design* **2011**, *43*, 1870–1878.
37. Wang, X.Y.; Li, S.J.; Kou, X.P. A self-adaptive three-term conjugate gradient method for monotone nonlinear equations with convex constraints. *Calcolo* **2016**, *53*, 133–145. [CrossRef]
38. La Cruz, W.; Martínez, J.; Raydan, M. Spectral residual method without gradient information for solving large-scale nonlinear systems of equations. *Math. Comput.* **2006**, *75*, 1429–1448. [CrossRef]
39. Bing, Y.; Lin, G. An Efficient Implementation of Merrills Method for Sparse or Partially Separable Systems of Nonlinear Equations. *SIAM J. Optim.* **1991**, *1*, 206–221, doi:10.1137/0801015. [CrossRef]
40. Yu, Z.; Lin, J.; Sun, J.; Xiao, Y.H.; Liu, L.Y.; Li, Z.H. Spectral gradient projection method for monotone nonlinear equations with convex constraints. *Appl. Numer. Math.* **2009**, *59*, 2416–2423. [CrossRef]

41. Dolan, E.D.; Moré, J.J. Benchmarking optimization software with performance profiles. *Math. Program.* **2002**, *91*, 201–213. [CrossRef]
42. Hale, E.T.; Yin, W.; Zhang, Y. A fixed-point continuation method for ℓ_1-regularized minimization with applications to compressed sensing. *CAAM TR07-07 Rice Univ.* **2007**, *43*, 44.
43. Beck, A.; Teboulle, M. A fast iterative shrinkage-thresholding algorithm for linear inverse problems. *SIAM J. Imaging Sci.* **2009**, *2*, 183–202. [CrossRef]
44. Figueiredo, M.A.; Nowak, R.D.; Wright, S.J. Gradient projection for sparse reconstruction: Application to compressed sensing and other inverse problems. *IEEE J. Sel. Top. Signal Process.* **2007**, *1*, 586–597. [CrossRef]
45. Birgin, E.G.; Martínez, J.M.; Raydan, M. Nonmonotone spectral projected gradient methods on convex sets. *SIAM J. Optim.* **2000**, *10*, 1196–1211. [CrossRef]
46. Xiao, Y.; Wang, Q.; Hu, Q. Non-smooth equations based method for ℓ_1-norm problems with applications to compressed sensing. *Nonlinear Anal. Theory Methods Appl.* **2011**, *74*, 3570–3577. [CrossRef]
47. Wang, Z.; Bovik, A.C.; Sheikh, H.R.; Simoncelli, E.P. Image quality assessment: From error visibility to structural similarity. *IEEE Trans. Image Process.* **2004**, *13*, 600–612. [CrossRef] [PubMed]

Permissions

All chapters in this book were first published by MDPI; hereby published with permission under the Creative Commons Attribution License or equivalent. Every chapter published in this book has been scrutinized by our experts. Their significance has been extensively debated. The topics covered herein carry significant findings which will fuel the growth of the discipline. They may even be implemented as practical applications or may be referred to as a beginning point for another development.

The contributors of this book come from diverse backgrounds, making this book a truly international effort. This book will bring forth new frontiers with its revolutionizing research information and detailed analysis of the nascent developments around the world.

We would like to thank all the contributing authors for lending their expertise to make the book truly unique. They have played a crucial role in the development of this book. Without their invaluable contributions this book wouldn't have been possible. They have made vital efforts to compile up to date information on the varied aspects of this subject to make this book a valuable addition to the collection of many professionals and students.

This book was conceptualized with the vision of imparting up-to-date information and advanced data in this field. To ensure the same, a matchless editorial board was set up. Every individual on the board went through rigorous rounds of assessment to prove their worth. After which they invested a large part of their time researching and compiling the most relevant data for our readers.

The editorial board has been involved in producing this book since its inception. They have spent rigorous hours researching and exploring the diverse topics which have resulted in the successful publishing of this book. They have passed on their knowledge of decades through this book. To expedite this challenging task, the publisher supported the team at every step. A small team of assistant editors was also appointed to further simplify the editing procedure and attain best results for the readers.

Apart from the editorial board, the designing team has also invested a significant amount of their time in understanding the subject and creating the most relevant covers. They scrutinized every image to scout for the most suitable representation of the subject and create an appropriate cover for the book.

The publishing team has been an ardent support to the editorial, designing and production team. Their endless efforts to recruit the best for this project, has resulted in the accomplishment of this book. They are a veteran in the field of academics and their pool of knowledge is as vast as their experience in printing. Their expertise and guidance has proved useful at every step. Their uncompromising quality standards have made this book an exceptional effort. Their encouragement from time to time has been an inspiration for everyone.

The publisher and the editorial board hope that this book will prove to be a valuable piece of knowledge for researchers, students, practitioners and scholars across the globe.

List of Contributors

Fiza Zafar
Centre for Advanced Studies in Pure and Applied Mathematics, Bahauddin Zakariya University, Multan 60800, Pakistan
Instituto de Matemáticas Multidisciplinar, Universitat Politenica de Valencia, 46022 Valencia, Spain

Alicia Cordero and Juan R. Torregrosa
Instituto de Matemáticas Multidisciplinar, Universitat Politenica de Valencia, 46022 Valencia, Spain

Zhang Yong
School of Mathematics and Physics, Changzhou University, Changzhou 213164, China

Neha Gupta and J. P. Jaiswal
Department of Mathematics, Maulana Azad National Institute of Technology, Bhopal 462003, India

Kalyanasundaram Madhu
Department of Mathematics, Saveetha Engineering College, Chennai 602105, India

Abdolreza Amiri and Mohammad Taghi Darvishi
Department of Mathematics, Razi University, Kermanshah 67149, Iran

Alicia Cordero and Juan Ramón Torregrosa
Institute for Multidisciplinary Mathematics, Universitat Politècnica de València, Camino de Vera s/n, 46022 Valencia, Spain

Anantachai Padcharoen
Department of Mathematics, Faculty of Science and Technology, Rambhai Barni Rajabhat University, Chanthaburi 22000, Thailand

Pakeeta Sukprasert
Department of Mathematics and Computer Science, Faculty of Science and Technology, Rajamangala University of Technology Thanyaburi (RMUTT), Thanyaburi, Pathumthani 12110, Thailand

Fuad W. Khdhr and Rostam K. Saeed
Department of Mathematics, College of Science, Salahaddin University, Erbil, Iraq

Fazlollah Soleymani
Department of Mathematics, Institute for Advanced Studies in Basic Sciences (IASBS), Zanjan 45137-66731, Iran

D. R. Sahu
Department of Mathematics, Banaras Hindu University, Varanasi-221005, India

Ravi P. Agarwal
Department of Mathematics, Texas A&M University-Kingsville, Kingsville, TX 78363-8202, USA

Vipin Kumar Singh
Department of Mathematics, Banaras Hindu University, Varanasi-221005, India

Mozafar Rostami, Taher Lotfi and Ali Brahmand
Department of Mathematics, Hamedan Branch, Islamic Azad University, Hamedan 15743-65181, Iran

Haifa Bin Jebreen
Mathematics Department, College of Science, King Saud University, Riyadh 11451, Saudi Arabia

Prem B. Chand and Pankaj Jain
Department of Mathematics, South Asian University, Akbar Bhawan, Chanakya Puri, New Delhi 110021, India

Francisco I. Chicharro
Escuela Superior de Ingeniería y Tecnología, Universidad Internacional de La Rioja, Av. La Paz 137, 26006 Logroño, Spain

Neus Garrido
Institute for Multidisciplinary Mathematics, Universitat Politècnica de València, Cno. de Vera s/n, 46022 València, Spain

Auwal Bala Abubakar
KMUTTFixed Point Research Laboratory, SCL 802 Fixed Point Laboratory, Science Laboratory Building, Department of Mathematics, Faculty of Science, King Mongkut's University of Technology Thonburi (KMUTT), 126 Pracha-Uthit Road, Bang Mod, Thrung Khru, Bangkok 10140, Thailand
Department of Mathematical Sciences, Faculty of Physical Sciences, Bayero University, Kano 700241, Nigeria

Poom Kumam
KMUTTFixed Point Research Laboratory, SCL 802 Fixed Point Laboratory, Science Laboratory Building, Department of Mathematics, Faculty of Science, King Mongkut's University of Technology Thonburi (KMUTT), 126 Pracha-Uthit Road, Bang Mod, Thrung Khru, Bangkok 10140, Thailand
Center of Excellence in Theoretical and Computational Science (TaCS-CoE), Science Laboratory Building, King Mongkut's University of Technology Thonburi (KMUTT), 126 Pracha-Uthit Road, Bang Mod, Thrung Khru, Bangkok 10140, Thailand
Department of Medical Research, China Medical University Hospital, China Medical University, Taichung 40402, Taiwan

Hassan Mohammad
Department of Mathematical Sciences, Faculty of Physical Sciences, Bayero University, Kano 700241, Nigeria

Aliyu Muhammed Awwal
KMUTTFixed Point Research Laboratory, SCL 802 Fixed Point Laboratory, Science Laboratory Building, Department of Mathematics, Faculty of Science, King Mongkut's University of Technology Thonburi (KMUTT), 126 Pracha-Uthit Road, Bang Mod, Thrung Khru, Bangkok 10140, Thailand
Department of Mathematics, Faculty of Science, Gombe State University, Gombe 760214, Nigeria

José Antonio Ezquerro and Miguel Ángel Hernández-Verón
Department of Mathematics and Computation, University of La Rioja, Calle Madre de Dios, 53, 26006 Logrono, Spain

Ioannis K. Argyros
Department of Mathematical Sciences, Cameron University, Lawton, OK 73505, USA

Stepan Shakhno
Department of Theory of Optimal Processes, Ivan Franko National University of Lviv, 79000 Lviv, Ukraine

Min-Young Lee and Young Ik Kim
Department of Applied Mathematics, Dankook University, Cheonan 330-714, Korea

Beny Neta
Naval Postgraduate School, Department of Applied Mathematics, Monterey, CA 93943, USA

Kanokwan Sitthithakerngkiet
Department of Mathematics, Faculty of Applied Science, King Mongkut's University of Technology North Bangkok, 1518 Pracharat 1 Road,Wongsawang, Bangsue, Bangkok 10800, Thailand

Index

www.ingramcontent.com/pod-product-compliance
Lightning Source LLC
LaVergne TN
LVHW060522210926
848485LV00005B/1129
* 9 7 8 1 6 3 9 8 7 7 2 6 3 *